BRICKWORK 1
AND ASSOCIATED STUDIES

Related volumes
Brickwork 2 and Associated Studies
Brickwork 3 and Associated Studies } Harold Bailey and David Hancock

Other title of interest
The Skills of Plastering, Mel Baker

BRICKWORK 1
AND ASSOCIATED STUDIES

Harold Bailey

Sometime Senior Lecturer
Stockport College of Technology

David Hancock

Senior Lecturer
Stockport College of Technology

Second Edition

MACMILLAN

Published by
MACMILLAN EDUCATION LTD
Houndmills, Basingstoke, Hampshire RG21 2XS
and London
Companies and representatives throughout the world

ISBN 0 – 333 – 51955 – 8

A catalogue record for this book
is available from the British Library

Printed in Hong Kong

First edition reprinted three times
Second edition 1990
10 9 8 7 6 5 4 3
00 99 98 97 96 95 94 93 92 91

CONTENTS

PREFACE

This series of three volumes is designed to provide an introduction to the brickwork craft and the construction industry for craft apprentices and all students involved in building. All too often, new entrants to the construction industry are expected to have a knowledge of calculations, geometry, science and technology irrespective of their previous education. It is the authors' aim to provide a course of study which is not only easily understood but is also able to show the relationship that exists between technology and associated studies.

The construction industry recognises that the modern craftsman, while maintaining a very high standard of skills, must be capable of accepting change — in methods, techniques and materials. Therefore it will be necessary for apprentices to develop new skills related to the constant advancements in technology.

The first volume deals with all work within the sub-structure of a building and the operations in which the bricklayer is involved. It will be appreciated that a knowledge of all related work is necessary if a successful start is to be made in this important area of construction.

The incentive at the beginning of an apprentice's career should be comparable to the craftsman's at the start of a building project — to reach a successful completion.

HAROLD BAILEY
DAVID HANCOCK

ACKNOWLEDGEMENTS

The authors wish to acknowledge the assistance and cooperation of J. & A. Jackson (Brick Manufacturers) Ltd.

FOREWORD: SAFETY IN BUILDING

There were almost 20,000 reportable injuries (those injuries involving more than three days' absence from work) in the Building Industry between April 1987 and 1988, including over 150 deaths.

The authors appreciate this opportunity to bring these appalling figures to the immediate attention of apprentice bricklayers, and at the same time to remind them of their responsibilities, as well as those of their employers.

The *Health and Safety at Work Act 1974* became effective in 1982. This Act made further provision to the existing *Construction Regulations* for ensuring the health, safety and welfare of persons at work, and may be briefly summarised as follows:

An employer must ensure as far as practicable
1. The health, safety and welfare of his employees while at work.
2. The provision, and maintenance, of safe plant and systems of work.
3. Information, instruction, training and supervision as necessary.
4. A safe place of work.

While at work an employee must
1. Take reasonable care of the health and safety of himself and all other persons who may be affected by his acts or omissions.
2. Co-operate fully with management in all health and safety matters.
3. Not interfere with, or misuse anything provided in the interests of health and safety.

TOOLS REQUIRED BY THE JOURNEYMAN BRICKLAYER

The journeyman bricklayer may be employed by a large national building company, a medium-sized firm or possibly a small employer requiring only one craftsman. The work of the bricklayer will obviously vary according to the type of work undertaken by his employer. It is therefore necessary for the bricklayer to possess a sound working knowledge of

(1) new work
(2) alterations and extensions
(3) maintenance and repair.

To be capable of exercising his skills on each of these types of work, the craftsman should possess the necessary tools for the situation.

The Working Rule Agreement, published by the National Joint Council for the Building Industry, lists a set of 20 tools which a bricklayer may be expected to provide in order to qualify for the weekly 'tool allowance', which at the time of writing is 72 pence/week. The list is as follows:

1 brick trowel	1 square
2 pointing trowels	1 bevel
1 lump hammer	1 pair of dividers
1 bolster	1 rule
1 brick hammer	1 boat level
1 scutch	1 hawk
3 cold chisels	1 pair of line pins
1 point	and line
1 plumb level	1 carborundum stone

These and other common handtools are illustrated and explained below.

The Brick Trowel

This tool is used for picking up and spreading mortar, and can also be used for obtaining a 'rough cut' when using soft or common bricks. The brick trowel can be obtained for a left or right-handed person, since when selecting the trowel it is essential to pick the type of trowel to suit the hand that will use it. Balance and lift too are important factors as are the size and width, which may vary from 225 mm to 350 mm in length, and from 112 mm to 138 mm in width. The brick trowel is also used for striking the mortar joints. Special, lightweight trowels are available for high-class work, but this type of trowel does not possess a cutting edge.

The Jobbing Trowel

This is a hand trowel, the blade being approximately 150 mm in length, usually a brick trowel which has been worn or cut down to a reduced size and is round nosed. It is extremely useful because of its reduced length, is able to pick up a reasonable amount of mortar, and is greatly appreciated when working in confined positions such as fireplace fixing, drain-laying and other situations which prevent the brick trowel from being used (figure 1.1).

The Brick Hammer

There are two types: the modern brick hammer has a

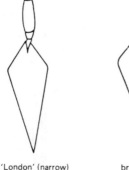  

'London' (narrow) brick trowel   broad trowel   jobbing trowel

Figure 1.1 Types of trowel

1

shorter, broader blade, the head being larger and heavier. While it is very good for cutting bricks, the shortened head length does not allow for plumbing while retaining the plumb level in its position. The long-headed type of brick hammer has both a longer blade and an increased length of hammer. Since the blade is narrower it is not as effective for cutting but the longer length of head allows the hammer to reach positions which would not be possible using the modern, short type of hammer.

The Lump or Club Hammer

The weight of this hammer varies between 1 and 2 kg, the lighter of the two being the most popular. The head of the hammer is flat and is also chamfered in order to reduce damage to the hands from careless blows. It is used in conjunction with all types of chisels.

The Bolster or Boaster Chisel

In some areas this chisel is also known as a blocking chisel. It is made of cold steel, with a 112 mm blade, and the overall length is 220 mm. It is used for accurate brick cutting and it is important that this tool is retained only for this purpose since if used for other work the blade can become distorted.

Note   The head of the bolster should not be allowed to become 'mushroomed' from excessive wear. This condition is dangerous to the hands and the head should be kept as shown in figure 1.2 by grinding as and when necessary on an abrasive wheel.

The Brick Scutch or Comb Hammer

This tool has slots cut in the end of the blades, into which combs or plain blades can be fitted. These can be removed and discarded, and replaced when worn out. The scutch is used when a rough-cut brick is required to be accurately dressed.

Raking-out Pick

This tool is shaped like a miniature pick, both ends being pointed and the head fixed to an ash handle. Its function is to 'pick out' or rake out mortar joints before pointing operations (figure 1.3).

Figure 1.3

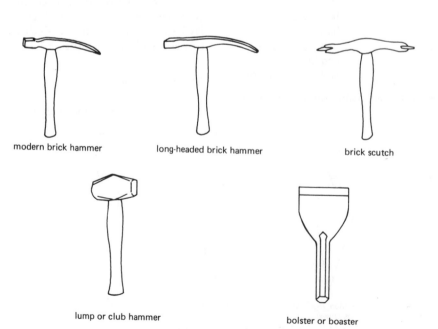

modern brick hammer   long-headed brick hammer   brick scutch

lump or club hammer   bolster or boaster

Figure 1.2 Types of hammer

Cutting Block or Board

This aid is often known as the bat and closer gauge and is usually made of hardwood. On this block are marked the lengths of quarters, halves and three quarter bats (figure 1.4).

Figure 1.4

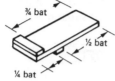

Spreading Trowel or Steel Float

Sometimes known as the laying-on trowel, the use of this tool is to spread and provide a surface when laying floor screeds. It is also used for applying renderings to walls (figure 1.5).

Figure 1.5

Wooden Float

This is a rectangular piece of timber about 250 x 125 x 20 mm It has a handle on top and is used in a circular motion to flatten out concrete and rendering surfaces. Nails can be knocked just through to the underside, when it is used to provide a scoured surface to rendering (figure 1.6).

Figure 1.6

Tingle Plate

This is made out of thin flat metal, usually copper or aluminium. Its size varies around 100 x 75 mm and its purpose is to prevent the occurrence of sagging in long lengths of line (figure 1.7).

Figure 1.7

Corner Blocks

These are preferably made from sections of hardwood and measure about 70 x 50 x 30 mm. A halving is cut out and a sawcut is made in the block, into which the line is inserted. The blocks are used in conjunction with the line and pins to build lengths of walling in between stopped ends or right-angle quoins. The line is pulled through the sawcut and wrapped round once or twice, and use of the corner blocks eliminates the necessity of pushing the line pin into a cross joint (figure 1.8).

Figure 1.8

Mortar Rake

This is a modern version of the raking-out pick, having only one pointed end in which a tungsten tip is fitted. Unlike the pick, this tool is pressed against the joint and pulled along by the handle while applying sufficient pressure to the other handle. The advantage of this tool is that if properly used it does not disfigure the brickwork and operations are considerably speeded up with the minimum of fatigue for the operative. The tip is replaceable (figure 1.9).

Figure 1.9

Rubber Hammer

This is a rubber-headed hammer, fitted into a metal head with a hardwood handle. It is very useful when laying and fixing thin cladding slabs of marble or slate. Also used for fixing glass blocks (figure 1.10).

Figure 1.10

Handsaw

This is a mason master handsaw. Its use is for cutting blocks of low density for exposed work when good quality facework is required. It can also be used to saw through mortar joints when forming toothings, etc. on existing work (figure 1.11).

Figure 1.11

Plumb Rule

This is made of seasoned timber, usually yellow pine.

The length is around 1200 mm and the width is 120–125 mm. A centre line is gouged into the face of the rule and a hole is formed near the bottom to allow the lead plumb bob to move with a 3 mm tolerance all round. The bob, weighing around 200 g, is attached to a line which is secured in saw-cuts formed in the top of the rule. A tapered piece of timber can be fixed to one side when it is required to build a battered wall. The tool is always accurate and is not affected by other influences, but its use in modern brickwork is limited because of the time required when an accurate vertical reading is sought (figure 1.12).

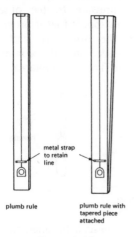

metal strap
to retain
line

plumb rule   plumb rule with
                          tapered piece
                          attached

Figure 1.12

Plumb Level

The modern plumb level is now available in shock-proof form. This requires the levels in the steel casing to be sealed and non-adjustable. It contains top and bottom levels for vertical adjustment and a centre level for horizontal levelling. Plumb levels are obtainable in various lengths, the most popular being 900 mm to 1200 mm (figure 1.13).

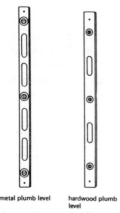

metal plumb level   hardwood plumb
                     level

Figure 1.13

Boat Level

This is a pocket level of wood or metal, 225 mm in length, containing horizontal and vertical bubbles. It is used for individual bricks, or short lengths in conjunction with a straightedge where a plumb level is too long. Also for plumbing brick on end (soldiers) and similar situations (figure 1.14).

metal boat   wooden boat
level   level

Figure 1.14

Line Levels

Small metal levels 100 mm to 120 mm in length containing a centre glass bubble and having two metal hooks formed on top of the tube. The hooks fit over the bricklayer's line and if placed exactly in the centre, horizontal levels can be obtained (figure 1.15).

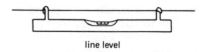

line level

Figure 1.15

Flexible Steel Measuring Tape

These tapes vary from 2 to 5 m in length and are extremely useful, the craftsman being able to bend the tape to obtain lengths of curves, etc. It can be inserted into openings and other positions which would not be possible using a rigid rule (figure 1.16).

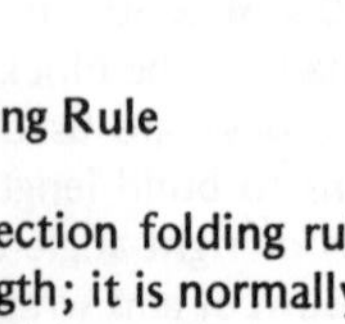

flexible steel
measuring tape

Figure 1.16

Boxwood Folding Rule

This is a four-section folding rule, each section being 250 mm in length; it is normally carried in the craftsman's overall (figure 1.17).

boxwood folding rule

Figure 1.17

Folding Steel Rule

This rule is 600 mm in length, folding in the middle. It is very useful because of its flexibility (figure 1.18).

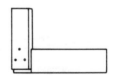

folding steel rule

Figure 1.18

Squares

Steel squares vary in size up to 600 x 450 mm and are useful for setting out quoins and junction walls, checking reveals etc. The try square has a wooden handle and metal blade and is useful for squaring bricks, marking gauge staffs, etc. (figures 1.19 and 1.20).

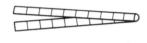

try square

Figure 1.19

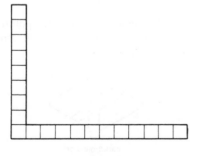

large steel square

Figure 1.20

Bevel

This tool consists of a wooden stock and adjustable steel blade, which is secured by a fly nut. Its use is to obtain any desired angle as may be required for skewbacks in arch construction or when squint quoins and splayed reveals are required (figure 1.21).

bevel

Figure 1.21

Cutting Tools

These consist of a variety of cold steel chisels of different widths and lengths according to use.

(1) The tiling chisel has a 6 mm blade and a length of 150 mm, and is used for cutting out ceramic wall tiles and floor tiles.
(2) The 12 mm blade chisel 200 mm in length is probably the most popular all-purpose chisel. Uses range from cutting drain pipes to cutting toothings in brickwork.
(3) The 18 and 25 mm bladed chisel is used extensively for cutting out defective brickwork or forming indents in existing walls (figure 1.22).
(4) The 38 mm chisel with a length of 450 mm is used for heavy work and when it is desirable to obtain a considerable depth.
(5) The raking-out or plugging chisel is angled to a point and is used for cutting out mortar joints for such purposes as re-pointing or fixing wooden plugs, etc. (figure 1.23).
(6) The seaming chisel (figure 1.24) is a smaller version of the brick bolster, having a 50 mm cutting blade width.
(7) The pitching tool is a cold steel chisel having a 50 mm blade with a thickness of 6 mm which is chamfered to a fine edge. It is often used by masons for stonework and is also required when paving slabs have to be reduced in size.
(8) Comb chisels are similar to seaming chisels but the blade is formed as a comb. It is used to remove rough projecting work and produces a dragged finish (figure 1.25).
(9) The point chisel is a chisel having a length of 225 mm and a diameter of 18 mm. It is reduced to a pointed end and is very often used to form an indentation in very hard surfaces (figure 1.26).

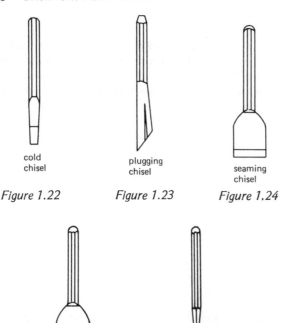

cold
chisel

Figure 1.22

plugging
chisel

Figure 1.23

seaming
chisel

Figure 1.24

comb
chisel

Figure 1.25

point
chisel

Figure 1.26

Pointing Trowel

Pointing trowels are similar in shape to brick trowels, but can vary in size from 100 to 150 mm in length. They are used to form struck joints and to place mortar in joints, including applying a finish to the joint (figure 1.27).

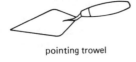

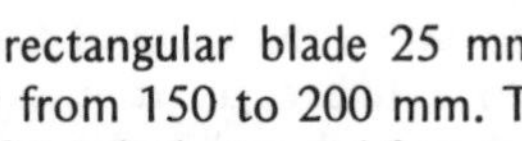

pointing trowel

Figure 1.27

Finger Trowels

This has a rectangular blade 25 mm in width, and lengths vary from 150 to 200 mm. The finger trowel can be used for pointing round frames and for forming joints which would be awkward or difficult using a pointing trowel (figure 1.28).

finger trowel

Figure 1.28

Jointing Tools

These are usually made from 12–18 mm diameter mild steel rod, and are used when a keyed joint is required. They produce a concave section for the mortar joint.

Recessed jointers are provided with a wooden handle and have a rectangular blade. The same type of tool can be used with varying forms of blade to produce external and internal V-joints (figure 1.29).

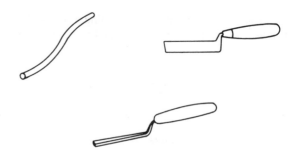

Figure 1.29 Alternative types of jointer available

Frenchman

This tool is formed by cutting and bending an old table knife to the shape shown. Its use is to cut off and remove mortar from the brickwork face when a joint is used to cover weathered arrises. This tool is always used in conjunction with a pointing rule (figure 1.30).

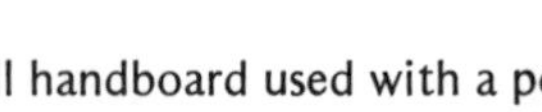

frenchman

Figure 1.30

Hawk

This is a small handboard used with a pointing trowel and its purpose is to support small amounts of pointing mortar (figure 1.31).

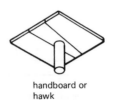

handboard or
hawk

Figure 1.31

Use of the Frenchman and Feather-edge

The bed and cross joints are filled and accurately cut off by drawing the frenchman across the top of the feather-edge. The wood pads pinned to the back of the edge allow any surplus mortar to fall clear (figures 1.32 and 1.33).

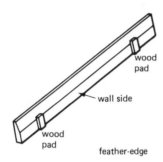

Figure 1.32

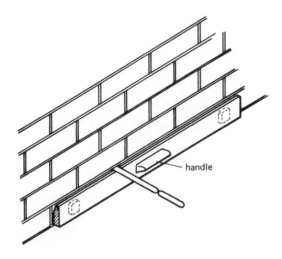

Figure 1.33 Use of the frenchman and feather-edge

Line and Pins

The pins are used in pairs and are made of steel with a circular head 25—28 mm in diameter, a shank and blade. The line is hemp and is obtained in balls or hanks 30—40 m in length. The line is wound round the shank of the pin, but to prevent line rot as a result of corrosion from the pin it is advisable to first place a wrapping of black adhesive tape around the shank. The use of a clove hitch every 5 m or so prevents all the line unwinding should the pin be dropped while working on a scaffold, for example (figure 1.34).

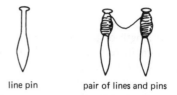

Figure 1.34

Dividers

These are similar to compasses but have two points. They are used, for example, for setting out voussoirs for an arch on a full-sized drawing (see figure 7.64, Volume 2). A wing nut is commonly used to tighten them up as necessary (figure 1.35).

Figure 1.35

Trammel Heads

These are used for setting out large span curves. A suitable batten, of the required length, is necessary. A hole for inserting a pencil is provided in one of the trammel heads (figure 1.36).

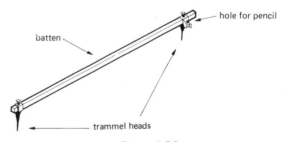

Figure 1.36

Carborundum Stone (approximate size 100 x 50 x 50)

This is an abrasive stone, nowadays more often associated with rubbing rough-cut edges of wall or floor tiles to a smooth surface prior to laying. It is also useful in producing a perfectly flat face of an arch built of red rubbers (figure 1.37).

Figure 1.37

Gauge Staff

A suitable, straight length of planed timber, approximately 50 x 20 mm in section, marked exactly at intervals of 75 mm. The bricklayer uses this constantly when building quoins and stopped ends in order to ensure that all corners are kept perfectly level. Where thicker bricks than the usual 65 mm are used, the gauge is adjusted accordingly. When building an extension to an existing dwelling, the gauge is marked off the existing work, and the new work is kept carefully to this. When building a quoin or stopped end, the order of working is always:

gauge, level, plumb, range

2
MEASUREMENT, SETTING OUT AND LEVELLING

The site manager for a new building project is usually invited by his contracts manager to attend a pre-construction meeting at head office. At this meeting the site manager will inspect and take charge of one set of the contract drawings, all the available detailed working drawings, and information on the new building project.

After the meeting, the contracts manager should introduce the new site manager to the project estimator and planning engineer who between them combine the most comprehensive knowledge in their building company of the proposed construction work. They should all visit the new building site together, to acquaint the site manager with the topography of

Table 2.1. Basic SI Metric Units

Measure	Units		Application
Length	km	— kilometre	journey, distance
	m	— metre	length generally
	mm	— millimetre	small dimensions
	1 km = 1000 m and 1 m = 1000 mm		
Area	km^2	— square kilometre	land area
	ha	— hectare	
	m^2	— square metre	superficial measure generally
	mm^2	— square millimetre	small area
	1 ha = 10 000 m^2		
	1 m^2 = 1 000 000 mm^2		
Volume	m^3	— cubic metre	cubic measure generally
	mm^3	— cubic millimetre	small volumes
Capacity	l	— litre	fluid measure
	ml	— millilitre	small fluid measure
	1 m^3 = 1 000 000 000 mm^3		
	1 litre = 1000 ml		
Mass	t	— tonne	large mass
	kg	— kilogram	all masses
	g	— gram	small masses
	1 tonne = 1000 kg		
	1 kg = 1000 g		
Density	kg/m^3	— kilograms per cubic metre	building materials
	g/cm^3	— grams per cubic centimetre	laboratory work
Force	kN	— kilonewton	large force
	N	— newton	small force
	mass in kg x 9.81 = force in N		
	1 kN = 1000 N		

the construction area. It is useful if the architect, together with his clerk of works, is also invited to attend this site meeting, in order to encourage all those responsible for the successful completion of the new structure, and for them to become acquainted with one another as soon as possible.

THE SI SYSTEM

SI units were evolved over a number of international conferences between 1889 and 1969. The building industry in the United Kingdom was the first to convert gradually from the imperial system, in 1969.

SITE PRELIMINARIES

Scale Drawing

Interpretation of Working Drawings

It is generally accepted that all building operatives should have little difficulty with the interpretation of an architect's drawings, provided that they have a sound knowledge of

(1) basic geometry
(2) basic mensuration
(3) the British Standard Specification for building drawing office practice, BS 1192:1987 Metric units.

Drawing Instruments

It is essential to use good instruments to produce accurate geometry and scale drawings. The basic requirements are indicated in figure 2.1.

Note   Drawing pins should never be used to secure the drawing paper to the board because their continual use would eventually damage the surface to such an extent that accurate drawing would be impossible.

Setting Out the Drawing Paper (BS 1192:1987)

Procedure

(1) The drawing paper should be squarely positioned, using the tee-square, on the drawing board.
(2) Secure the paper to the board using two clips.
(3) Draw a margin round the edges of the paper, 20 mm wide on the left-hand side only, using

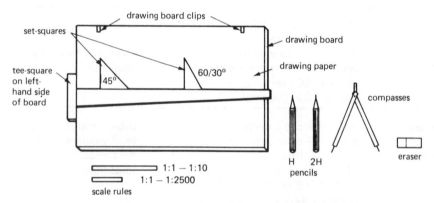

Figure 2.1

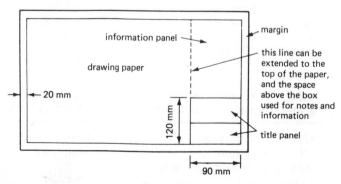

Figure 2.2

(a) the tee-square for the two lines running across the paper from side to side
(b) one of the set-squares for the two lines running from the top to the bottom of the paper.

Note   The tee-square should never be removed from the left-hand side of the board to be used to draw lines running from the top to the bottom of the paper because it cannot be guaranteed that the board corners are square.

(4) Provide a title panel, 90 x 120 mm, in the bottom right-hand corner of the paper (figure 2.2).

Dimension Lines

Dimensions should always be drawn in a position in which they cannot be confused with the subject, and the points to which they relate must be clearly indicated. All dimensions should be written above the line as near to the centre as possible and be arranged so that they can be read from the bottom or from the right of the drawing (see figure 2.3).

Lettering

The aim of lettering any part of a drawing is to provide information necessary to clarify important

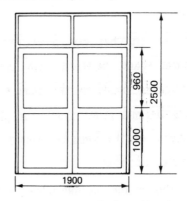

Figure 2.3 (dimensions in millimetres)

features of the work. The recommendations are

(1) print — do not write
(2) letters and figures should be uniform
(3) letters should be carefully formed and well spaced
(4) flowery styles are not recommended
(5) notes should be grouped together and placed near the item to which they refer, but they must not obscure any part of the drawing
(6) nothing is to be placed in any of the margins
(7) punctuation is not to be used unless the notes would be ambiguous without it
(8) always use pencil, never ink or ballpoint pen.

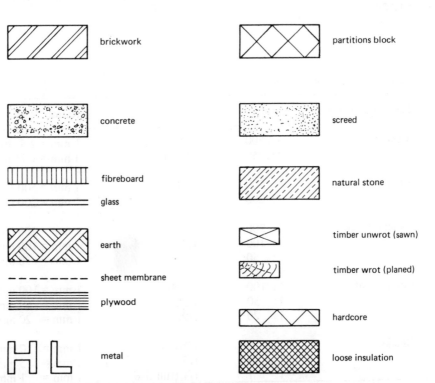

Figure 2.4

Example 2.1

> Form the letters between 3mm lines
> with a 3mm space between each line of
> lettering. Figures should be the same size as
> the letters. Decimal points should be placed
> in the centre of the 3mm lettering lines,
> for example 12·650

Hatching This is the recommended method of indicating materials, usually on plans and sections. Thin lines are used and where diagonal lines are shown they should be at 45°. Where large areas of hatching are required, especially for concrete, only a small area near the end or edges of the concrete should be drawn (see figure 2.4).

Scales

The architect's drawings should communicate to the building site every architectural detail he wishes to incorporate into the finished building. To assist the builder, the drawings are drawn to scale. This means that the land, buildings and constructional details are not represented full size, but are usually reduced in proportion, to enable the necessary information to be presented on drawing paper.

Scale rules are available either 150 or 300 mm long and it is necessary to use two rules which should between them be graduated so as to indicate the popular scales used.

Table 2.3. Scale Rules

Rule No.	Side 1	Scales Indicated Side 2
No. 1 V	1:1	1:5
	1:2	1:10
No. 3 V	1:1 and 1:100	1:2500 and 1:1250
	1:20 and 1:200	1:5 and 1:50

The plans, elevations and sections on a builder's drawing are the representation of a three-dimensional object on a flat plane. It is essential that all operatives who have to interpret the working drawings understand how the plans and elevations are positioned on the drawing paper.

Projections

In the United Kingdom the first-angle method is recommended for building drawing (see figures 2.5 and 2.6).

Table 2.2. Scales in Common Use

Type of Drawing	Scales		Ratio
Block plan	1:2500	1/2500	1 mm = 2.5 m
	1:1250	1/1250	1 mm = 1.25 m
Site plan	1: 500	1/500	1 mm = 500 mm
	1: 200	1/200	1 mm = 200 mm
General location	1: 200	1/200	1 mm = 200 mm
	1: 100	1/100	1 mm = 100 mm
	1: 50	1/50	1 mm = 50 mm
Component	1: 100	1/100	1 mm = 100 mm
	1: 50	1/50	1 mm = 50 mm
	1: 20	1/20	1 mm = 20 mm
Details	1: 10	1/10	1 mm = 10 mm
	1: 5	1/5	1 mm = 5 mm
	1: 1	1/1 (full size)	1 mm = 1 mm

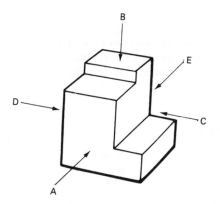

B

E

D

C

A

Figure 2.5 View in direction of: A, front elevation or view; B, plan or view from the top; C, end elevation or view from the right; D, end elevation or view from the left; E, rear elevation or view

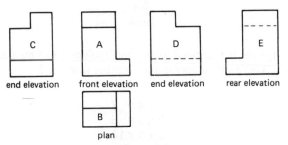

C — end elevation

A — front elevation

D — end elevation

E — rear elevation

B — plan

Figure 2.6

Isometric Projection This is a method of indicating three faces of an object on a flat plane. Usually the front and one side elevation are shown, together with the plan (see figure 2.7). All vertical lines remain

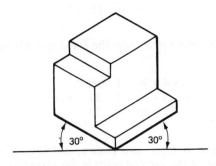

30°   30°

Figure 2.7

vertical, but horizontal lines on the front and end views are at 30° to the horizontal. It is drawn using the 30°–60° set-square. The ratio of the length, height and width is 1:1:1.

Oblique Projection This is another method of indicating three elevations on a single plane (see figure 2.8). All vertical lines remain vertical, and the

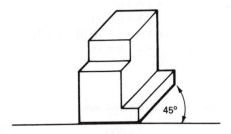

45°

Figure 2.8

horizontal lines on the front view remain horizontal, but the horizontal lines on the end view are drawn at 45°, using the 45° set-square. The ratio of the length, height and width is 1:1:½.

Calculations

Calculations using Decimals

The site manager and every craftsman in the building industry are constantly manipulating decimal figures. They are adding, subtracting, multiplying or dividing, usually linear or length dimensions. Decimals should not prove difficult provided the decimal point is placed accurately.

$$1 \text{ m} = 1000 \text{ mm}$$

therefore

$$1000 \text{ mm} = 1.000 \text{ m}$$
$$900 \text{ mm} = 0.900 \text{ m}$$
$$90 \text{ mm} = 0.090 \text{ m}$$
$$9 \text{ mm} = 0.009 \text{ m}$$

Addition (+)

Example 2.2

6.100 + 4.350 + 0.625 m

```
 6.100
 4.350
 0.625
------
11.075
```

Answer   11.075 m

Method   As above, list the numbers with their decimal points directly below one another. Place the decimal point for the answer directly below the points above. Add together the four columns, starting on the right-hand side

$$5 + 0 + 0 = 5$$
$$2 + 5 + 0 = 7$$
$$6 + 3 + 1 = 10 \text{ (0 carry 1)}$$
carry $\quad 1 + 4 + 6 = 11$

Example 2.3

6 m + 705 mm + 6.5 m + 964 mm

$$
\begin{array}{r}
6.000 \\
0.705 \\
6.500 \\
0.964 \\
\hline
14.169 \\
\hline
\end{array}
$$

Answer $\quad$ 14.169 m

Method $\quad$ As above, list the numbers again, but this time we must convert any millimetres into metres. Place the decimal point for the answer and add, starting from the right-hand side of the numbers

$$4 + 0 + 5 + 0 = 9$$
$$6 + 0 + 0 + 0 = 6$$
$$9 + 5 + 7 + 0 = 21 \text{ (1 carry 2)}$$
carry $\quad 2 + 6 + 0 + 6 = 14$

Subtraction (−)

Example 2.4

5.697 − 2.473 m

$$
\begin{array}{r}
5.697 \\
2.473 \\
\hline
3.224 \\
\hline
\end{array}
$$

Answer $\quad$ 3.224 m

Method $\quad$ As above, place the decimal point for the answer and subtract the figure on the bottom line from the figure above it, starting from the right-hand side of the numbers

from $\quad$ 7 take 3 = 4
$\quad\quad\quad$ 9 $\;-\;$ 7 = 2
$\quad\quad\quad$ 6 $\;-\;$ 4 = 2
$\quad\quad\quad$ 5 $\;-\;$ 2 = 3

Check $\quad$
$$
\begin{array}{r}
3.224 \\
2.473 \\
\hline
5.697 \\
\hline
\end{array}
$$ Check

Check the answer by adding the answer 3.224 to 2.473.

Multiplication (×)

More than one method can be used when multiplying two decimals together. Always use the method taught to you during your formal schooling.

Example 2.5

5.34 × 6.2

$$
\begin{array}{r}
5.34 \\
6.2 \\
\hline
\end{array}
$$

List the two numbers with their decimal points directly below one another, as above.

$$
\begin{array}{r}
5.34 \\
6.2 \\
\hline
1068 \\
\end{array}
$$

Ignore the decimal point and multiply 534 by 62, taking each number in turn starting with the 2

$$2 \times 4 = 8$$
$$2 \times 3 = 6$$
$$2 \times 5 = 10$$

$$
\begin{array}{r}
5.34 \\
6.2 \\
\hline
1068 \\
32040 \\
\hline
33108 \\
\hline
\end{array}
$$

Answer 33.108

534 × 60 place zero under the 8 and multiply by 6

$$6 \times 4 = 24 = 4 \text{ carry } 2$$
$$6 \times 3 = 18 + 2 = 20 = 0 \text{ carry } 2$$
$$6 \times 5 = 30 + 2 = 32$$

Add the bottom two figures together. Fix the position of the decimal point in the answer by counting the number of figures after the decimal points in the sum. There will be the same number of decimal places in the answer. In this case 3. Therefore count 3 figures from the right of 33108 and place the point.

Calculations on Perimeters, Areas and Volumes

The perimeter of a figure is the distance, or the length around its boundary.

Example 2.6

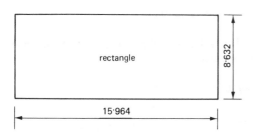

Figure 2.9

Figure 2.9 has four sides, and because it is a rectangle the sides opposite one another will have the same linear dimension.

Method 1   Add the four sides together.

$$
\begin{array}{r}
15.964 \text{ m} \\
8.632 \text{ m} \\
15.964 \text{ m} \\
8.632 \text{ m} \\
\hline
49.192 \text{ m} \\
\end{array}
$$

Method 2   Multiply the length by 2 and the width by 2 and add together.

$$
\begin{array}{l}
2 \times 15.964 \text{ m} = 31.928 \text{ m} \\
2 \times 8.632 \text{ m} = 17.264 \text{ m} \\
\hline
\phantom{2 \times 8.632 \text{ m} = 1}49.192 \text{ m} \\
\end{array}
$$

Perimeters are the same in A and B (figure 2.10).

Perimeter = (2 x maximum length)
       + (2 x maximum width)
       = (2 x 7.800) + (2 x 4.500)
       =   15.600 +   9.000
       =   24.600 m

In all cases where rectangular corner insets occur in a figure, the total perimeter will be equal to that of the rectangle with the maximum dimensions of the figure. This principle will not apply where any inset is not positioned at a corner.

Example 2.7

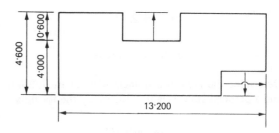

Figure 2.11

In figure 2.11

perimeter = (2 x 13.200) + (2 x 4.600) + (2 x 0.600)
       =   26.400 +   9.200 +   1.200
       =   36.800 m

To determine the area of a figure

(1) On plan, multiply the length by its width or breadth, using the formula

area = length x width

(2) On elevation, multiply the length by its height, using the formula

area = length x height

The triangle in figure 2.12 is half the area of the rectangle

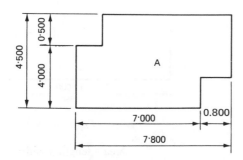

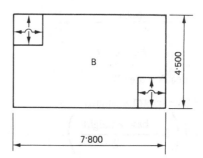

Figure 2.10

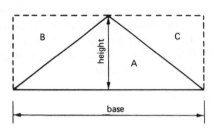

Figure 2.12

area B + C = area A

Determine the area of a triangle with a base 7.000 m and a perpendicular height 3.000 m, by the formula

$$area = \frac{base \times vertical\ height}{2}$$

$$= \frac{7.000 \times 3.000}{2}$$

$$= \frac{21.000}{2}$$

$$= 10.500\ m^2$$

Example 2.8

Determine the area of the gable end of a bungalow, illustrated in figure 2.13.

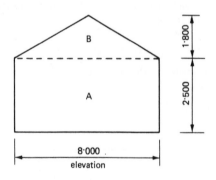

elevation

Figure 2.13

Total area = area A + area B

= (A = length x height)

$$+ \left(B = \frac{base \times height}{2} \right)$$

$$= (8.000 \times 2.500) + \frac{(8.000 \times 1.800)}{2}$$

= 20.000 m² + 7.200 m²

= 27.200 m2

The volume is the amount of three-dimensional space a figure occupies.

Example 2.9

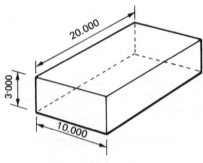

Figure 2.14

See figure 2.14.

Volume = length x width x height or area x height

= 20.000 x 10.000 x 3.000

= 200.000 m² x 3.000

= 600.000 m³

The square is a four-sided figure, where all four sides are of equal length and all four angles formed by the sides are at 90° (see figure 2.15).

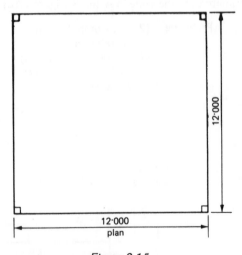

plan

Figure 2.15

Area = length x width

= 12.000 x 12.000

= 144.000 m²

Where the area of a square is known the length of one side of the square can be found by finding the square root of the area ($\sqrt{}$).

Example 2.10

Determine the length of one side of a square having an area of 81.000 m².

Length of one side = $\sqrt{81.000}$ m

$\qquad\qquad\qquad$ = 9.000 m (because 9 x 9 = 81)

Right-angled Triangle　Any triangle containing a right-angle is termed a right-angled triangle, and the line opposite the right-angle is termed the hypotenuse. The area (figure 2.16) is found exactly as for figure 2.12, that is

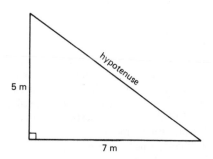

5 m

7 m

Figure 2.16

$$\text{area} = \frac{\text{base x vertical height}}{2}$$

$$= \frac{7 \times 5}{2}$$

$$= \frac{35}{2}$$

$$= 17.5 \text{ m}^2$$

Use of Log Tables

The following is a very brief explanation of working with logs and is intended to form revision notes only. As the student will know, log tables can turn a tedious process into a quick operation. For example, to multiply such numbers as 0.009 164 by 7663 by 22.98 would take nearly half a page using long multiplication. By logs this can be solved in four lines.

　Logs consists of two parts, a characteristic and a mantissa. The characteristic must be arrived at by the student and goes before the point and the mantissa is obtained from the log tables and goes after the point,

for example, the log of 25.25 is 1.4023. The 1 is the characteristic and the .4023 is the mantissa.

The Characteristic　That of any number greater than unity is positive and is less by one than the number of figures to the left of the decimal point. That of any number less than unity is negative and is greater by one than the number of noughts following the decimal point.

The Mantissa　Open the log books at logarithms since many mistakes are made by using the wrong page. Locate the first two figures of your number in the left-hand column and put a rule across the page below these two figures. The third figure is found across the top of the page and a cross check is taken. The fourth figure, if one exists, is found at the top of the page in the right-hand columns, known as mean differences. Take a further cross check and add the figure indicated to your previous cross check.

Antilogs　To convert logs back to numbers the table of antilogs must be used. Look up the mantissa only in the tables and place the point as the characteristic indicates.

Using Logs　To multiply numbers, find the logs of the numbers, add them and antilog the result. For example

(1) $\qquad\qquad$ 2.114 x 87.24 x 666.6

No.	Log
2.114	0.3251
87.24	1.9407
666.6	2.8239
1229	5.0897

Answer 122900.0
(add noughts as necessary to place the point)

(2) $\qquad\qquad$ 0.0561 x 0.000724 x 118.8

No.	Log	
0.0561	$\bar{2}$.7490	
0.000724	$\bar{4}$.8597	(referred to as bar 4.8597)
118.8	2.0749	
4826	$\bar{3}$.6836	(referred to as bar 3.6836)

Answer 0.004826
('bar 3' means two noughts after the point)

　For long division, find the logs of the numbers concerned, subtract that of the denominator from that of the numerator, and antilog the result. For example

(1)

$$\frac{684.2}{77.91}$$

No.	Log
684.2	2.8352
77.91	1.8916
8782	0.9436

Answer   8.782

(2)

$$\frac{3.212}{0.00671}$$

No.	Log
3.212	0.5068
0.00671	$\bar{3}.8267$
4787	2.6801

Answer   478.7

In the last example a bar number had to be subtracted. Where this is necessary, the easiest way is to follow the simple rhyme — 'change the sign of the bottom line and add'. In this case we are subtracting bar three from bar one. Using the rhyme this becomes plus 3 added to bar 1, which gives 2.

Areas and volumes can be quickly calculated using the methods described.

Square Roots   There are several methods for finding the square root of a number, but possibly one of the surest is to use log tables in the following way. Find the log of the number, divide it by 2 and antilog the result.

For example, to find the square root of 28

$$\log 28 = 1.4472$$

$$1.4472 \div 2 = 0.7236$$

$$\text{antilog } 0.7236 = 5.291$$

To find the square root of a number the log of which has a negative characteristic, the characteristic must be made divisible by 2, and it may be necessary to add bar 1 to the characteristic and plus 1 to the mantissa.

For example, to find the square root of 0.4634

$$\log 0.4634 = \bar{1}.6660$$

This becomes

$\bar{2} + 1.6660$ (now the characteristic is divisible by 2)

$$\bar{2} + 1.6660 \div 2 = \bar{1}.8330$$

$$\text{antilog } \bar{1}.8330 = 0.6808$$

Squares   In this case find the log of the number, multiply it by 2 and antilog the result.

For example, to find the square of 7.139

$$\log 7.139 = 0.8536$$

$$0.8536 \times 2 = 1.7072$$

$$\text{antilog } 1.7072 = 50.95$$

One of the first practical operations on site is to mark accurately the position of the proposed building in relation to its site boundaries. Both vertical and horizontal setting out are necessary.

Use of Calculators

Calculators are now freely accepted in Craft and Advanced Craft examinations, and with a little practice students can rapidly arrive at an answer. However, when using a calculator it is still necessary to write a full answer to a question, that is

1. State the formula.
2. Change this to figures.
3. Write down your answer, not forgetting to state the units used (m, m^2, m^3, etc.)

It is also good practice, if time allows, to check through your answer again on the calculator since silly mistakes are easily made pressing small buttons!

VERTICAL SETTING OUT

If possible this task should be accomplished first by establishing a site datum (figure 2.18) to which all site levels are related. This should be set up in a readily accessible position, preferably in the vicinity of the site office but well out of the way of site traffic. This datum may be related to the nearest Ordnance Bench Mark (O.B.M.) (figure 2.17) which marks the height above mean sea level at Newlyn in Cornwall. O.B.M.s are levels established by the Government Ordnance Survey of Great Britain, and are permanently incised into walls, usually of public buildings.

Alternatively an assumed datum may be used, known as a Temporary Bench Mark (T.B.M.) which is given a sufficient value to ensure that the lowest point on site will still be above zero since negative values should be avoided where possible.

Datum level generally corresponds with the finished ground floor of the proposed building and

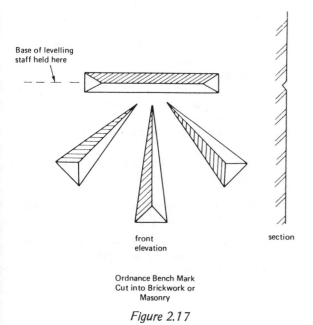

Base of levelling
staff held here

front elevation   section

Ordnance Bench Mark
Cut into Brickwork or
Masonry

Figure 2.17

can take the form of a timber or, preferably, steel peg, driven into the ground. To give it protection and added stability, it should be surrounded with concrete. The peg should be fenced round to give further protection from site traffic (figure 2.18).

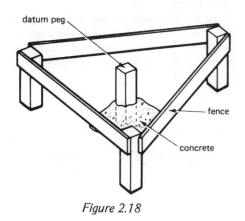

datum peg

fence

concrete

Figure 2.18

If the proposed building is to be constructed in a built-up area, datum can be the top of an adjacent permanent object, for example, an inspection chamber cover, road kerb, or the top surface of a coping stone on a boundary wall. This type of datum does not usually correspond with the ground floor of the new building.

When datum has been established on the site, its height in relation to an O.B.M. may be too cumbersome for site use so the architect may designate a value of 100.00 to this level and relate all other heights or depths of the proposed building to this figure.

For example

datum	= 100.00
basement floor level	= 3.00 below datum

Therefore

finished floor level, basement = 97.00

datum	= 100.00
1st floor level	= 3.00 above datum

Therefore

finished floor level, 1st floor = 103.00

LEVELLING

Transferring levels across a site must be done with care and accuracy. There are several traditional methods in use, and the most common are

(1) Levelling board with spirit levels
(2) Boning rods
(3) Water level

Levelling Board Method

A levelling board is a long timber rule with its longest edges straight and parallel with each other. It is used in conjunction with a spirit level.

Procedure

(1) Ensure that the rule is clean.
(2) Sight with one eye along one edge to check straightness.
(3) Measure the length of the rule and mark the centre. The spirit level should be placed in the centre on the top edge of the rule (figure 2.19).

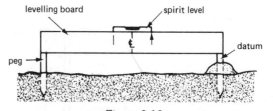

levelling board   spirit level

datum

peg

Figure 2.19

(4) Place one end of the rule on datum and drive a temporary peg into the ground the same distance

away from datum as the length of the rule.
(5) Place the rule on the two pegs, and position the spirit level on the rule.
(6) Check the position of the bubble in relation to the two graduation marks on the spirit-filled glass tube (figure 2.20).

high this ← end   level   high this → end

Figure 2.20

(7) Adjust the temporary peg until the spirit level indicates that the board is horizontal.
(8) Turn the level end for end to check its accuracy.
(9) Repeat the procedure, moving the board in the direction required.

Notes

(1) Minimise levelling errors by reversing the board each time a new temporary peg is driven into the ground (figure 2.21).
(2) Where it is impossible to drive pegs into hard ground, packings can be used.

Boning Rods Method

Boning rods are used in sets of three. Each rod is made up of two pieces of timber, one vertical and one horizontal, in the shape of a T. The vertical pieces of timber for each rod must be of equal length (figure 2.22).

Procedure

(1) Provide a temporary level point from datum, using the levelling board, in the direction of the proposed boned level required.
(2) Start to drive the proposed boned level peg into the ground, and stand one boning rod on the top of the peg.
(3) Stand another rod on the top of the temporary levelled point.
(4) The boner stands the third rod on the top of datum, and sights the tops of the rods.
(5) The proposed level is adjusted until the top surfaces of the three rods are in coincidence with the line of sight (figure 2.23).

Intermediate level points, between two previously established levels, can also be determined using the boning rods, as follows.

Procedure

(1) Establish two main level points, using the procedure described previously.
(2) Stand one boning rod on each of the two master points.
(3) Start to drive the first intermediate level point and stand the third boning rod, the traveller, on top of the peg.
(4) Adjust the peg until the top surfaces of the three boning rods are in coincidence with each other.
(5) Repeat this procedure until the required number of intermediate level points has been established (figure 2.24).

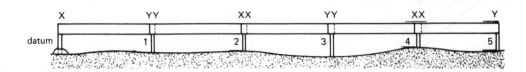

Figure 2.21

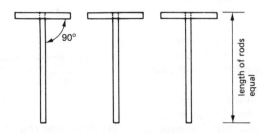

Figure 2.22

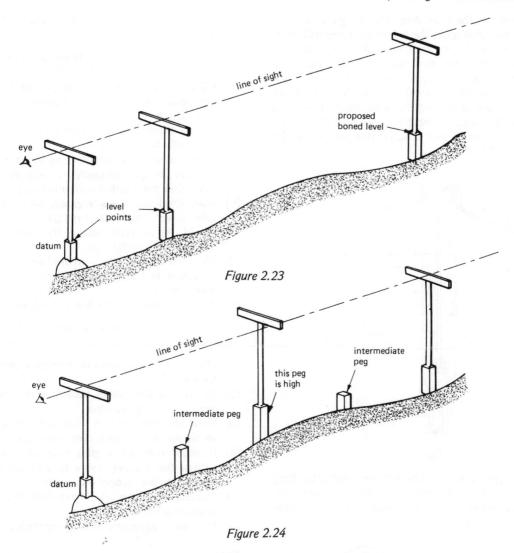

Figure 2.23

Figure 2.24

Only operatives with a good trained eye should consider using the boning rods for levelling purposes and it is most important that each rod is perfectly perpendicular when being held on a levelling point. To reduce errors it is recommended that the vertical leg of the rod be made as a plumb rule (figure 2.25).

Water Level Method

The modern water level comprises a length of small-bore rubber tube with a glass tube attached at each end (figure 2.26). The tube is carefully filled with water at one end only, to ensure that no air bubbles

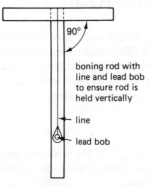

Figure 2.25

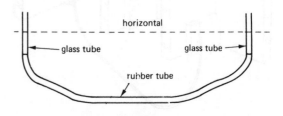

Figure 2.26

are trapped in the tube. When the two glass tubes are held at the same level, the height of water will be at the same level in each. If one tube is significantly lower than the other, then water will pour out of the lower tube, until the liquid once again finds its own level. To stop this happening in practice, especially when carrying or storing the level, both glass tubes are provided with a rubber cork with a hole pierced through its centre and a screw stopper to seal the hole (figure 2.27).

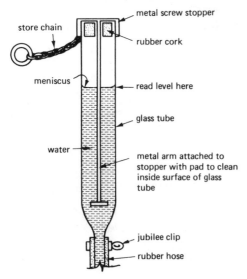

Figure 2.27

This type of level can be used for general levelling purposes, but it is particularly useful when determining level points which are accessible, but not visible from one another (figure 2.28).

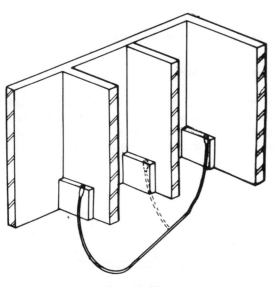

Figure 2.28

Essential precautions when preparing the level are as follows.

(1) Unscrew the metal cap from the glass tube at both ends before starting to fill the tube.
(2) Fill the tube with water from one end only.
(3) Both glass tubes should be approximately at the same height.
(4) Try to ensure that the glass tube being filled is constantly full of water.
(5) When the water is seen rising up the other end of the tube, at approximately half way up the glass stop filling and tighten both metal stoppers.
(6) Store the level in a safe place, with the rubber tube lower than the two glass tubes, which should be approximately the same height. Unscrew both metal caps.
(7) The level must not be used until air bubbles stop appearing in the glass tubes, when the rubber tube is moved.
(8) Screw down metal caps before the level is moved.

Essential precautions when using the level are as follows.

(1) There must be no kinks in the tube when reading the level.
(2) No part of the rubber tube must be above the bottom of the glass tubes.
(3) Metal stoppers must be released to allow atmospheric pressure to enter both glass tubes.
(4) If water rises up a glass tube and overflows, tighten the stopper, move to a higher position and release the stopper once again.
(5) The level reading is taken from the bottom of the meniscus (figure 2.27).
(6) Close the stoppers before moving the level.

Note

The water must be removed from the level during freezing conditions. A solution of car antifreeze and water can be used in an emergency.

Adjustment of Spirit Levels

Modern levels have a metal alloy body and completely enclosed spirit-filled glass tubes. They are shock-proof and not capable of being adjusted. Traditional levels have a hardwood body and spirit tubes that can be adjusted when errors are evident.

Procedure

(1) Mark the length of the level on a length of timber which has been placed on a bench or table (figure 2.29).

(2) Check the position of the bubble in relation to the two graduation marks on the glass tube (figure 2.30).
(3) Reverse the level, ensuring that it is replaced to the marks on the timber.
(4) Re-check the spirit bubble. The bubble should occupy the same position as it did before reversing, if the level is reading correctly (figure 2.31).

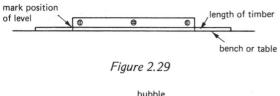

Figure 2.29

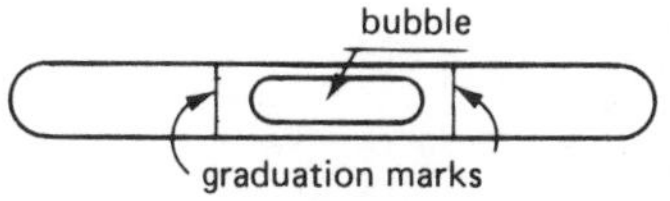

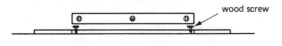

Figure 2.30

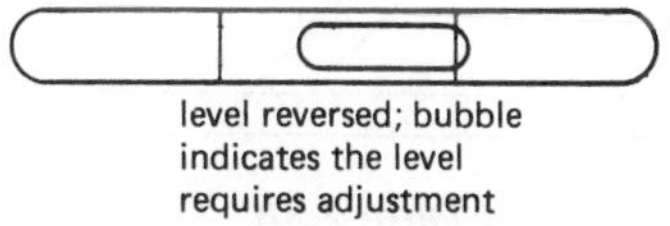

Figure 2.31

(5) Using a screwdriver, drive two wood screws into the length of timber, with each screw just inside the position marks (figure 2.32).

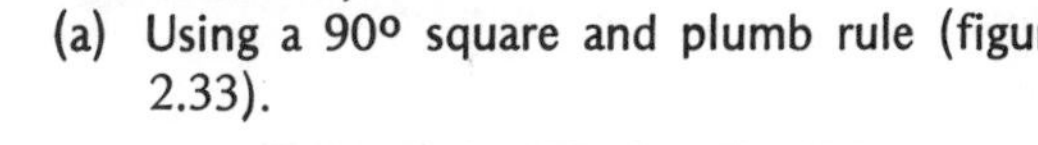

Figure 2.32

(6) Adjust the depth of penetration of the wood screws until their heads are perfectly level with one another by either
(a) Using a 90° square and plumb rule (figure 2.33).

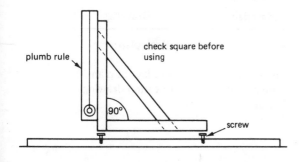

Figure 2.33

(b) Continual reversing of the level until the bubble is in error an equal distance to the graduation marks (figure 2.34).

Figure 2.34

(7) Replace the level on the screws and adjust the spirit tube to a position that brings the air bubble into a central position with the graduation marks on the glass.
(8) Reverse again and check.

Care of Spirit Levels

(1) Never drop a level.
(2) When levelling work is in progress never strike the level with any other tool. It is the work that is in error, not the level.
(3) Keep the level clean.
(4) Never immerse hardwood levels in water; they should be wiped clean with a cloth. They can be periodically varnished or given a coat of linseed oil.
(5) Never use it as a lever or packing.
(6) Never leave the level lying on the ground where it can be run over by passing site traffic.

HORIZONTAL SETTING OUT

The plan shape of the proposed building must be accurately marked on the ground in relation to the site boundaries.

Profiles

The face side of every loadbearing wall in the structure is determined and marked by ranging lines secured to temporary timber structures termed profiles. From the wall positions the foundation trenches can be determined and marked on the profiles, using ranging lines, one for the front and one for the back of the trench (figure 2.35). The ranging

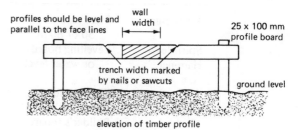

Figure 2.35

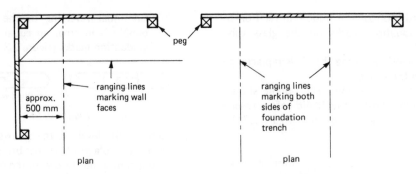

Figure 2.36   Figure 2.37

lines are heavy gauge builder's line, which is much thicker than bricklayer's chalk line.

There are two types of timber profile in use

(1) return profile containing three pegs and two profile boards (figure 2.36).
(2) single profile containing two pegs and one profile board (figure 2.37).

The tools, equipment and materials required are listed in table 2.4.

Steel measuring tapes are preferred where accuracy is required because linen tapes tend to stretch with continual use. It is essential when using the steel tape to measure from the end of the ring, and never allow the tape to kink, otherwise it will crack and snap at the kink point. When measuring operations are complete, the tape should be fully withdrawn from its case and cleaned as it is rewound back into the case.

Builder's Square

A builder's square is usually made on site by a carpenter. It comprises three pieces of timber fastened together to form a 90^0 angle and can be checked by the 3:4:5 method before it is used (figure 2.38).

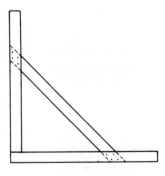

Figure 2.38

To check for squareness: measure 3 units along the outside of one 'leg', 4 units along the outside of the other 'leg' and the diagonal length between these points should be exactly 5 units.

Suitable units for checking a builder's square are

300 mm	400 mm	500 mm
600 mm	800 mm	1000 mm
900 mm	1200 mm	1500 mm

The reason why these numbers are suitable for checking a builder's square is explained by Pythagoras'

Table 2.4

Tools	Equipment	Materials	Drawings
Metre rule	Steel tape 30 or 60 m	Nails	Block plan
Spirit level	Levelling board or water level	Timber pegs	Site layout plan
Hammer		Timber profile	Location drawing
Axe		Boards	
Saw	Boning rods		
Pencil	Builder's line		
	Builder's square		
	Mawl or 2 kg hammer		

theorem, which states that

'in any right-angled triangle, the square on the hypotenuse is equal to the sum of the squares on the other two sides'

Right-angled Triangle A triangle having one of its angles equal to 90°.
Square The value obtained when a number is multiplied by itself, for example, 3 squared (written 3^2) = 9.
Hypotenuse The longest side in a right-angled triangle. It is always opposite to the right-angle.
Sum The value obtained when numbers are added together.

Example 2.11

See figure 2.39. To find AC.
By Pythagoras' theorem

$$AC^2 = AB^2 + BC^2$$
$$= 12^2 + 5^2$$
$$= 144 + 25$$
$$= 169$$

Therefore

$$AC = \sqrt{169}$$
$$= 13 \text{ m}$$

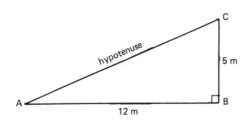

Figure 2.39

If we are to position the building accurately, for both horizontal and vertical setting out, then the top surface of every profile board must be level within itself, and level with every other profile board used. The top surface of the board should be set at the height of the damp-proof course above ground level.

The Building Line

This is an imaginary line on site, set by the local authority's Planning Officer. No part of the main structure should be in front of the line. It can be found on the site layout plan.

The procedure for a reasonably level site is as follows.

(1) Determine from the site layout plan the position of the building line, and mark its position on the site with a ranging line attached to pegs or profiles (figure 2.40).
(2) Determine from the site layout plan the main setting out quoin, or corner, of the structure and approximately mark the position using a peg directly under the ranging line. The top of the peg can be levelled using boning rods from the top of the building line profiles. Accurately mark the point with a nail in the top of the peg (figure 2.40).
(3) Hook the tape on this nail and approximately mark the length of the front of the building by driving peg 2 under the ranging line and level with peg 1. Pull the tape taut and drive the second nail half-way into the peg.
(4) Set up a builder's square on packings so that one of its legs is parallel to the building line and its corner is positioned at the main setting-out point nail (peg 1).
(5) Attach a line to the nail at peg 1, and mark the position of peg 3. The line should run parallel with the other leg of the builder's square. Accurately mark the position of the end elevation with a nail in peg 3 (figure 2.40).
(6) Measure the width of the building from the nail in peg 1, and fix peg 4. Accurately mark this back quoin with a nail.
(7) Mark the position of the remaining quoin peg 5, by using two steel tapes, one for the length, and one for the width (figure 2.40).
(8) Check the setting out for squareness by measuring the diagonals. The distance from nail to nail on pages 1 and 5 should be the same as the distance from nail to nail on pegs 2 and 4 (figure 2.40).

Note Where the two diagonal measurements do not correspond with one another, all the setting out must be checked again. For example, the lineal measurements 1–2 must be the same as 4–5 and 1–4 must equal 2–5. Also, angle 2–1–4 must be 90°.

(9) Adjust setting out, if required, and check again.
(10) Erect temporary timber profiles approximately 500 mm outside the setting-out lines. Return profiles for the quoins and single profiles for the internal loadbearing walls (figure 2.41).
(11) Transfer the ranging lines to the profiles by unhooking the line from the first set of pegs, and mark each wall in turn by projecting through the nails, with a line on to the profile boards (figure 2.42).

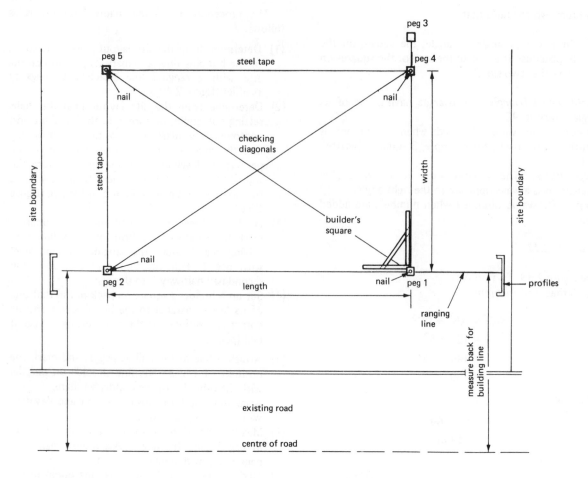

plan of first set of pegs

Figure 2.40

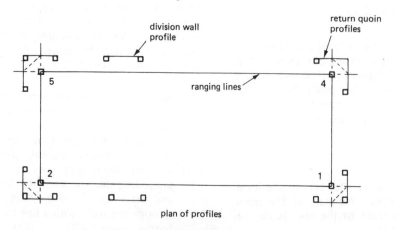

plan of profiles

Figure 2.41

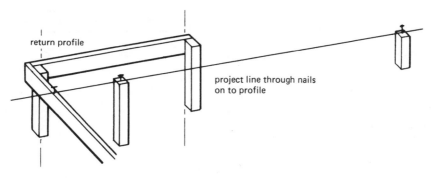

Figure 2.42

(12) Stretch the ranging lines around all the nails on the profile boards, marking the external face of the shell walls of the structure.
(13) Check the setting out again, and if satisfactory, remove pegs 1, 2, 4, 5.
(14) Mark on the profile board the width of the walls.
(15) Determine the width and projection of the concrete foundations and mark both front and back of the trench with a V saw cut in the profile board (figure 2.35).

SETTING OUT ON SLOPING SITES

The frontage line is located and the two front corners marked as described, but it is vital that the tape is held horizontally if the measurements are to be accurate (figure 2.43).

It will be obvious from figure 2.43 that the greater the slope, the greater the inaccuracy if the tape is held out of level; but with even a slight slope the

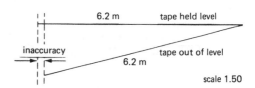

Figure 2.43

measurements would be wrong. For example, if the ground slopes only 1 m in a distance of 10 m and the tape is allowed to follow the ground level the measurement will be almost 50 mm out.

If the dimension to be set out is short and the fall not too great, this can be set out using pegs of different lengths as shown in figure 2.44a. If, however, the distance or the fall is considerable, setting out may have to be carried out as shown in figure 2.44b.

Having accurately pegged out the frontage line, the flank walls must now be squared back. This operation is usually carried out in one of the following three ways

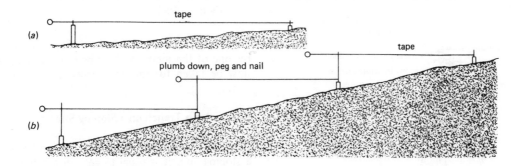

Figure 2.44

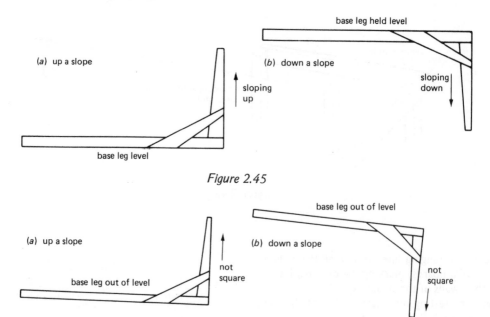

Figure 2.45

Figure 2.46

Using the Builder's Square

This method will be found to be quite accurate if the base 'leg' of the square is held level and the other 'leg' sloped up or down hill as required (figure 2.45). If the base 'leg' is held out of level, the line obtained will be out of square (see figure 2.46).

Using the 3—4—5 Method

This method is effective provided the tops of any pegs used for this purpose are kept level.

Using the Sitesquare

The instrument requires setting up roughly over the nail in the corner peg and will accurately set out an angle of 90° whether the ground is level or sloping (figure 2.47).

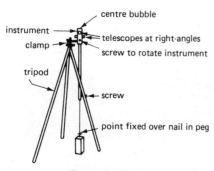

Figure 2.47

Setting up and using the sitesquare presents little difficulty; the method is as follows.

(1) Set up the tripod firmly over the setting-out peg, which must be a little off centre and not co-incident with the legs of the tripod.

(2) Clamp the instrument on the spike and drop the telescopic point over the setting-out nail; tighten up the screw.

(3) Undo the instrument clamp and slide this back-wards or forwards, left or right as required until the centre bubble shows correct; clamp up.

(4) Turn the rotating screw until the top telescope is pointing directly over the setting-out line (figure 2.48a).

(5) Look through the bottom telescope and the centre of the cross depicts a line exactly square off the first. The ranging line is adjusted until coincident with this cross and then pegged (figure 2.48c).

Having squared back from the two front corners by one of the methods described, measurements can now be taken for the flank walls as shown in figure 2.40.

Excavating a Trench on a Sloping Site

Whether excavating is carried out by hand or plant on a sloping site, it is usual to get out a rough slope in the first place and arrange the steps by hand (figure 2.49).

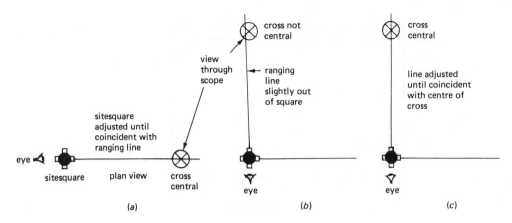

Figure 2.48

Stage 1

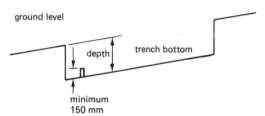

Figure 2.49

Stage 2

A peg is placed in the lowest end to give the required concrete thickness (minimum 150 mm) and the operative must excavate by hand from this peg to provide a level base for the concrete until a step is necessary. This will depend largely on the slope of the ground, and the height of each step is required to work in to the gauge set for the brickwork. A peg is inserted level with the first (figure 2.50).

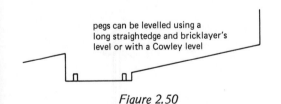

Figure 2.50

Stage 3

Put in a peg on the step, carefully to gauge and minimum 150 mm above peg 2 and repeat stage 2 (figure 2.51).

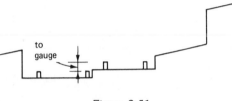

Figure 2.51

The rest of the trenches are excavated similarly until all necessary pegs are in place.

Note It is important on completion to count the courses from top to bottom in both directions around the trenches. If the number of courses is not equal the result may be a 'course in pig' (figure 2.52).

To explain, a course in pig is what could occur when bricklayers set up opposite corners in their own way without first coming to an agreement. For example, if the total height from concrete foundation to d.p.c. height is 720 mm and the bricks are 65 mm thick, what would the gauge be? We could use nine courses at a gauge of 80 mm, totalling 720 mm, or ten courses at a gauge of 72 mm, again totalling 720 mm. Figure 2.53 shows the result of setting up using different gauges.

Further Levelling

The use of the spirit level and long straightedge and water level have been discussed previously; we will now consider the Cowley level and the tilting level.

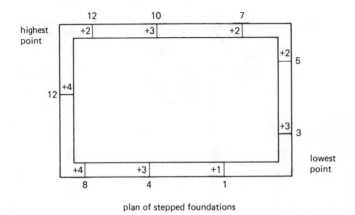

plan of stepped foundations

Figure 2.52

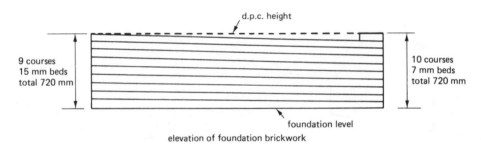

elevation of foundation brickwork

Figure 2.53

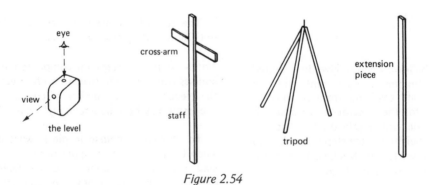

Figure 2.54

The Cowley Level

The level itself is a compact metal case containing a dual system of mirrors. This is placed on top of the tripod and directed at the staff, for which there is an extension piece if required (figure 2.54).

When the level is placed on top of the tripod a clamp is released within the level and it is ready for use; there are no adjustments on the instrument. To take a reading, look into the aperture on top of the instrument, which is rotated in the direction of the staff (figure 2.55).

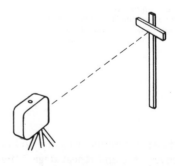

Figure 2.55

When viewing the staff, the cross-arm usually appears as in figure 2.56 *a* or *b*. If this is the case the cross-arm is not level with the instrument and must be moved up or down until it appears horizontal as in *c*. The circle may be distorted as in *d, e* or *f*, but this does not make sightings impossible and the view in *f* is level with the instrument. When a level has been obtained the staff man clamps the cross-arm in place and this level can be transferred where required. If a rise or fall is wanted the cross-arm is raised or lowered

by the difference in level and clamped in place. A peg is now hammered in where required so that when the staff is placed on top, the cross-arm appears horizontal when sighted through the instrument.

The Cowley level is ideal for levelling the tops of pegs in a foundation or as a check on d.p.c. height, for example. It is said to be accurate to 3 mm in 30 m. The staff is marked every 5 mm on the back so that the staff man can easily read the difference between two or more points (figure 2.57).

Note   It is important not to move the Cowley level while the instrument is located on the tripod since this will damage the mechanism. Always separate these two before moving.

The Tilting Level (figure 2.58)

This is a relatively modern instrument sometimes known as the quickset owing to the speed of setting

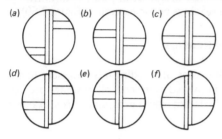

Figure 2.56

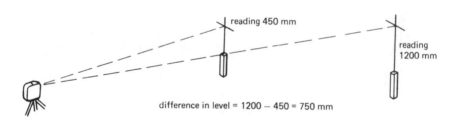

Figure 2.57

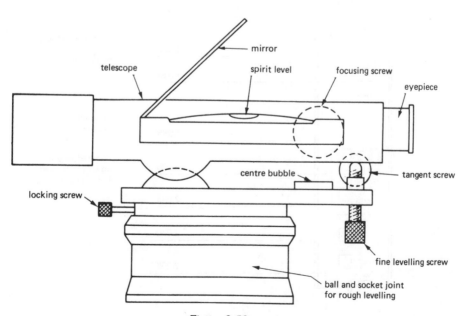

Figure 2.58

up. It is usually available mounted on a ball-and-socket joint on a tripod and is used in conjunction with a folding or telescopic staff.

The tripod is firmly set up where required so that the eyepiece will be approximately level with the user's eye. He should not have to stoop to take readings. The level is placed on top of the tripod and tightened up from underneath when the centre bubble shows correct. Sight on to the staff held about 40 m away and rotate the eyepiece until the cross-hairs are at their sharpest. Focus as necessary and, using the tangent screw, move the instrument so that it is looking at the centre of the staff. Next adjust the spirit level by rotating the fine levelling screw.

Note   The spirit level must be checked and adjusted as necessary before each reading.

Before using the quickset level it is necessary to practise reading the staff, which is approximately 4 m long. Alternate metres may be marked in black and red, and these are split into tenths and hundredths. The hundredths have to be split up by eye, but this is a fairly easy task. It is important that the staff is held in an upright position or mistakes in readings will occur as shown in figure 2.59.

When viewed through the level the staff will be seen to read upside down, but, with a little practice, taking readings becomes simple (figure 2.60).

Taking Readings

For all classes of levelling it is necessary to start from some known datum and all readings are related to this. A datum can be assumed; for example, some

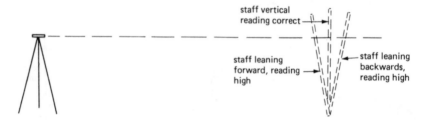

Figure 2.59

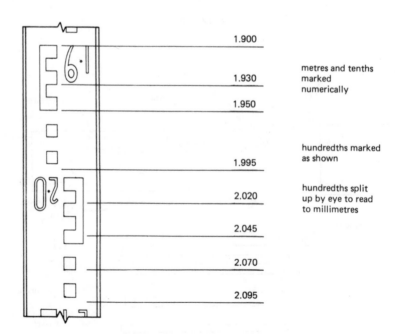

Figure 2.60

immovable object at or near the starting point such as a manhole cover is given a value of say 10.000. It is important never to start from zero since if the ground falls below this a minus reading can be recorded.

The method is as follows.

(1) Set up the level so that all the points to be 'read' can be seen if possible.
(2) Hold the staff on datum; take a reading.
(3) Move the staff on to the other points in turn and record the readings.

The first reading taken from an instrument position is called a *backsight* (B.S.), the last reading from an instrument position is called a *foresight* (F.S.) and all readings in between are *intermediate sights* (I.S.).
Readings are 'booked' as they are taken and may be reduced in one of two ways: the 'collimation height' method or the 'rise and fall method', details of both of these methods being given in advanced texts.

Using the tilting level is the best method of determining the difference in level of two points some distance apart. While the Cowley level may be used

for this purpose, it has neither the accuracy nor the range. For example, if it should be required to find the difference in level between a rainwater gully and a main sewer some 150 m apart, simply set up the instrument approximately mid-way between the two points and hold the staff on each point in turn (figure 2.61).

As a second example, consider a datum peg, fixed close to the entrance to a large site and this level has to be transferred to the other end of the site, a distance of 100 m. Again, set up the instrument about mid-way between these two points and take a reading on the datum peg. Assume this is 1.650. Next sight the instrument in the direction where the new peg is required and adjust the height of this until the staff reads 1.650 (figure 2.62).

If the ground rises or falls to any extent between these two points it would be simpler to adjust the height of peg 2 in relation to peg 1. For example, assuming peg 1 has a value of 10.00 and the ground falls 2 m in 100 m, peg 2 could have a value of 8.00, in which case the reading at peg 2 would be 1.65 + 2.00 = 3.65 (figure 2.63).

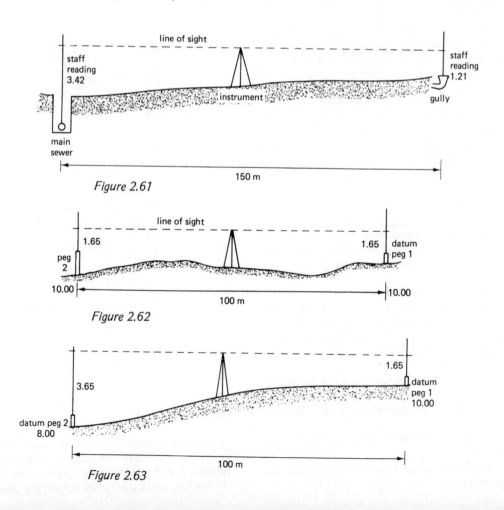

Figure 2.61

Figure 2.62

Figure 2.63

SETTING OUT PROJECTIONS

When setting out trenches for buildings that have projections, it is usual for the type of foundation structure to determine the method of setting out. Most strip foundations are now excavated by mechanical plant, that is, diggers, mechanical shovels, etc. Excavation of small, circular or splayed projections cannot often be carried out efficiently with mechanised equipment, therefore hand excavation is required.

Projections are set out by using one of the following methods

(1)   trammel
(2)   templet
(3)   triangulated frame.

The trammel or radius rod method can be used accurately for any circular work having a radius not exceeding 3 m. When it is used to describe an arc in excess of 3 m the trammel rod becomes unwieldy and the pivot needs to be increased in diameter and substantially stabilised.

The templet method has the advantage that it can be used efficiently for setting out splayed, segmental or semicircular projections; because it is not a fixed position method it is most commonly used.

The triangulated frame method is usually reserved for circular work when the radius point is inaccessible. Because of the length of framework required and the skill needed for its application it is not often used; natural site hazards also prevent its use.

The trammel method of setting out is based on a knowledge of the circle and its application, recognising that the trammel is the radius of a circle.

Definitions and Terms

The circle is a plane figure within the confines of a curved line termed the circumference; all points on the circumference are equidistant to the centre of the circle.

Arc   This is part of the circumference.
Chord   This is a straight line, less than the diameter, terminating at both ends on the circumference.
Diameter   A straight line, terminating at both ends on the circumference and which passes through the centre of the circle.
Normal   A straight line from any point on the circumference which is radial to the centre of the circle.
Quadrant   This is the quarter of a circle both in shape and area.
Radius   A straight line from the centre point to the circumference.
Sector   A part of a circle bounded by two radii and the circumference.

Semicircle   Half a circle in shape and area.
Segment   This is the part of a circle contained between a chord and an arc.
Tangent   A straight line meeting the circumference at one point and also at right-angles to the normal at that point. The circumference is approximately 3.142 or $3\frac{1}{7}$ times the diameter in length. See figures 2.64 or 2.66.

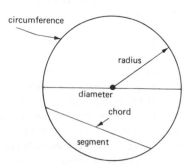

Figure 2.64

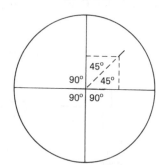

Figure 2.65

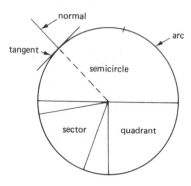

Figure 2.66

Trammel Method

When setting out segmental bay windows using the trammel, the length of span and the radius are required. For the semicircular bay only the span length is needed, since the radius is half the diameter (see figure 2.64). Segmental projections can be set out when the span length is provided and the radius is omitted but the amount of rise is indicated; this is achieved by using the theorem of intersecting chords (see figure 2.67). That is

$$a \times b = c \times d$$

In figure 2.68, calculate the length of the radius rod

$$0.5 \times x = 1.5 \times 1.5$$

Therefore

$$0.5 \times x = 2.25$$

$$x = \frac{2.25}{0.5} = 4.5$$

diameter $0.5 + 4.5 = 5$ m

radius rod $4.5 \div 2 = 2.25$ m

In figure 2.69, the span of the bay window is 2.4 m and the rise 0.6 m; calculate the length of the trammel rod required.

$$0.6 \times x = 1.2 \times 1.2$$

Therefore

$$x = \frac{1.2 \times 1.2}{0.6} = 2.4$$

diameter $= 0.6 + 2.4 = 3.0$ m

radius $= 1.5$ m (length of trammel)

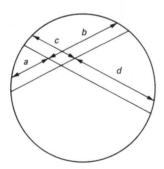

Figure 2.67

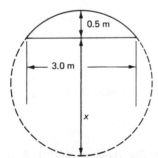

Figure 2.68

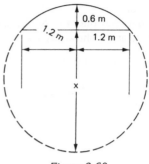

Figure 2.69

Another method of finding the length of a trammel is to use the formula

$$\text{radius} = \frac{(\tfrac{1}{2}\,\text{chord})^2 + \text{rise}^2}{2 \times \text{rise}}$$

For example, in figure 2.69

$$\text{radius} = \frac{(\tfrac{1}{2} \times 2.4)^2 + 0.6^2}{2 \times 0.6}$$

$$= \frac{1.2^2 + 0.36}{1.2}$$

$$= \frac{1.8}{1.2}$$

$$= 1.5 \text{ m (required trammel length)}$$

Setting out is illustrated in figure 2.70.

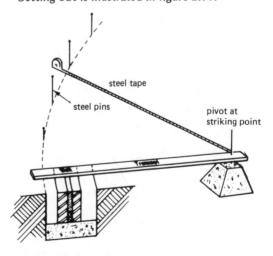

Figure 2.70 Setting out using trammel and measuring tape

Measurement of curves is often necessary when setting out buildings or when quantities are required. The following factors must first be determined

(1) length of radius
(2) the angle between the radii of the sector (this is termed theta and is shown by the symbol θ).

To obtain length of arc (curve) (figure 2.71):

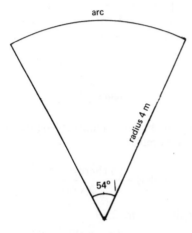

arc

radius 4 m

54°

Figure 2.71

length of arc $= 2\pi R \times \dfrac{\theta}{360}$

$= 2 \times 3.142 \times 4 \times \dfrac{54}{360}$

$= 3.77$ m

Calculate the length of the arc in figure 2.72.

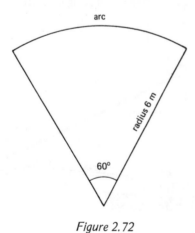

arc

radius 6 m

60°

Figure 2.72

length of arc $= 2\pi R \dfrac{\theta}{360}$

$= 2 \times 3.142 \times 6 \times \dfrac{60}{360}$

$= 6.284$ m

SETTING OUT ANGLES

When setting out splayed projections it is necessary to have some knowledge of acute and obtuse angles and how these can be produced.

An acute angle lies between 0° and 90°; an obtuse angle between 90° and 180° and a reflex angle between 180° and 360° (see figures 2.73 to 2.75).

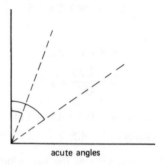

acute angles

Figure 2.73

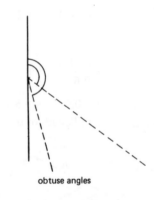

obtuse angles

Figure 2.74

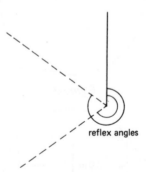

reflex angles

Figure 2.75

An Angle of 45°/135°

Measure past the required corner A for a known distance, say 600 mm, to B. Square off from this point for the same distance to C and join AC (figure 2.76).

An angle of 60°/120°

Measure past the required corner A for a known distance, say 600 mm, to B. Describe arcs intersecting at C using A and B as the striking points and with the radius equal to AB (600 mm). Joining these points will produce an equilateral triangle (all internal angles 60°) (figure 2.77).

Angles of 22½°, 30°, 67½° and 75°

These can be obtained simply by bisecting the appropriate one of the above angles (figure 2.78).

Any angle

Any angle can be set out using the tangent method (see an advanced text).

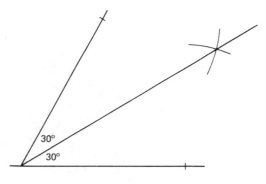

Figure 2.78

The previous work has described the traditional methods that can be used to set out projections, but modern setting-out techniques require economy in labour and the use of excavating equipment, which results in the templet method being used whenever possible. A modern method often used on site is the construction of a rectangular frame; this is fixed in position and the area within its confines is completely excavated, the concrete foundations are placed and the brickwork is then set out with another templet designed to the actual size and shape of the required projection.

The rectangular templet can also incorporate the actual shape of the projection as shown in figure 2.79.

When building firms are engaged in the erection of a considerable number of houses, each having the same type of projection, the usual practice is to obtain an actual size templet from the manufacturer or stockist supplying the window frames (figure 2.80).

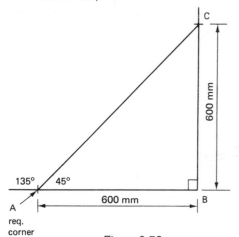

Figure 2.76

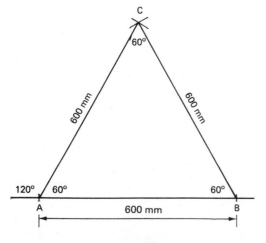

Figure 2.77

The Triangulated Frame

This method is used when the radius length is excessive or when the radius point is inaccessible. Pegs are fixed on the building line indicating the starting points on each side of the curve; another peg is also fixed at the centre of the curve and at right-angles to the building line. A lightweight timber frame is then constructed as shown in figure 2.81. The frame is then placed against the three pegs, peg A is removed, and the frame can then be moved across while keeping it pressed against the two side pegs, using the point of peg A to describe the required arc.

This method is difficult to operate single-handed since the length of the framework and the unevenness of many sites often cause considerable problems for the operators of the frame.

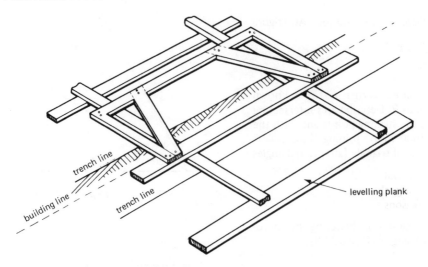

Figure 2.79 Rectangular templet incorporating splay

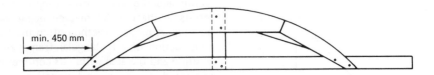

Figure 2.80 Segmental bay templet

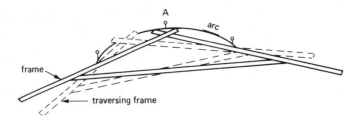

Figure 2.81 Using the triangulated frame

3
EXCAVATING AND TIMBERING

TRENCHES

When trenches are excavated for works such as foundations or drainlaying, the sides may have to be supported. The main reasons for carrying out timbering to trench sides are

(1) to provide safe working conditions for operatives
(2) to prevent disruption of work due to collapse of the trench sides
(3) to prevent damage to adjacent property which may occur if trenches are got out too close to

structures such as boundary walls, etc.; if the trench bottom is below the foundation the wall may become unsafe and will possibly overturn.

It is obvious from figure 3.1 that if the trench shown is not timbered the boundary wall will almost certainly collapse in the direction of the trench. (Note the direction of the load dispersal lines.)

Timbering is usually carried out by operatives as work proceeds and must be done under the supervision of a competent person. Different methods are necessary for different subsoils and the following figures show common methods of timbering down to a depth of approximately 1.5 m.

Firm Ground

A system of open timbering will normally be sufficient; either of the two examples shown in figures 3.2 and 3.3 would be adequate.

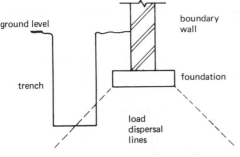

Figure 3.1

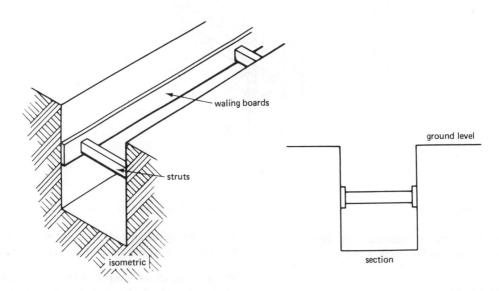

Figure 3.2

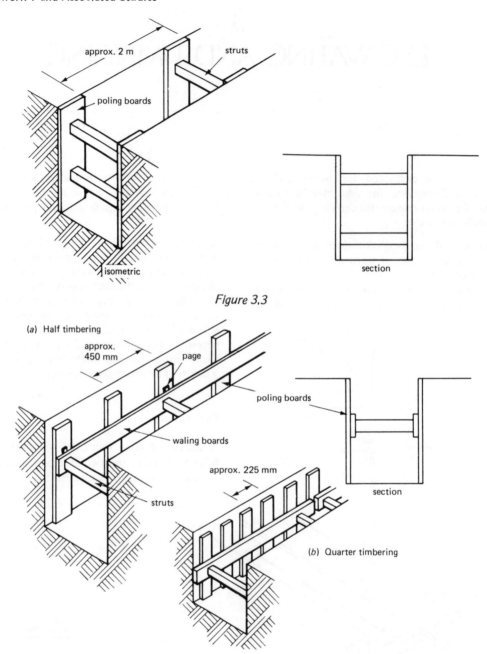

Figure 3.3

Figure 3.4

Moderately Firm Ground

Either of the methods illustrated in figure 3.4 is suitable, depending on the circumstances.

Loose, Dry Ground

Close timbering is required, that is, poling boards close together, walings and struts, or a system known as sheeting as shown in figure 3.5.

Note   An adjustable strut as shown in figure 3.6 provides a useful alternative to using timber struts.

Where trenches are deeper, many other members are used, some of which are shown in figure 3.7. Again, pages would be used to tighten the runners to the waling boards and folding wedges can be used as required to tighten up the puncheons.

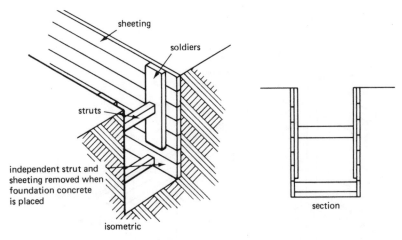

Figure 3.5

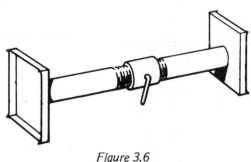

Figure 3.6

A brief description of the members mentioned so far is given below.

Poling boards　Short, vertical members in direct contact with the trench sides. May be close or open depending on the type of subsoil. Sectional size usually about 225 x 38—50.

Waling boards　Long, horizontal members placed up against the polings to prevent their movement. Sizes vary from 100 x 75 upwards.

Struts　Strong, horizontal members at right-angles to and in contact with the walings, cut to length to hold these apart at the required distance. Sizes vary from 100 x 100 upwards.

Trench props　Resemble small Acrow props and can be used in place of struts.

Strutlips　Short timber sections fastened to the tops of the struts and resting on the walings to prevent the struts from falling into the trench should the system become loose (figure 3.7).

Puncheons　Vertical members fastened between walings to prevent their downward movement.

Sheeting　Horizontal boarding which may be used

in loose soil for example. These are placed next to the trench sides and are supported by soldiers and struts.

Soldiers　Short vertical supports for sheeting. These are held apart by struts or trench props.

Runners　Long, vertical boards not less than 50 mm thick, which may be used in place of polings in unstable ground for deeper trenches. Their lower edges are chisel-shaped and they are driven downwards as excavations proceed.

Wedges　Tapered timber sections used in pairs for tightening up a system, for example, between struts and walings (at one end only) or at the base of puncheons to tighten these between walings.

Pages　Narrow wedges, preferably of hardwood, used singly, for example to tighten poling boards to walings.

Safety in Excavations

The Construction Regulations include the following rules for work in excavations

(1) An adequate supply of timber of suitable quality or other suitable support to be provided.

(2) Every part of an excavation to be examined at least once a day while persons are employed there. The working end of any trench to be inspected by a competent person before the beginning of each shift.

(3) All timbering to be carried out by a competent person and any alterations to be supervised by a competent person.

(4) A suitable barrier or guard rail to be provided to any excavation over 2 m deep.

(5) Materials must not be stacked near the edges of excavations if they endanger men working below.

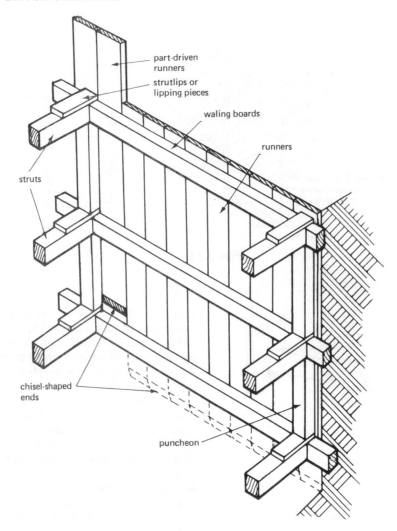

part-driven runners

strutlips or lipping pieces

waling boards

runners

struts

chisel-shaped ends

puncheon

Figure 3.7

(6) Plant and equipment not to be in the vicinity of an excavation if there is any likelihood of collapse.

Some of the main points to be considered when a trench is to be excavated are as follows.

Nature of the Ground

Rock may need no support at all down to considerable depths whereas loose sandy soil will need support almost from the start. Table 3.1 (from CP 2003) gives an indication of the timbering required at different depths, where the nature of the ground is uniform.

Depth of Trench

Many subsoils are self-supporting down to normal strip foundation depths (approximately 750 mm) but the deeper the trench the more likely the need for support.

Presence of Ground Water

When excavating below the ground-water table and water is standing in the trench the likelihood of collapse is much greater in many subsoils, especially sands and clays, etc.

Table 3.1

Type of Soil	Depth of Excavation		
	To 1.5 m deep	1.5 to 4.5 m	over 4.5 m
Soft peat	C	C	C
Firm peat	A	C	C
Soft clays and silts	C	C	C
Firm and stiff clays	A*	A*	C
Loose gravels and sands	C	C	C
Packed gravels and sands	A	B	C
Gravels and sands below water table	C	C	C
Fissured rocks, shales, etc.	A	A*	B
Sound rock	A	A	A

A. No support required.
B. Open timbering.
C. Close sheeting or similar.
* Depends on site conditions.

Proximity of Buildings

All structures exert pressure on the subsoil whether it is a free-standing boundary wall or a dwelling house. This pressure is exerted downwards and outwards at 45° from the foundation base and any trench within this pressure zone will require stronger timbering than usual.

Presence of Spoil Heaps

It is recommended that spoil heaps and materials, etc., are kept at least 1200 mm away from trench sides because of the pressure they will exert on the trench sides.

Can Trench Sides be Splayed?

The need for timbering can be completely obviated if this is the case but it will entail much more excavation, storage and backfilling on completion.

How Long is the Trench to Remain Open?

Many subsoils are self-supporting to some extent and if the excavation is to be filled in the same day little support may be necessary, for example, pinchers at 2 m centres. If, on the other hand, it is to remain open much longer than this, collapse of the sides can be expected unless further timbering is provided.

Prevailing Weather Conditions

Wet weather will often bring down trench sides much more quickly than dry conditions. It is therefore important that this factor is taken into consideration when timbering. Here again clay subsoils are a problem since clays swell when wet and shrink when dry. Thus timbering may become loose in hot weather unless regularly tightened, whereas in wet weather members may fail owing to increased pressure.

Table 3.2 gives some indication of the type of timbering required in different soils and the length of time a vertical face may be expected to remain standing without support.

Stacking of Timber

When not in use, timber for trenches should be carefully stored and not simply thrown in an untidy heap. Figure 3.8 shows an adequate method of stacking. This should be kept clear of the ground and covered against the elements by a waterproof cover.

Table 3.2

Type of Ground	Length of time an *Unsupported* Face may be Expected to Remain Standing	Type of Timbering Required
Very hard chalk and rock	Indefinitely	None or open
Hard chalk and firm ground	10–15 h	Half
Soft chalk, sand and gravel and clayey ground	3–8 h	Quarter
Loose soil, made-up ground, soft clay and earth	1–3 h	Close
Waterlogged ground	None	Close

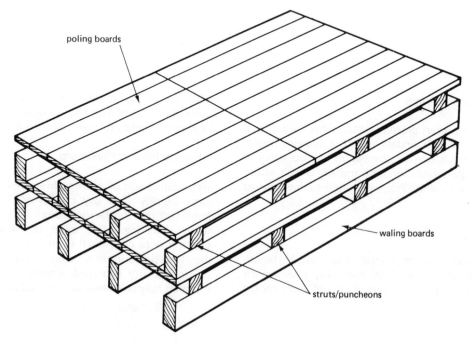

Figure 3.8

SMALL BASEMENTS

When carrying out excavating and timbering for small basements, the choice is normally between the following four methods, each method having its advantages and disadvantages.

(1) Excavate the basement and slope the sides out-wards to completely avoid the need for propping.

(2) Get out the whole of the excavation and support the sides from the bottom.

(3) Excavate the trenches only, build the external walls and when these have 'set' take out the rest of the spoil.

(4) Use sheet piles and support with raking shores.

Sloping Sides (figure 3.9)

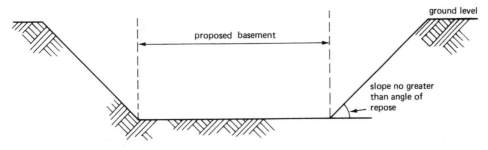

Figure 3.9

The angle at which the sides may be sloped will depend on

(1) The type and condition of the subsoil and its angle of repose. When a material is tipped or excavated the sides will eventually settle to a certain angle depending on the type of material and its condition (wet or dry). This is known as the angle of repose.
(2) The area of land available to accommodate the slope. Advantages include large savings in timbering and freedom from props, but a great deal of extra excavating is required, coupled with the need for a large storage area for the spoil; then the backfilling often rules out this method unless the sides are stable at greater angles, for example, rock or similar subsoil.

Excavate Completely and Support the Sides from the Bottom

This method can be carried out as follows.

(1) Excavate as much spoil as possible down to the required depth, leaving a 'wedge' of earth, known as a berm, supporting the sides (figure 3.10).
(2) Excavate vertically from ground level as far as possible, depending on the subsoil, leaving a smaller berm to support the rest. Fix the top set of polings, walings and props, etc. (figure 3.11).

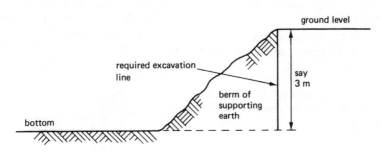

Figure 3.10

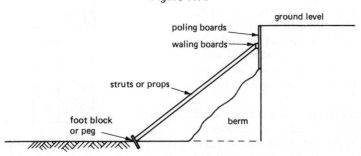

Figure 3.11

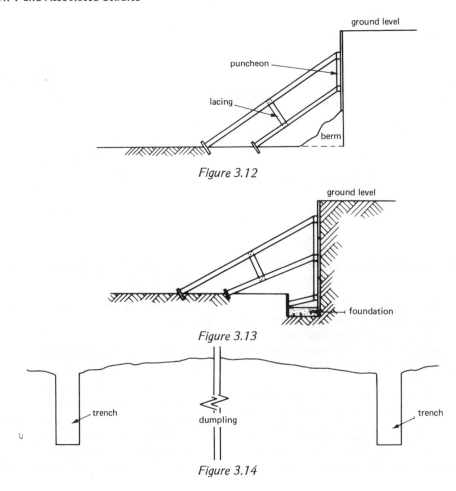

Figure 3.12

Figure 3.13

Figure 3.14

(3) Excavate vertically as in (2) and fix the second set of timbers (figure 3.12).

(4) Excavate the remainder, including the foundation, complete the propping and place the foundation concrete (figure 3.13).

As the walls are built the timbers can be removed.
Note   The timbering must not be removed while the brickwork is still 'green'.

Dumpling Method

Ordinary trench timbering is all that is required using this method. Poling boards are used in preference to runners since this facilitates removal as the walls are erected. The periphery trenches are excavated wider than usual and a dumpling of earth is left inside the proposed basement area, which is got out later (figure 3.14).

The foundation, including part of the base slab, is cast, leaving reinforcement projecting on the inside to provide continuity with the rest of the slab, which is cast when the substructure brickwork is completed

and fully hardened, and the dumpling removed.

The lower timbering is removed and backfilling on the outside is begun. To even the pressure the walls can be strutted off on the inside until they gain sufficient strength (figure 3.15).

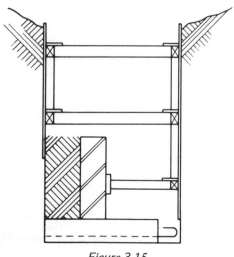

Figure 3.15

Sheet Piling

Interlocking rolled steel sections known as sheet piles are driven into the ground to a depth somewhere below that of the basement floor level, and in advance of any excavating work. Further details of this method will be found in an advanced text.

Excavating Equipment

Hand excavation is out of the question for small basement work — unless a machine cannot get to the job because of nearby buildings, etc. Probably the most common excavator in use today is the combined digger and loading shovel such as the J.C.B.3, etc. This is a rubber-tyred machine which can be brought to site under its own power and has a limit of reach of 4 m, getting out about 9 m^3/h, depending on subsoil type. The J.C.B.6 is a larger machine on crawler tracks and is brought to site on a low loader. It can work in very bad ground conditions, has a reach limit of 7 m and a capacity of up to 40 m^3/h. Other excavators include the face shovel, which works from the bottom of the excavation, the back actor, the dragline — for large areas, the crane and grab and the multi-bucket trench digger.

'BULKING' OF EXCAVATED MATERIAL

When a trench is excavated for a 100 mm drain and backfilled on completion, there is invariably a certain amount of spoil to be disposed of. Subsoil which has been *in situ* for many years will have become compacted, especially the harder strata such as rock or chalk. When it is excavated it tends to 'bulk' as it is broken up and despite hand or mechanical compaction rarely becomes as compact as before. Obviously different subsoils will bulk to different extents and these differences have been calculated and tabulated as follows

1 m^3	sound rock becomes	1.75 m^3
1 m^3	chalk becomes	1.33 m^3
1 m^3	earth becomes	1.25 m^3
1 m^3	clay becomes	1.25 m^3
1 m^3	sand becomes	1.1 m^3
1 m^3	gravel becomes	1.1 m^3

For example, if 2 m^3 earth are excavated, the amount to be carted away will be

$$2 \times 1.25 = 2.5 \ m^3$$

Experiment to Show that Subsoils 'Bulk' on Excavation

Requirements

Flat tray about 450 x 300 with 50 mm sides; piece of polythene, sufficient to line the tray; stone, sand and cement, trowel and hammer; two short straight-edges, one to fit the tray as shown in figure 3.16, the other to span the tray and rest on the sides.

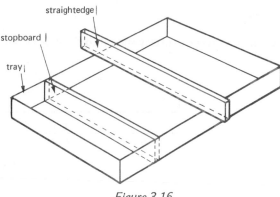

Figure 3.16

Method

Make a weak mix of stone, sand and cement, fill the tray and level with the longer straightedge. When set and hardened, tip out and smash into lumps, taking care not to lose any. Refill the tray with the broken pieces, making sure none protrudes above the top of the tray, that is, using the longer straightedge as a screedboard. Tip out and put in the remainder, using the small straightedge as a stopboard, once again keeping the top level with the sides of the tray. This simple experiment can be likened to excavating sound rock or similar and placing in a lorry for carting away. It will be obvious that the volume increases considerably, depending to some extent on how small the pieces were.

Example 3.1

If 20 m^3 chalk is excavated from a foundation, how much cart-away volume will this produce assuming none is to be backfilled?

$$\text{Cart-away volume} = 20 \times 1.33$$
$$= 26.6 \ m^3$$

Example 3.2

Same problem in sound rock.

$$\text{Cart-away volume} = 20 \times 1.75$$
$$= 35 \ m^3$$

Example 3.3

Same problem in sand.

$$\text{Cart-away volume} = 20 \times 1.1$$
$$= 22 \text{ m}^3$$

Obviously to calculate the cart-away volume for any subsoil, first find the amount of excavated material and multiply this by the correct bulking factor from the table given.

Example 3.4

A trench is excavated in clay subsoil to the dimensions given in figure 3.17. Calculate the cart-away volume.

Example 3.5

Where the trench bottom slopes, the simplest method is to obtain the average depth and the formula is as shown below (figure 3.18).

A drainage trench is excavated to the dimensions given in figure 3.18. Subsoil is chalk. Calculate the cart-away volume to two places of decimals.

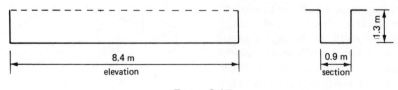

Figure 3.17

$$\text{Excavated volume} = l \times b \times d \qquad \text{(formula)}$$
$$= 8.4 \times 0.9 \times 1.3 \quad \text{(figures)}$$
$$= 9.828 \text{ m}^3 \qquad \text{(answer)}$$

Cart-away
$$\text{volume} = 9.828 \times 1.25 \text{ (answer x multiplication factor)}$$
$$= 12.285 \text{ m}^3 \qquad \text{(final answer)}$$

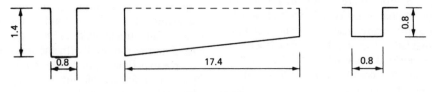

Figure 3.18

$$\text{Excavated volume} = l \times b \times \text{average } d$$

$$= 17.4 \times 0.8 \times \frac{(1.4 + 0.8)}{2}$$

$$= 17.4 \times 0.8 \times \frac{2.2}{2}$$

$$= 17.4 \times 0.8 \times 1.1$$

$$= 15.31 \text{ m}^3$$

$$\text{Cart-away volume} = \text{excavated volume x multiplication factor}$$

$$= 15.31 \times 1.33$$
$$= 20.36 \text{ m}^3$$

Example 3.6

A basement excavation measures 17 m x 9 m x 8 m deep. If the subsoil is clay calculate the cart-away volume.

$$\text{Volume of basement} = l \times b \times d$$
$$= 17 \times 9 \times 8$$
$$= 1224 \text{ m}^3$$
$$\text{Cart-away volume} = 1224 \times 1.25$$
$$= 1530 \text{ m}^3$$

When it is required to calculate the volume of spoil to be excavated from such works as a house foundation, the best method is first to find the length of centre line of the trench and multiply this by the depth and the width. The centre-line length is calculated as follows.

Where the External Dimensions are Given

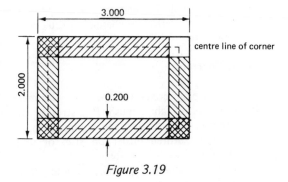

Figure 3.19

Length of centre line = 2 x length + 2 x breadth
− 4 x trench width
(2*l* + 2*b* − 4 x trench width)

Figure 3.19 shows that twice the length plus twice the breadth will give the corners twice and therefore the four corners must be deducted. To obtain the length of the corner is simple: looking at the top right-hand corner in figure 3.19 the length of the centre line at this point is equal to half the trench width plus half the trench width, which is of course equal to the trench width.
In the above case

$$\text{Centre line} = 2l + 2b - 4 \times \text{trench width}$$
$$= 2 \times 3 + 2 \times 2 - 4 \times 0.2$$
$$= 6 + 4 - 0.8$$
$$= 10 - 0.8$$
$$= 9.2 \text{ m}$$

Where the Internal Dimensions are Given

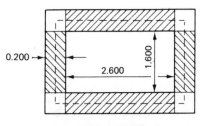

Figure 3.20

Length of centre line = 2 x length + 2 x breadth
+ 4 x trench width

Figure 3.20 shows that twice the length plus twice the breadth will not give the overall centre-line and therefore the four corners must be added this time. Finding the centre line of the corner is performed exactly as explained previously.
In the above case

$$\text{Centre line} = 2l + 2b + 4 \times \text{trench width}$$
$$= 2 \times 2.6 + 2 \times 1.6 + 4 \times 0.2$$
$$= 5.2 + 3.2 + 0.8$$
$$= 9.2 \text{ m}$$

Not surprisingly the answer is the same as in the first example, since it is the same trench.

Note The centre line of brickwork is obtained in this manner.

4

CONCRETE

Most of the materials used in the construction industry today are manufactured under factory conditions. This implies that the majority of manufacturers exercise strict quality control in the production of their goods, which are tested to conform to a British Standard Specification. Concrete is one of the very few materials manufactured in any great quantity on site, very often under adverse weather conditions, and therefore it is essential that building operatives concerned with its production are knowledgeable about the recommended principles of manufacture.

GENERAL PRODUCTION

Concrete is produced by mixing together

(1) cement powder
(2) water
(3) mineral aggregates.

Admixtures are available to improve the properties of the concrete in various ways if carefully used. The mixture is placed in a trench, container or mould, compacted and allowed to harden. The mix can be regarded as being made up of two parts

(1) the aggregates — stone and sand
(2) cement paste — water and cement

The cement paste covers the surface area of every stone and sand particle in the mix, binding them together when the concrete hardens. The aggregates are not altered in any way, but firmly set into a rock-like hardened cement paste.

The cementing material is formed by the water and cement powder combining together chemically by a process termed 'hydration', during which heat is evolved. The chemical reaction takes place quite slowly and the cement paste becomes harder and stronger as it matures, provided that the temperature and moisture conditions are favourable. If the concrete is allowed to dry out completely, hydration will cease and the strength increase will cease. The longer the concrete remains in a damp condition, up to 28 days depending on the type of cement used, the stronger it will become. Disintegration of the cement paste at any age can be brought about by atmospheric conditions or chemical action and it is therefore necessary to ensure that it is protected and of sufficient strength to withstand such attacks.

TYPES OF CONCRETE

Plain Concrete

This consists solely of cement powder, water, and graded coarse and fine aggregates. No reinforcement is used, and it can be manufactured on site, or can be purchased from a ready-mixed concrete company.

Uses include simple foundations, garden paths and drives, paving slabs, kerbs and channels, protection of drainage pipes, etc.

Reinforced Concrete

This consists of plain concrete reinforced with metal, usually steel bars or fabric mesh. It is stronger than plain concrete in both tension and compression, and it can be manufactured either on site, or under factory conditions away from site.

Uses include foundations, walls, columns, lintels and beams, floors, roofs, etc.

Precast Concrete

This is usually in the form of some kind of unit which can be manufactured either on or away from the site. The unit is made in some other place than that which it is to permanently occupy. It can be plain, reinforced or prestressed.

Uses include bricks, blocks, cladding panels, pad stones, copings, window sills, canopies, chimney caps, flue liners and all types of structural unit, etc.

Note

When the concrete is placed on site in the position where it is to remain permanently, it is termed *in situ* concrete.

Prestressed Concrete

High-strength concrete can be precast, or cast *in situ*,

cast on site or in a concrete works. The concrete, usually a structural unit, is given a very high tensile strength by tensioning high-tensile wire or cables while the concrete is either (1) fresh – pretensioned or (2) hardened – post-tensioned.

Uses include large span beams, bridges and elevated motorways, water-retaining structures, etc.

No-fines Concrete

This is concrete composed of coarse aggregate only, cement and water. It is usually made on site.

Many large national building companies are now using this technique for the production of pre-cast or cast *in situ* structural units.

Lightweight Concrete

This is made with lightweight aggregates and cement paste. It is used when it is necessary to produce concrete of a low thermal conductivity, and low density.

Uses include lightweight concrete blocks, flue blocks, lining existing flues and chimneys, floor and roof screeds.

Note

Lightweight concrete blocks can be manufactured using foamed concrete. The concrete mix is made up of pulverised fuel ash (P.F.A.) and sand to form the aggregates, with cement paste and a foaming agent. Aluminium powder can be introduced to the mix, causing it to rise, and creating minute air bubbles which form a cellular structure in the hardening concrete cake.

Heavyweight Concrete

This is concrete made of heavy aggregate, for example, barytes, pig iron, steel punchings, etc.

Uses include protection from radioactivity, since the material acts as a shield against radiation.

Flow Concrete

This is a concrete with an additive included in the mix, which causes the resultant concrete to have a very high workability. 'Mighty 150' is one additive that can be used to produce a very wet concrete with no apparent loss of strength in the hardened concrete.

Flow concrete is used for thin deep heavily reinforced sections.

PROPERTIES OF CONCRETE

In its fresh state, concrete should

(1) be composed of accurately batched (measured) proportions of cement, fine and coarse aggregate, and water
(2) have the correct cement–aggregate ratio
(3) have the required water–cement ratio. This refers to the proportion of water to cement by weight and it is an essential factor in the production of concrete. Actually a 0.3 water–cement ratio is all that is required to hydrate the cement, but more water than this is necessary for workability. Usually 0.45 is approximately the minimum water cement ratio that can be used to provide sufficient workability, depending on the method of compaction and the use to which the concrete is being put
(4) be well mixed
(5) not segregate during transportation and placing
(6) be fully compacted since voids cause weakness
(7) be provided with the specified finish
(8) be well cured, in other words kept in a damp condition as long as specified since premature drying out can cause deterioration.

When concrete is completely hardened it should be

(1) durable
(2) of the required strength
(3) of the correct density
(4) impermeable to water
(5) resistant to friction
(6) free from shrinkage cracks.

The ensuing information attempts to explain how to achieve all the above requirements.

MATERIALS FOR CONCRETE

Cement

Several types of cement powder have been developed since Joseph Aspdin patented Portland cement in 1824. Each type produced is required to conform to a British Standard Specification, and is for use in a particular situation.

Generally a site operative cannot distinguish by visual inspection between the various types unless the cement being used is white, coloured or high alumina, which is blue–black in colour.

Cement powder is the most expensive material used in the production of concrete and the most common type is ordinary Portland cement (O.P.C.), which is manufactured to conform to BS 12.

Ordinary Portland Cement

There are two methods of manufacture: the wet process, used where the basic materials are wet as in the case of clay; and the dry process, which uses shale in place of clay. An outline of the wet process is as follows.

Calcium carbonate in the form of chalk or limestone is crushed and mixed with washed clay to form a slurry, which is ground in a ballmill. This is then sieved and any coarse particles are returned to the ballmill for re-grinding. The slurry is then fed into large storage tanks where it is kept agitated to prevent settlement. From here it is fed into the top end of a long, sloping, cylindrical kiln which rotates slowly on its axis. As the slurry passes down the kiln it is gradually heated and all the water is evaporated. The hottest part of the kiln is about 1500 °C, where it is heated to such a degree that a chemical reaction takes place and cement clinker is formed. From the kiln it is transferred to a clinker store from which it is fed into a ballmill where it is finely ground and 3 to 7 per cent gypsum is added. The process is complete at this point and the cement powder is transferred to storage silos, then bagged and discharged or sent out in bulk.

The process is simply illustrated in figure 4.1.

Note   The addition of gypsum is to control setting time and is now explained in detail.

Testing

O.P.C. has to conform to the requirements of BS12, which includes the following tests

(1) fineness
(2) chemical composition
(3) strength development
(4) setting time
(5) soundness.

While tests 1, 2 and 3 are usually considered the prerogative of laboratories with considerable experience in this type of work, the tests for setting time and soundness can be carried out with a little care by students.

Setting Time   The setting time of concrete must be sufficient to allow for mixing, transporting, placing, compacting and finishing before any stiffening takes place. Ground cement clinker normally sets very rapidly and it is for this reason that gypsum is added at the dry grinding stage since it has a retarding action on the setting process.

Setting and hardening are part of a continuous process, setting referring to the stiffening of the

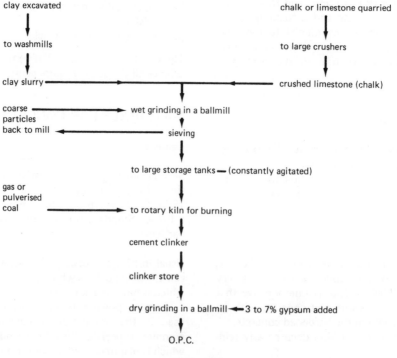

Figure 4.1

cement paste while hardening results in the development of strength which, though rapid at first, may continue slowly for a number of years.

Cement is said to set in two stages, the initial and the final set. BS 12 requires O.P.C. to have an initial setting time of not less than 45 min, and a final set of not more than 10 hours. The test is carried out using the Vicat apparatus (figures 4.2 and 4.3).

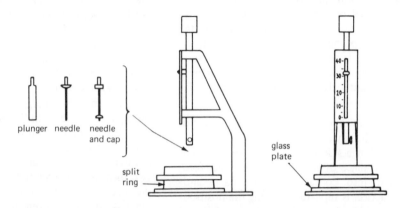

Figure 4.2 The Vicat apparatus

Figure 4.3 The Vicat apparatus

To carry out the test it is first necessary to determine the amount of water to mix with the cement powder since a standard paste must be used. To do this a 10 mm dia. plunger is fitted in the Vicat and a trial paste is made and quickly filled into the split ring, which is placed on a glass base. The plunger is brought into contact with the surface of the paste and released. The plunger should settle to within 5–7 mm of the base. If penetration is too deep or too shallow, the amount of water is adjusted and further pastes are made until the depth of penetration is correct.

Note   It is usual to mix cement and water to a ratio of 3:1 in the first instance.

Having arrived at a standard paste, a needle 1 mm square is used in place of the plunger, and as before this is brought into contact with the paste and released. This action is repeated, moving the position of the split ring each time, until the needle settles to approximately 5 mm of the base. The period between the time when the water was first added to the cement to the time when the needle settles to the required depth is the initial setting time. It may be well over the permitted minimum of 45 min but this is considered quite satisfactory.

For the final setting time test the needle and cap are fitted in the Vicat in place of the needle. The needle is again 1 mm square and projects 0.5 mm below the cap. The needle is brought into contact with the surface and the final set is said to have taken place when the needle makes an impression but the cap does not. The time between adding the water to the cement and this occurring must not exceed 10 hours.

Soundness   If too much lime is used in the manufacture of cement it may not all combine with the other constituents and a small amount of free lime may be present in the cement powder. When water is added this lime is liable to slake slowly and the resulting expansion may damage the concrete. The soundness test can be carried out to determine whether the cement contains any particles of free lime by hastening the slaking process.

The Le Chatelier apparatus (figure 4.4) is used, consisting of a split metal ring to the sides of which two pointers are welded. This is stood on a glass plate, filled with a paste of standard consistency taking care not to force open the split, and covered with another glass plate which is weighted down with a suitable lead weight.

The apparatus is then submerged in water at a temperature of $19\,^\circ C + 1\,^\circ C$ for 24 hours, after which time it is removed and the distance between the two pointers is measured.

It is then submerged again and the water is brought to the boil in 25 to 30 min. Boiling is continued for 1 hour, then the apparatus is removed and allowed to cool. Once again the distance between the pointers is measured and the difference between the two measurements should not exceed 10 mm. If this did occur a further test should be made from the same sample after seven days' aeration of the cement powder, whereupon the expansion should not exceed 5 mm.

Testing Cement on Site   Lumpy cement can be tested on site by crushing in the hand. If the lumps crush easily the cement is satisfactory for normal use; if not it should be discarded.

Storage

On large construction sites 10 tonne loads may be delivered in tankers and then blown into storage silos. Where bagged cement is being used it should be stored in a damp-proof, draught-free shed having a raised wooden floor. To avoid 'warehouse set' the bags should not be stacked any higher than 1.5 m, and to avoid the situation of the lower layers of bags becoming rock hard it is preferable to keep to the rule: 'first in, first out'. Four months is considered the maximum possible storage period.

If a shed is not available, the bags of cement should be stacked on a raised wooden platform and securely covered with tarpaulins or polythene sheeting, which should be weighted down to prevent penetration by wind and rain.

Other Cements

Rapid-hardening Cement   This is similar in composition to O.P.C., but is more finely ground. It does not have a quicker setting time but it gains strength more rapidly. Use of this cement in concrete enables earlier striking of formwork, and the increased heat of hydration makes it useful when concreting in cold weather.

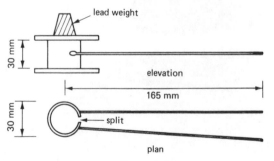

Figure 4.4

High-alumina Cement This is unlike other types in composition and properties. The basic materials from which it is made are bauxite (aluminium ore) and limestone or chalk. These are melted in a furnace, the resultant mass being allowed to cool before grinding down to a fine powder. High-alumina cement is comparatively slow setting, but hardens very quickly, giving off considerable heat in the process. It should be kept continuously damp for a period of at least 24 hours. Concrete made using this type of cement has the same strength at 24 hours as concrete made with O.P.C. at 28 days.

While high-alumina cement is highly resistant to the action of sulphates and is therefore an alternative to sulphate-resistant cement and super-sulphated cement where sulphate attack can be expected, if it is used in concrete where humidity and high temperatures are present simultaneously, a serious loss of strength may result.

Sulphate-resistant Cement O.P.C. contains a percentage of tri-calcium aluminate (C3A) which can be attacked by soluble sulphates present in certain sub-soils, and foundations may eventually disintegrate. Sulphate-resisting cement contains less C3A and can therefore be used safely where normal concentrations of sulphates exist.

Super-sulphated Cement This cement contains even less C3A, and can be used in foundation concrete where high concentrations of soluble sulphates are present.

To test the ability of cements to withstand sulphate attack, small cubes of cement and sand should be made using different cements, carefully marking each cube. The cubes are then placed in large jars containing, for example, calcium, magnesium, sodium or potassium sulphate solutions. A regular check on the cubes should be kept and any deterioration noted.

Aggregates

The term aggregate refers to the sands, gravels, crushed stone, etc. that are mixed with cement and water to produce concrete, and since usually at least 85 per cent of most concretes consist of aggregates, it is essential that the correct material is selected.

Desirable properties of aggregates for the production of good quality concrete are as follows.

(1) Cleanliness is essential since any dust, clay or other impurities may prevent proper bonding together of the material. Sands and gravels should be washed before delivery, and any excess dust in crushed stone should be used sparingly.

(2) The crushing strength of the aggregate should be at least equal to that of the cement paste that binds the particles together.

(3) Aggregates must be durable and not liable to any form of shrinkage, swelling or decomposing.

(4) Organic impurities should not be present; their inclusion often affects the setting and hardening of the concrete.

(5) Because of their salt content unwashed seashore sands and gravels often promote efflorescence on finished work, also causing corrosion of reinforcing steel.

Types of Aggregate

Aggregates can be classified as normal, heavy or light-weight; some examples of each type are given in table 4.1.

Table 4.1

Normal	Heavy	Light
Sand	Barytes	Clinker
Gravel	Iron and	Pumice
Crushed stone	Steel punchings	Foamed slag
Broken brick		Expanded shale
Blast furnace slag		

Aggregates are further divided into fine and coarse. *Fine aggregate* is material mainly passing a 5 mm sieve, usually consisting of sand, stone dust or crushed gravel. *Coarse aggregate* is mainly retained on a 5 mm sieve, the most commonly used being gravel or limestone chippings.

All-in aggregate can be obtained from certain suppliers. This consists of pre-mixed aggregate from the largest size specified down to dust, simplifying mixing on site; owing to possible variation in the grading this aggregate would not be used for the production of high-strength concrete.

Testing Aggregates

Sampling When a sample of aggregate is taken for testing it should be representative of the whole of the load or stockpile, therefore it is necessary to collect a little aggregate from different places rather than a sample from one place only. This should then be thoroughly mixed and the amount required for testing arrived at by quartering or by use of the riffle box (figure 4.5).

Quartering   The well-mixed sample should be tipped on to a clean, hard, dry surface. It is then quartered (see figures 4.6 and 4.7) and opposite quarters are selected, re-mixed, and the process continued until the correctly sized sample is obtained. The unwanted aggregate can be returned to the stockpile.

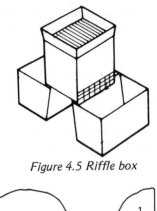

Figure 4.5 Riffle box

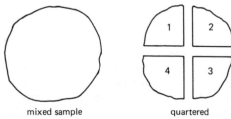

Figure 4.6

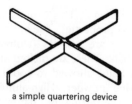

a simple quartering device

Figure 4.7

The Riffle Box   This is carefully designed to divide the sample in half each time it is tipped into the top container. One of the containers is selected and tipped again into the top container, the process continuing until the required amount is obtained.

Sieve Test   An aggregate should be well graded, that is, it should consist of all sizes of sand or gravel from the largest specified down to dust. Figure 4.8 shows examples of poorly-graded and well-graded aggregates.

To ascertain if an aggregate is well graded the sieve test is used. The sieve sizes in general use are 75 mm, 63 mm, 37.5 mm, 20 mm, 14 mm, 10 mm, and 5 mm for coarse aggregate; and 2.36 mm, 1.18 mm, 600 μm, 300 μm and 150 μm for fine aggregate. The minimum sizes of sample for this test are 250 g for fine aggregate and 5 kg for 19 mm coarse aggregate. If the samples are considered too large for the sieves, the test can be carried out in two or more sievings as necessary.

The sample is dried, weighed carefully, then passed through the sieves, starting with that having the largest mesh. If sieving is being carried out by hand, each sieve should be shaken separately for at least 2 min. If, however, mechanical shaking is available, then the sieves are arranged in the correct order, that is, on top of one another in a 'nest', then agitated for 10 to 15 min.

After sieving the amount retained on each sieve is carefully weighed and the percentage passing each sieve is calculated and recorded in the form of a graph

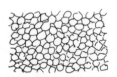

well graded   poorly graded

Figure 4.8

Silt Test for Sand (figure 4.9)   As explained previously, silt or fine clay particles are detrimental to concrete and the amount contained can be checked in the following way.

Place 50 ml of a 1 per cent salt solution in a 250 ml measuring cylinder. Add the sand until the height of the sand in the cylinder is 100 ml. Add further solution until up to the 150 ml mark and shake vigorously, allowing 3 hours to settle. The grains of sand being heavier will settle first, and the layer of silt that settles on top can be measured and should not exceed 6 per cent.

Bulking of Sand   Perfectly dry sand occupies the same volume as saturated sand, but when sand becomes damp it increases in volume by up to 35 per cent. This increase is known as bulking and if a concrete mix was specified as 1 cement, 3 sand and 6 gravel by volume, and the sand was bulking 33 per cent, the mix would be undersanded. Instead of 3 parts of sand going into the mix, because of the bulking only 2 would be included.

For this reason the amount by which a sand is bulking must be determined in order that additional sand can be added as required. The percentage bulking can be calculated by the following method.

A beaker or glass jar is loosely filled to about two-thirds of its height with the damp sand and the exact depth measured with a steel rule. The sand is then emptied out and the jar is half-filled with water. Replace the sand slowly, re-measuring the depth of this on completion (figure 4.10). To calculate the bulking factor, take amount (c) from amount (a) and express as a percentage of amount (a).

Organic Impurities Test for Sand   Fill a 300 ml medicine bottle to the 110 ml mark with a sample of the sand to be tested. Add a 3 per cent solution of caustic soda up to the 175 ml mark and shake thoroughly. After 24 hours check the colour of the liquid and if it is darker than a light straw colour it would be advisable to consult the site engineer before proceeding further.

Note   A 12 oz bottle was originally used, which is 340 ml. The nearest metric size is 300 cm^3, the other figures having been adjusted accordingly.

Mixing Shrinkage   When materials for concrete are mixed together with water their volume will shrink by up to approximately 35 per cent and therefore when ordering materials for a particular purpose it is important to allow for this shrinkage. For example,

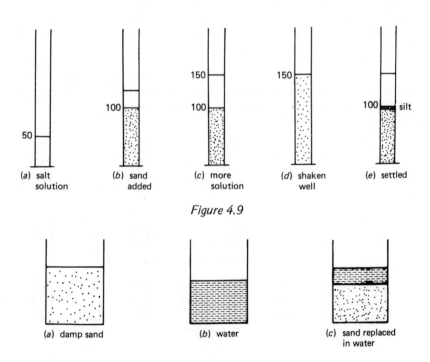

Figure 4.9

Figure 4.10

1 m^3 cement, 3 m^3 sand and 6 m^3 gravel will not produce 10 m^3 of concrete. The sand will merge into the voids in the gravel, as will the cement, and when the water is added the volume will shrink still further to produce about 7 m^3 of concrete. A simple test to demonstrate mixing shrinkage is as follows (figure 4.11).

Take ten jars or suitable containers, fill six with gravel, three with sand and one with cement. Tip out on to a clean, hard, dry surface and mix thoroughly dry, filling as many jars as possible and noting how many are left empty. Tip out once more, add water to produce a workable mix and return to the jars, compacting as far as possible. Again note the number of jars left unfilled.

Testing Sand on Site for Cleanliness   A sample should be rubbed between the palms of the hands. If a stain is left on the palm, the sand may be dirty and should be further tested as described previously. If no stain is left the sand may be deemed suitable for use.

Storage of Aggregates

The practice of tipping aggregates directly on to the ground is wasteful since the bottom 150 mm or so will become contaminated with material such as mud and weeds. Aggregates should be stored on a clean concrete base laid to falls to prevent the lower part of the stockpiles from becoming damp or saturated.

Where required, different sizes can be kept apart by dwarf walls or sleepers and, if possible, duplicate stockpiles should be available to allow any excess moisture content to drain away for at least 12 hours before using.

Aggregate from the bottom of stockpiles should be left as a base for the next load since this is where

contamination may exist owing to impurities percolating through the upper layers. If possible cover aggregates, particularly sand, with tarpaulins in order to exclude rain, snow and leaves, and also to provide some protection against frost.

Admixtures

These are materials other than cement, aggregate and water that may be added to the concrete during mixing in order to improve or modify its properties. Where admixtures are being used it is important that the manufacturer's instructions are strictly adhered to since the addition of too little may not have the desired effect, and too much may cause deterioration of the finished concrete.

Air-entraining Agents

Adding these to the mixer not only improves the workability of the concrete but the provision of millions of tiny bubbles tends to improve the frost resistance. The pores produced provide spaces into which water can expand harmlessly during freezing weather.

Accelerators

Probably the best known accelerator is calcium chloride but several other types are available. Accelerators, as the name implies, cause an increase in the heat of hydration and may be used for example when concreting at low temperatures or to promote rapid hardening of *in-situ* concrete members to allow early striking of formwork. It must be remembered, however, that accelerators are not antifreezes, and do not provide protection for concrete at very low temperatures.

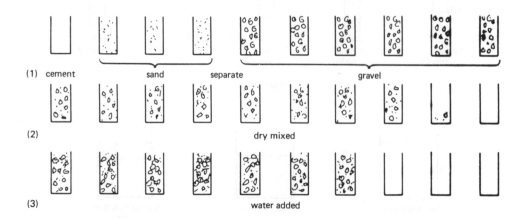

Figure 4.11

Plasticisers

These have the effect of adding workability and reducing the amount of water required, the latter being desirable when a high-strength concrete is necessary. Approximately 0.2 per cent of the weight of cement is added, calling for accurate dispensing at the mixer. (This is a 100 g per bag of cement.)

Retarders

These are based on such materials as sugars, starches and gums, and tend to slow down the setting process. They are useful in hot weather to counteract premature stiffening and also when transporting concrete over long distances.

Damp-proofers

Used integrally to decrease permeability. They should be used with a dense mix, which requires a low water content to be effective.

Water

As a general rule, water for use in concreting work should be fit for drinking. If drinking water is not available and water has to be used from a natural source, that is, a river or lake, for example, a preliminary treatment such as settlement or filtration may be required. Sea water will normally produce a good strength concrete but may cause efflorescence on the finished unit; it may also cause rusting of any steel reinforcement.

To test if the water is suitable, standard pastes should be made and initial and final setting times tested. These should not differ from the usual figures by more than 30 min. Furthermore the compressive cube strength must be at least 80 per cent of that of cubes made using distilled water.

Water—Cement Ratio

One of the most important factors in the production of concrete is the water content. Too much water will result in the finished product being full of pores which are left behind when the water evaporates. For a 50 kg bag of cement, only 15 litres of water are necessary for hydration to take place, but, for workability, extra water must be added to the mix. Generally speaking, the best concrete for any job is the driest possible coupled with sufficient workability for the circumstances.

CONCRETE PRODUCTION

Batching

This refers to the measuring out of materials, which may be carried out by weight or by volume.

By Weight

This is the only accurate method of measuring quantities of materials and is becoming common practice on all but the smallest jobs.

Proportions are normally based on the 50 kg bag of cement, fine and coarse aggregate too being given by weight, for example 50 kg cement, 150 kg sand and 300 kg stone. The weight of the sand is affected very little by water content, the possible error being about 5 per cent.

Most machines have a dial gauge and once the machine has been set up level the accuracy can be checked by placing a known weight such as a bag of cement on to the hopper. Loading the hopper with the correct amounts for discharge into the mixer is simple since the dial usually has markers to show the weights of each material required (see figure 4.12).

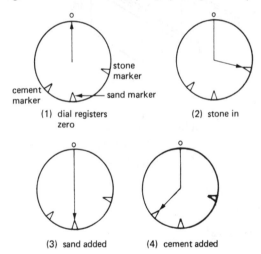

(1) dial registers zero   (2) stone in

(3) sand added   (4) cement added

Figure 4.12

Water can be measured by weight or volume as 1 litre of water weighs 1 kg. If the mixer does not have a water-measuring tank, a container such as a bucket of known volume should be used.

By Volume

On small jobs batching is still carried out using the volume method and with a little care this can be reasonably successful. Batching of cement by volume should never be permitted for, once out of the bag, cement becomes aerated. The mix should therefore be based on a whole bag (0.035 m³) or a number of whole bags. Batching of aggregates by the shovelful is a poor method since no two shovelfuls are the same.

Gauge boxes should be used for aggregates after first calculating the percentage bulking of the sand and taking the extra amount to be added into consideration. These should be deep and narrow and have two handles (figure 4.13). They must be strongly made, not too large and cumbersome, and may be bottomless. A suitable sized gauge box is about 450 x 300 x 300. For example, assume a mix was required to be batched by volume in the ratio of 1 cement, 2 sand and 4 stone. Since 1 bag of cement contains 0.035 m³, then the gauge box should be made to accommodate 0.07 m³. This is then filled twice with coarse aggregate and once with sand to 1 bag of cement. The dimensions could therefore be 500 x 350 x 400 (see figure 4.13).

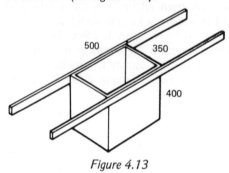

Figure 4.13

Since

$$volume = l \times b \times d$$
$$= 0.5 \times 0.35 \times 0.4$$
$$= 0.07 \text{ m}^3$$

Mixing

By Machine

Large mixers are usually loaded in the sequence: coarse aggregate, cement, fine aggregate and water; but when batching by volume, having accurately measured out the materials it is usual practice to put half the water in first or the mixer may tend to clog up. Mixing times range from 30 s to 3 min, depending on type and size.

By Hand

The general rule to be observed is: turn thrice dry and thrice wet or until the concrete is of an even colour and consistency throughout. Hand mixing should be carried out on a clean, hard base, after measuring out the proportions, and these are turned completely three times. The heap is then hollowed out into a circle and half the mixing water is added. The material is brought in from the sides until the water is used up and the rest of the water is added as required, preferably through a rose while keeping the materials turning until the concrete is ready for transporting.

Transporting

Transporting the concrete can be by wheelbarrow, dumper, crane and skip, conveyor belt, monorail, pump, lorry, etc., and it is important that the journey from the mixing area to the point of pour is smooth or segregation may occur. It occurs when heavier aggregate falls to the bottom and the fines, cement and water come to the top. Not only is the concrete difficult to pour but its condition may be well below the desired standard.

Placing

Concrete should be placed as near as possible to its final position and before any stiffening occurs. Water should not be added once the concrete leaves the mixing area since this will increase the water—cement ratio and thus decrease the final strength.

Compacting

Wet mixes are easily compacted since they will flow readily into the required position, but where high-strength concrete is to be made dry mixes are mostly used and compaction is more difficult. Compaction, however, is vital since voids left in the concrete can considerably reduce the final strength. For example, 5 per cent voids may reduce the strength by up to 30 per cent.

Internal Compaction

Most commonly used for this purpose is the poker vibrator, which is placed into the concrete to its full length and will compact a layer up to a maximum of 600 mm in depth. Use of the poker vibrator is as follows.

(1) Place the poker into the concrete and allow it to run, pulling it out slowly to allow the hole to close up after it as much as possible.
(2) Move the poker frequently, replacing sufficiently close to each preceding position so as to completely close this up.
(3) When compacting a layer on top of the previous

one, vibrate 100 mm into the previous layer to ensure an effective joint.
(4) The poker should not be allowed to touch any formwork, otherwise a mark may be visible on the finished concrete.
(5) The poker should not be allowed to come into contact with reinforcement, otherwise displacement may occur.
(6) Over-vibration can cause segregation, and a layer of laitance may form on the surface.
(7) Do not attempt to move concrete laterally with the poker since this too tends to cause segregation.
(8) Vibrate until air bubbles cease to come to the surface, or the tone of the poker becomes constant.

External Compaction

Where congested reinforcing steel or lack of depth prevents the use of a poker vibrator, external vibrators can be used. These are fastened to the formwork at sufficiency close centres and will cause compaction. Thin slabs too are best compacted using a tamping beam to which a vibrator is fastened. The beam rests on the side forms, concrete is placed loosely in position with approximately 1/5 surcharge and the vibrator will compact the concrete as the beam is moved slowly across.

For cubes, precast beams and paving slabs a vibrating table can be used but these are usually seen in precast works and laboratories rather than on site.

A Kango hammer too will provide efficient vibration for cubes, small lintels etc. A short length of timber about 100 x 50 mm in section is placed across the mould and the hammer is allowed to run while in contact with the timber.

Hand Compaction

For smaller jobs where a vibrator may not be available, concrete can be compacted by hand in any of the following ways.

(1) A short length of 50 x 50 mm timber can be used internally with a ramming action.
(2) The sides of the formwork can be hammered using a hand hammer.
(3) For slabs a heavy plank with two raised handles at the ends can be operated by a man at each side (figure 4.14). This is known as a tamping board.

Finishing

For exposed horizontal surfaces, finishing should take place immediately after compacting and the surface

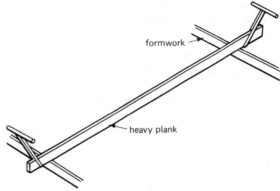

Figure 4.14

should be trowelled over with a steel float, but not over-trowelled so as to bring to the surface a layer of laitance. Wood-float the surface after 2 to 3 hours, depending on the drying rate, in a circular motion to flatten any lumps and fill any hollows and, after a further drying period, lightly run a steel float over the surface.

Curing

Concrete hardens by the interaction of cement and water, and the longer the period that water is present the better will be the quality of the concrete. Curing refers to the process of keeping concrete damp, which is preferably carried out for up to 28 days, during which time the concrete gains most of its strength. After this period strength gain is slow. Some of the methods of curing concrete are

(1) cover as soon as possible with waterproof sheeting tied down at the edges
(2) cover the surface with damp sand, sacking or sawdust and ensure that this is kept damp
(3) continuous spraying with water: the spray must be fine or the surface can be damaged
(4) ponding, that is, heaping sand around the periphery and filling in with water
(5) spraying on a sealing membrane.

Some of the advantages of curing are

(1) elimination of surface cracking
(2) greater final strength
(3) improved resistance to abrasion
(4) less likelihood of surface dusting
(5) an increase in impermeability
(6) improved weather resistance and durability.

Concreting in Cold Weather

At 0 °C concrete ceases to gain strength and, if the mixing water in the concrete freezes, irreparable damage can be caused. If concreting is to be carried out in frosty conditions or when a night frost is expected the following points will assist in preventing damage

(1) heat the water before mixing
(2) heat aggregates with blowers, steam lances or braziers
(3) add an accelerator if permissible or use a rapid-hardening cement
(4) heat the surface to receive the concrete if possible
(5) cover the completed work with insulating quilts on top of waterproof sheeting.

Concreting in Hot Weather

Ensure that aggregate heaps and mixing water are not hot before mixing since this tends to hasten hydration of the cement. Protect completed work against drying winds and strong sunlight since these dry out the surface too quickly, often causing multiple surface cracks. Treat the surface with a fine cold water spray when the surface is sufficiently hard.

TESTING OF CONCRETE

Slump Test

Except for drier mixes this test is a measure of workability. It is the simplest method of ensuring that the consistency of the concrete does not alter throughout the job. The equipment required consists of a slump cone (figure 4.15), a 15 mm diameter steel rod 600 mm in length, a trowel, straightedge and rule.

The test is carried out in the following way.

(1) Make sure the cone is clean and stand it on a smooth, hard, surface, preferably a sheet of metal.

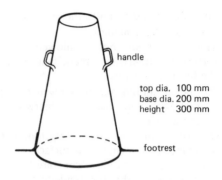

top dia. 100 mm
base dia. 200 mm
height   300 mm

Figure 4.15

(2) Stand on the footrests and fill the cone in three layers, rodding each layer 25 times. Do not rod to the bottom each time or the first layer will have been rodded 75 times on completion.
(3) Top up as necessary and smooth off the top.
(4) Clean round the base and lift the cone vertically, placing it upside down beside the resulting mound of concrete.
(5) Place the straightedge across the cone and measure down to the topmost point with the rule. This dimension is termed 'the slump' and it should be reasonably constant throughout the job (figure 4.16).

Three forms of slump are possible

(1) natural or true slump: the mound sinks but keeps its shape more or less (figure 4.17)
(2) shear slump: the mound falls away sideways (figure 4.18)
(3) collapsed slump: the mound pancakes (figure 4.19).

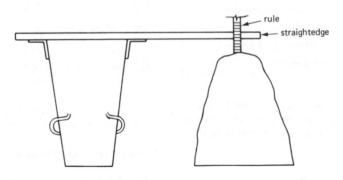

Figure 4.16

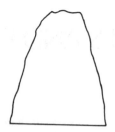

Figure 4.17

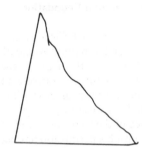

Figure 4.18

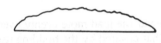

Figure 4.19

If a true slump is obtained, it should be measured and recorded. A shear slump should be re-tested and if it occurs again, measure it and record it as a shear. A collapsed slump should be recorded as such.

Making Test Cubes

Compression tests are made on cubes to maintain a check on the quality of the concrete being produced. For concrete with a maximum aggregate size of up to 40 mm, the cubes are usually 150 mm. These are made and tested in the following way.

(1) The steel moulds are assembled, cleaned and oiled prior to use.
(2) A representative sample of the concrete is taken for testing. If the concrete is being discharged from a mixer, the sample should be collected in at least three parts at different discharge times. If taken from a heap, the sample should be collected from at least five different places.
(3) The sample is thoroughly mixed and filled into the moulds in three layers, each layer being tamped 35 times with a special bar weighing 1.8 kg, 375 mm in length and 25 mm square.
(4) Smooth off the top and store under damp sacking for 24 hours at a temperature of 20 °C ± 1 °C.

Mark on identification as required.
(5) The cubes are then removed from the moulds and stored under water at between 20 °C ± 1 °C until required for testing, which is usually carried out at 7 and/or 28 days.
(6) Just before testing the cubes are removed from the water, wiped dry and weighed. They are then crushed in a compression testing machine, the results being expressed in N/mm^2.

To calculate the crushing strength, the load at failure is divided by the area of one face.

Example 4.1

A cube fails at 500 kN. Calculate the crushing strength.

$$\text{Crushing strength} = \frac{\text{load in N}}{\text{area of one face}}$$

$$= \frac{500 \times 1000 \quad (N)}{150 \times 150 \quad (mm)}$$

$$= \frac{500\,000}{22\,500}$$

$$= 22.22 \text{ N/mm}^2$$

For aggregates with a maximum size of up to 20 mm, 100 mm cubes may be used. These are filled in 2 layers, each layer being tamped 25 times with the same bar as the larger cubes. The compressive strength is found as follows:
For example, a 100 mm cube fails at 340 kN.

$$\text{Compressive strength} = \frac{\text{load in newtons}}{\text{area of one face}}$$

$$= \frac{340 \times 1000}{100 \times 100}$$

$$= 34 \text{ N/mm}^2$$

5
FOUNDATIONS FOR WALLS AND PIERS

All loadbearing walls and piers are required to stand on a concrete foundation, and the concrete must be capable of receiving and transmitting the loads placed upon it to the natural foundation below. The depth of the natural foundation below the ground surface must be at the line of saturation.

NATURAL FOUNDATION

The sub-soil on which the concrete foundation is placed is termed the natural foundation, and it may be in the form of rock, clay, gravel, sand, or even waterlogged and reclaimed ground. Each type of sub-soil has its own safe bearing pressure (S.B.P.), usually expressed in kN/m^2.

Classification of Sub-soil

(1) Cohesive soils are fine grained, and when they absorb moisture they become plastic and sticky, for example, clays, peats and silts.
(2) Non-cohesive soils are coarse grained and are not cohesive when moist, for example, rocks, boulders, gravels and sands.

Line of Saturation

Can be any depth below the vegetable soil to which ground water will not penetrate, or which will be unaffected by changes in atmospheric conditions. Natural foundations which do not retain moisture can be used when they occur immediately below the vegetable soil (figure 5.1).

Inspection of Natural Foundation

The Building Control Officer is required to visit the site and inspect the natural foundation. He may test the sub-soil, measure the depth of the natural foundation, also the depth of concrete foundation which may be indicated by pegs driven into the base of the excavation. It is necessary to give the inspector 24 hours' notice of the inspection and concrete must not be poured until his permission has been obtained.

PURPOSE OF FOUNDATIONS

(1) To spread the load more evenly over a larger area than that covered by the building itself.
(2) To prevent walls from tilting over because of undue settlement.
(3) They help to bridge over any soft spots which may occur in the natural foundation.
(4) They form a level base from which all building operations can start.

BUILDING REGULATIONS

These require that foundations shall

(1) safely receive and transmit loads upon them to

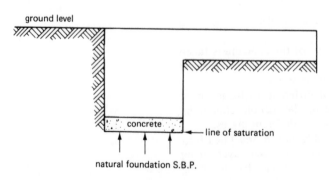

Figure 5.1

64

the ground below, so that no settlement will occur which can cause damage
(2) be taken down to such a depth that they are safeguarded against damage from swelling, shrinkage or freezing of the sub-soil
(3) resist attack by sulphates and noxious matter in the ground.

DAMAGING FORCES

Concrete foundations can be weakened or damaged as a result of any of the following factors.

(1) *Inequality of settlement*   This is caused by the unequal resistance of the sub-soil or by unequal loading on the foundation.
(2) *Lateral escape*   This failure is mainly due to sandy sub-soils, which are not contained, and 'leak' out from under the concrete foundation.
(3) *Sliding*   On sloping sites, structural forces can cause sliding if they are in the direction of the slope.
(4) *Withdrawal of water*   When clayey sub-soils lose their moisture content, they dry out and shrinkage occurs. This in turn results in unequal settlement.
(5) *Atmospheric action*   When sub-soils close to ground level accept an excessive amount of water, and are then subjected to frost action, the result is an upheaval of the sub-soil, and fracturing of the concrete foundation.
(6) *Soluble sulphates*   These are often found in sub-soils, and if excessive they can cause lamination or disintegration of ordinary Portland cement concrete.

Frost Heave

In any type of soil that retains water instead of allowing it to drain away quickly, ice may form during freezing conditions and lift the ground surface. The forces produced may be intense enough to lift and buckle concrete foundations and walls.

(1) It is water which changes into ice.
(2) Expansion occurs when the water changes from liquid to solid.

TYPES OF CONCRETE FOUNDATION

There are five main types of foundation

(1) strip or continuous foundation
(2) pad foundation
(3) raft or slab foundation
(4) stepped strip foundation
(5) pile foundation

All the above types of concrete foundation can be strengthened by steel reinforcement.

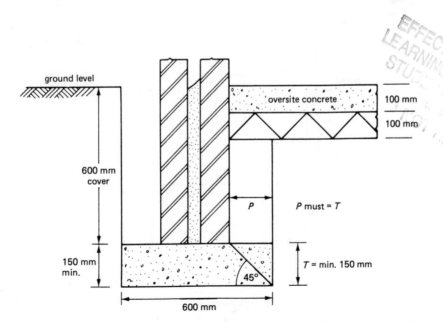

Figure 5.2 Standard strip foundation on normal load bearing soil (vertical direction)

Concrete Strip Foundation

This type of foundation is a continuous strip of concrete, minimum thickness 150 mm, placed centrally under all the loadbearing walls of the structure. It is recommended for use where the loads produced by the building are not excessive, and when there is no variation in the soil structure under the foundation concrete (figure 5.2).

Uses

(1) For houses, bungalows, and similar structures, the width and depth of the concrete strip is required to provide stability. It can be ascertained from a table in the Building Regulations.
(2) For larger and heavier structures the width and depth of the concrete strip is determined by calculated design.

Attached Structural Projections

The concrete foundation under a pier, buttress or chimney forming part of a wall, must project on all sides to at least the same extent as they project beyond the face of the wall (figure 5.3).

Depth of Foundations

The local authority will generally require 600 mm of ground cover for normal concrete foundations.

Wide Strip Concrete Foundation

This type of concrete strip foundation is used when the natural foundation is relatively weak; the width of the concrete strip is increased to allow distribution of the load over a greater area. Wide concrete strip foundations are designed by calculation (figure 5.4).

The additional advantage of this type of foundation is that substructure productivity is considerably increased; this is achieved because of the increased amount of working space within the foundation trench, and although the amount of concrete required for the foundation is greater than normal strip, the increased productivity usually offsets this cost.

Figure 5.3

Figure 5.4

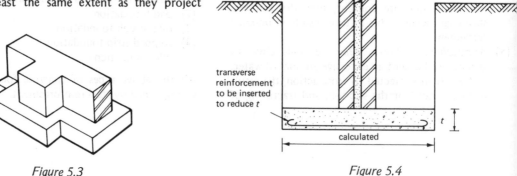

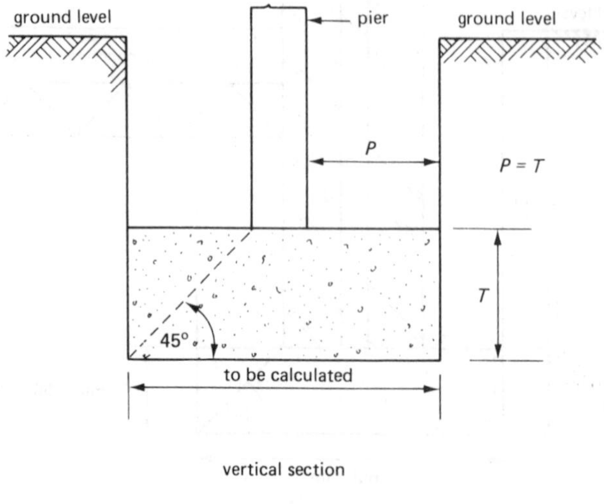

vertical section

Figure 5.5 Concrete pad foundation

Concrete Pad Foundation

Pad foundations are isolated blocks of concrete supporting concentrated loads which the pad must accept and transmit to the natural foundation below. They are used to support isolated

(1) brick or masonry piers
(2) concrete columns
(3) steel stanchions.

The size of the pad is determined by calculation, and the thickness of the concrete must equal the projection from the face of the pier (figure 5.5).

Raft Foundation

A raft foundation consists of a concrete slab formed at ground surface level; they cover the area beneath the building and often extend beyond the external walls, forming a protective apron around the base of the building. Concrete rafts should be strong enough to support all intended loads; the overburden must be removed to a depth of at least 300 mm and a layer of well-consolidated hardcore formed as a base for the concrete (figure 5.6).

Uses

A raft foundation is used for light structures where a sound natural foundation is obtained close to ground level (figure 5.6).

Reinforced Concrete Raft Foundation

This is used when forming a slab on reclaimed ground, or situations where the natural foundation is at an uneconomical depth (figure 5.7).

Aprons

Where the natural foundation below concrete rafts could be damaged by moisture penetration, a protective apron should be constructed. This extends the raft beyond the effective ground bearing.

For two-storey buildings, where the ground is of a soft nature, or on reclaimed ground, a semi-raft construction can be used. The reinforced concrete is thickened out under all loadbearing walls. This treatment can also be used to prevent lateral escape of sub-soils (figure 5.8). This type of raft is extensively

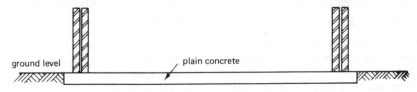

Figure 5.6 Plain concrete raft foundation (not reinforced)

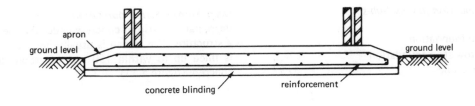

Figure 5.7 Reinforced concrete raft

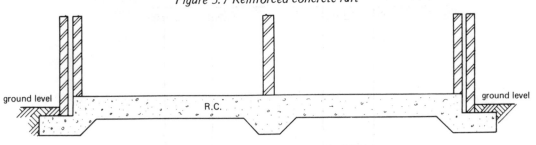

Figure 5.8 Concrete semi-raft construction

used on reclaimed ground in towns and cities after the demolition of slum areas.

Problems of Clay Soil

The strength and durability of clays are affected by water content because they

(1) shrink on drying
(2) swell again when wetted.

The disruptive effect is most marked around the outer perimeter of the building, and the movement of the clay may cause the concrete foundations and walls to buckle and crack as a result of the differential movement between the dry clay under the building, and the wet clay around the perimeter (figure 5.9).

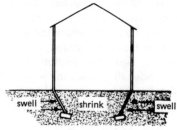

Figure 5.9

The volume changes become less with depth and, below 1.00 m, ground movement is slight and should not affect the substructure.

A depth of 1.00 m may not be effective against severe drying shrinkage caused by tree roots. In this case it will be necessary to excavate to a greater depth.

Types of Foundation in Clay Sub-soils

(1) deep strip foundation
(2) deep narrow strip foundation
(3) short bored piles

Deep Strip Foundation   This is similar in construction to the normal concrete strip foundation, but the natural foundation is at a depth of at least 1.00 m. The width of the strip is usually 600 mm (figure 5.10).

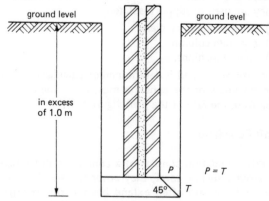

Figure 5.10 Deep strip foundation

Advantages are

(1) site personnel are familiar with this method of construction
(2) special plant is not required.

Disadvantages are

(1) extra costs because of increased
    (a) excavation and spoil
    (b) brickwork
    (c) cavity fill
    (d) hardcore, and return fill and ram
(2) restricted working space for bricklayers working in a 600 mm wide trench at such a depth.

Deep Narrow Strip Foundation   This is a continuous, deep, narrow strip of concrete under the shell walls of a structure. The trench is usually 380 mm wide, and is filled up to within 150 mm of ground level with concrete (figure 5.11).

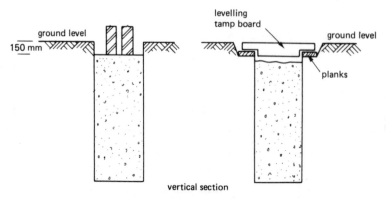

Figure 5.11 Method of levelling concrete

The advantage is that it overcomes the difficulty of bricklayers working in restricted conditions.

The narrow strip foundation is often termed trench fill. Bricklayers constructing the substructure walls on the concrete foundations work in very restricted conditions. Some builders increase their productivity when working on foundations by reducing the amount of brickwork in the substructure; this is achieved by increasing the depth of concrete in the foundation. The trench is filled with concrete up to within 150 mm of the ground surface. Builders who are proficient in the use of concrete report a reduction in cost over the more traditional methods used. Facing brickwork is set out two courses below ground level, therefore it is essential to determine the correct bonding arrangement immediately off the concrete foundations.

Short Bored Piles A continuous reinforced concrete ring beam is supported by circular concrete columns. These piles are placed under all loadbearing walls of the structure. The top of the ring beam is at ground level, and is cast upon a layer of ash or clinker blinding, which is used to take up any differential movement in the ground (figure 5.12).

The pile holes can be bored using a hand or machine auger (figure 5.13) and are filled with concrete immediately after boring (figure 5.14).

Diameters of piles may be

(1) 250 mm for normal loading
(2) 350 mm if placed under walls where greater forces are expected, for example, chimney breasts, etc.

Depths of piles are

(1) 2.00 m for the internal walls
(2) 4.00 m for shell walls and under chimney breasts.

Positions of piles are

(1) at corners of the structure
(2) wall junctions
(3) under chimney breasts; intermediate piles may be spaced from 1.00 m to 2.00 m, in sufficient numbers to adequately support the load. Ground floor door and window openings are usually between the planned positions of the piles.

Advantages are

(1) working space is not restricted
(2) excavation work is reduced.

Disadvantages are

(1) site personnel are not familiar with this method of construction
(2) special mechanical plant may be necessary, together with a trained operator.

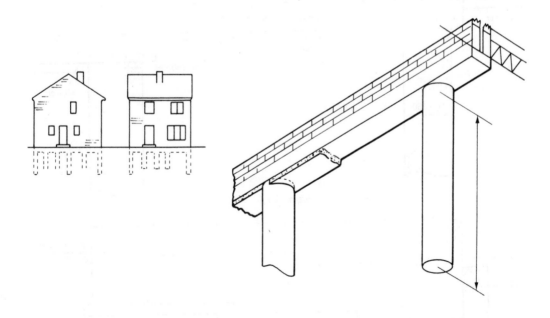

Figure 5.12

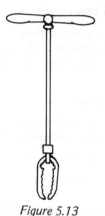

Figure 5.13

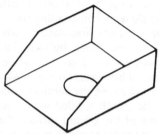

Figure 5.14 Metal tray for filling pile holes

It is possible to use the small bored pile method on sloping sites. The concrete beams are stepped with at least 150 mm lap beyond the beam and pile below;

the same length of pile can be used throughout (figure 5.15).

Stepped Strip Foundation

On sloping sites, where the natural foundation is found to be running parallel to the ground surface, it is good building practice to construct a foundation in a series of concrete steps (figure 5.16).

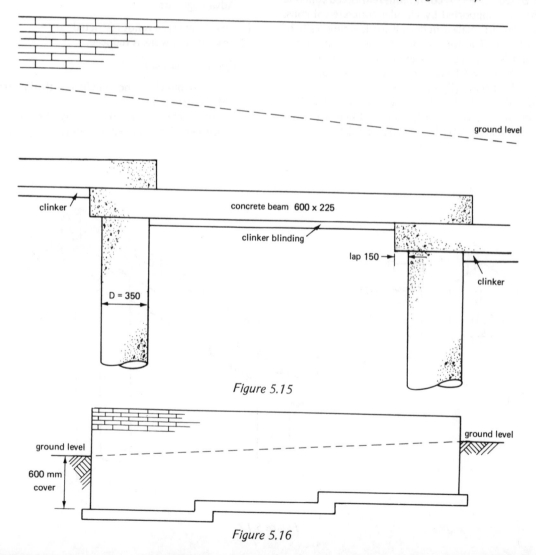

Figure 5.15

Figure 5.16

Advantages of the stepped foundation are

(1) It is more economical because it reduces the quantity of excavation.
(2) The concrete foundation is constructed on the same class of sub-soil and this will reduce the possibility of differential settlement.

Requirements for Stepped Strip Foundations

(1) The concrete foundation should rest on the same class of sub-soil throughout its length.
(2) The top surfaces of the concrete steps must have the minimum amount of cover required by the Building Regulations.
(3) The lengths of all concrete steps should be regular and not exceed 2.450 m nor be less than 1.000 m.
(4) A step in a concrete foundation shall not be of greater height than the thickness of the foundation.
(5) The higher concrete strip must lap over and combine with the lower strip by

    (a) twice the height of the step
    (b) the thickness of the foundation
    (c) 300 mm

whichever is greater.
For example (see figure 5.17):

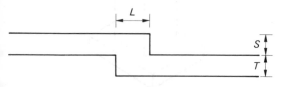

Figure 5.17

minimum overlap L = twice the height of the step, or the thickness of the foundation, or 300 mm, whichever is greater.
and S = not greater than T.

THE NEWTON

In the SI system forces are measured in newtons (N) and, due to the force of gravity at the Earth's surface, every mass exerts a force. For example, a bag of cement has a mass of 50 kg and due to gravitational force it exerts a force of nearly 500 N (0.5 kN).

The speed of a free-falling body, drawn towards the Earth by gravity, is 9.81 m/s^2. In practice, when calculating forces the figure 10 may normally be used. This not only simplifies problems with calculations, but also increases the factor of safety.

Example 5.1

Determine the magnitude of the force produced at the base of a brick pier with a mass of 3000 kg. Express the answer in kN.

$$\text{Force} = \text{mass} \times 9.81 \quad (\text{say } 10)$$
$$= 3000 \text{ kg} \times 10$$
$$= 30\,000 \text{ N}$$

There are 1000 N in 1 kN, therefore

$$30\,000 \text{ N} = 30 \text{ kN}$$

LOADS ON FOUNDATIONS

Controlling Factors

These determine the type and size of concrete foundation

(1) the structure to be placed on the concrete foundation
(2) the type and magnitude of the loads placed on it
(3) the safe bearing pressure of the sub-soil.

Safe Bearing Pressure (S.B.P.)

Each type of soil has its own S.B.P., that is, the magnitude of the force which it will safely accept, usually expressed in kN/m^2.

Factor of Safety (F.O.S.)

Structural engineers test samples of soil to failure and allow a factor of safety.
For example, assume a bridge will carry an ultimate load of 300 kN and it is decided to allow a working stress of 100 kN/m^2. To find the F.O.S.

$$\frac{\text{ultimate stress}}{\text{working stress}} = \frac{300}{100} = 3$$

$$\text{F.O.S.} = 3$$

F.O.S. is necessary because

(1) the material may be weaker than expected
(2) the load may be higher than calculated
(3) workmanship may be poor.

For example, calculate the working stress of a concrete foundation allowing a factor of safety of 3 where the ultimate stress is 510 kN/m^2.
    Since F.O.S. = ultimate stress ÷ working stress, then

$$\text{working stress} = \frac{\text{ultimate stress}}{\text{F.O.S.}}$$

$$= \frac{510}{3} = 170 \text{ kN/m}^2$$

This will ensure that the foundation is three times stronger than required.

Classification of Loads

(1) *Dead load* is the force produced by the fabric of a building, for example, foundation, walls, floors, roof, etc.
(2) *Imposed load* is the forces produced by other elements introduced after the completion of the building, for example, tables, chairs, machinery, people, etc.

Types of Load

(1) *Evenly distributed*, for example, walls on a strip foundation.
(2) *Isolated, that is, concentrated*, for example, pier on a pad foundation.

DESIGN OF CONCRETE FOUNDATIONS

This is arrived at by calculation, and the following information is required

(1) load to be supported
(2) the S.B.P. of the subsoil.

Width of Strip Foundations

This can be determined using one of two methods

(1) Table in Building Regulations. Widths are given for various loads on seven classes of soil, but for loads above 30 kN/m run, there are no widths indicated for soil types 5, 6 or 7. These will require calculation.
(2) By calculation. The formula is

$$\text{width/metre run} = \frac{\text{load/m run}}{\text{S.B.P.}}$$

Depth of Concrete

The Building Regulations require the depth of concrete strip foundations to be equal to the projection, and in no case to be less than 150 mm.

Pad Foundations

Calculations are always required for pad foundations. The formula is

$$\text{area} = \frac{\text{load}}{\text{S.B.P.}}$$

Example 5.2

An isolated brick pier, its supporting pad foundation and the superimposed load have a total mass of 7500 kg. If the S.B.P. of the natural foundation is 300 kN/m^2, calculate the area of the pad foundation using log tables (figure 5.18).

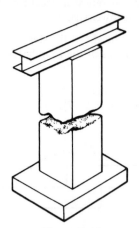

Figure 5.18

$$\text{Area} = \frac{\text{force}}{\text{S.B.P.}} \qquad \text{(always state formula)}$$

$$= \frac{7500 \times 9.81 \ (N)}{300 \ (kN)} = \frac{7500 \times 9.81 \ (N)}{300 \times 1000 \ (N)}$$

No.	Log
7500	3.8751
9.81	0.9917
	4.8668
300	2.4771
1000	3.0000
	5.4771
2453	$\overline{1}$.3897

$$\text{Area} = 0.2453 \text{ m}^2$$

or

$$\text{Area} = \frac{7500 \times 10}{300 \times 1000}$$

$$= \frac{75}{300}$$

$$= 0.25 \text{ m}^2$$

(slightly increased F.O.S.)

To obtain the depth of the concrete, the dimensions of the pier must be known since the depth must equal the spread or 150 mm, whichever is greater. Let the pier be 340 x 340 mm on plan (figure 5.19). Using the increased F.O.S. figure (0.25 m), since

$$\text{area of fdn} = 0.25 \text{ m}^2$$

$$\text{length of one side} = \sqrt{0.25}$$

$$= 0.5 \text{ m}$$

$$= 500 \text{ mm}$$

Therefore

$$\text{spread} = \frac{\text{length of fdn} - \text{length of pier}}{2}$$

$$= \frac{500 - 340}{2}$$

$$= \frac{160}{2}$$

$$= 80 \text{ mm}$$

depth of concrete = spread or 150 mm, whichever is greater, therefore

$$\text{depth} = 150 \text{ mm}$$

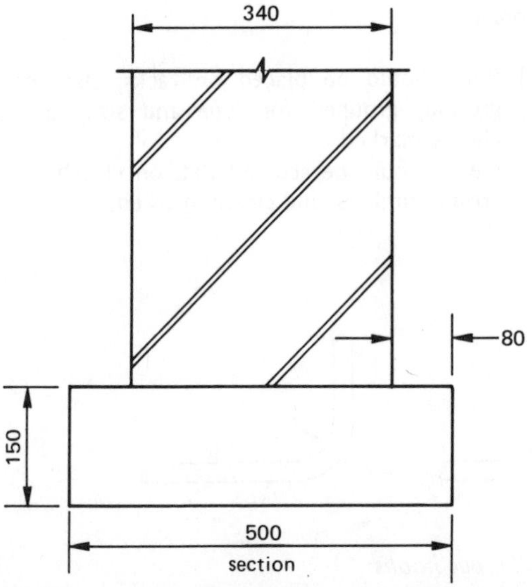

Figure 5.19 1:10

Example 5.3

A 215 mm wall exerts a pressure of 100 kN/m run on a natural foundation of firm clay. Calculate the width of strip foundation and the depth of concrete required. The S.B.P. = 214 kN/m².

$$\text{Width} = \frac{\text{load/m run}}{\text{S.B.P.}}$$

No.	Log
100.00	2.0000
214.00	2.3304
4673	$\overline{1}$.6696

Therefore

$$\text{width of strip} = 0.4673 \text{ m}$$

$$\text{depth of concrete} = \frac{\text{width of fdn} - \text{width of wall}}{2}$$

$$= \frac{467 - 215}{2}$$

$$= \frac{252}{2}$$

Therefore

$$\text{depth of concrete} = 126 \text{ mm}$$

But the Building Regulations require a minimum thickness of 150 mm, therefore

$$\text{width of strip required} = 467 \text{ mm}$$

$$\text{depth of concrete required} = 150 \text{ mm}$$

PRESSURE ZONES (figure 5.20)

Plain concrete is strong when resisting compressional forces, but weak in the resistance of tensional forces. Steel reinforcement must be provided when the thickness of foundation concrete is reduced, to counteract shear and tensional forces. Steel must always be positioned accurately in the tension zone.

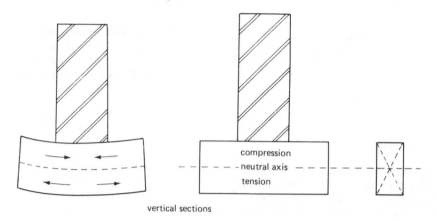

vertical sections

Figure 5.20 Pressure zones

Blinding

This is a 50 to 75 mm layer of plain concrete 1:3:6 mix placed next to the ground. Its purpose is to provide a flat, level base for

(1) correct cover to be maintained
(2) steel fixers to work accurately
(3) protection of the steel from mud coatings.

REINFORCEMENT OF CONCRETE FOUNDATIONS

Steel must be free from loose mill scale, loose rust, grease, oil, paint and mud, etc., which impairs its bond strength to the concrete.

Types of Steel

Usually mild steel is used, but medium tensile and high tensile bars are used when the loading is excessive.

Types of Bar

All are available in various diameters

(1) the plain round bar, which is the most common
(2) deformed bars, which give greater frictional resistance
   (a) twin round bars twisted together
   (b) twisted square bars
   (c) twisted ribbed bars
   (d) ribbed bar
   (e) welded fabric mesh.

Hooked Ends

Bars should be provided with hooked ends since this prevents their being forced out of the concrete when it takes up its full loading (figure 5.21). Three types are illustrated in figure 5.21.

Fabrication

Storage

(1) Bars should be placed on racks, clear of the ground, grouped for type and size, and also clearly marked.
(2) Mesh should be stacked flat, on closely spaced ground timbers, and clearly marked.

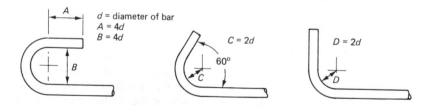

Figure 5.21 Shapes of anchorage hooks

Cutting and Bending

This must be done accurately and must follow the requirements of the Reinforcement Schedule. The prepared reinforcement must be grouped according to type and size, tied together, marked and stored in readiness for the fixer.

Fixing

The bars must be positioned accurately; at intersections the bars should be securely tied using 'black' iron wire. Cover is maintained by using plastic or concrete spacing chairs (see Volume 2).

Cover

Steel reinforcement below ground level must have a minimum cover of 50 mm concrete; it is recommended that 75 mm protection should be given when the concrete is cast against rough timbering or is in direct contact with the soil.

Batching

This must be accurate (by weight). The leanest permissible mix is 1:2:4. The water—cement ratio should provide a workable mix, and produce a concrete of the required density.

Placing

This should be undertaken with care to prevent displacement of the steel. Compaction should be carried out to remove all voids (trapped air) in the mix. Vibrating equipment must not be allowed to come into contact with reinforcing steel, otherwise displacement may occur.

6

BRICKS

DEFINITION

A brick is defined as a walling unit not exceeding 337.5 mm in length, 225 mm in width and 112.5 mm in height. Apart from metric modular bricks, which are discussed later, it is usual for the length of a brick to equal twice the width plus one joint and three times the height plus two joints (see figure 6.1).

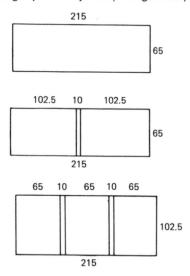

Figure 6.1

Brick size   215 x 102.5 x 65

Format (brick plus joint)   225 x 112.5 x 75

These dimensions have evolved over about 6000 years as the most convenient for handling, the easiest for bonding and provide the best aesthetic appearance.

REQUIREMENTS

Bricks should

1. Be well-burnt
Underburnt bricks are not durable and tend to disintegrate under adverse weather conditions. A well-burnt brick emits a clear ringing sound when struck with a trowel, and often has shiny surfaces. The underburnt variety gives a dull sound and often has a drab appearance.

2. Have good arrises
Well-defined, straight arrises are important, especially when erecting quoins and stopped ends. Some bricks distort slightly when fired, and if one of these is used as a quoin brick it may be laid level or plumb, but not both.

3. Have an even colour
Separate deliveries of bricks occasionally tend to vary in colour, making it very difficult for the bricklayer to maintain a wall of an even, attractive appearance. One load of red facings, for example, may be true to colour but the next load of the same bricks may have darker patches — these usually occur on the stretcher face. Domestic buildings are sometimes seen built in buffs up to chamber height, then darker buffs to completion. If colour cannot be guaranteed, all the bricks for a job should be delivered simultaneously, then the different colours — if any — can be distributed about the building.

4. Be easily cut
Some types of bricks are very brittle and cutting the awkward shapes that are often necessary (see figure 6.7) can be very wasteful in time and materials. Some of the perforated bricks available nowadays are difficult to use, for example, on cut-up gables, without the provision of an electric bench saw (again, causing much waste of time).

5. Have a regular size
Most bricks are of regular size, conforming to BS 3921, but occasionally variations in length or thickness can be experienced, and a length of wall set out with bricks from one delivery can produce tight joints with those from another.

6. Have no lime blows
Lime is sometimes combined with the clay during the manufacturing process and particles of free lime that have not all combined with the clay may still be present in the finished article. If these particles are near the face, they will expand in contact with rainwater. This will blow a small, unsightly, conical hole in the face of the brick, which can be identified by the small white speck at the centre. A completed wall suffering from this problem has a very shoddy appearance.

7. Be salt-free
Salts occur in some clays used for the manufacture of bricks, and this is one of the causes of efflorescence

76

in brickwork (see page 80). Rainwater soaks into the bricks and dissolves the salts, and in dry weather the rain evaporates, leaving a dirty white scum on the face.

8. *Be unmarked*

Rough handling is the main cause of unsightly marks on brick faces. This can be caused at the brickworks, while loading on to lorries, or at the builder's merchants, but most often it is caused on the building site itself. Bricks may be handled many times before being laid in their final position, and those with chipped faces should be used where they will not be seen.

9. *Have adequate strength and density*

It is important that bricks are suitable for the situation in which they are laid. For example, bricks below d.p.c. need to be dense and well-burnt, or they may suffer from spalling due to damp conditions and frost action. Bricks for manhole construction should preferably be class B engineerings, and for brick-on-edge copings a smooth engineering is to be preferred.

CLASSIFICATION

There are numerous methods by which bricks are classified, for example

Place of origin Leicester reds, Accringtons, Staffordshire blues, London stocks, etc.
Method of manufacture hand-made, wire-cut, pressed
Uses commons, facings, engineerings
Colour blues, reds, buffs, etc.

Bricks are generally classified under three headings.

(1) Commons: these are suitable for general building but have no special claim to give an attractive appearance. They are manufactured by most brickyards throughout the country and are normally used below ground level externally and internally for housing and for garages, boundary walls, inspection chambers, etc., depending on quality.

(2) Facings: these are specially made or selected to give an attractive appearance. Any type of brick which has this quality and a reasonable resistance to exposure falls into this category. Used externally above ground level or internally unplastered as a feature.

(3) Engineerings: these have a dense, strong, semi-vitreous body conforming to defined limits for strength and absorption. Owing to their high crushing strength and low porosity they have been successfully used for bridges, sewers, engine pits, power houses and damp-proof courses.

BS 1192 classifies clay bricks with regard to their frost resistance:

(1) *Frost resistant*. Bricks durable in all building situations, including those where they are in a saturated condition and subjected to repeated freezing and thawing.

(2) *Moderately frost resistant*. Bricks durable except when in a saturated condition and subjected to repeated freezing and thawing.

(3) *Not frost resistant*. Bricks liable to be damaged by freezing and thawing if not protected as recommended in BS 5628. These bricks are considered suitable for internal use.

The difference between solid, perforated, frogged and cellular bricks is defined in BS 3921.

(1) Solid bricks shall not have holes, cavities or depressions, a cavity being a hole closed at one end.

(2) Cellular bricks shall not have holes but may have frogs or cavities not exceeding 20 per cent of the gross volume of the brick.

(3) Perforated bricks shall have holes not exceeding 25 per cent of the gross volume of the brick and the area of any one hole shall not exceed 10 per cent of the gross area of the brick.

(4) Frogged bricks shall have depressions in one or more bed faces but their total volume shall not exceed 20 per cent of the gross volume of the brick.

The above details are summarised in figure 6.2.

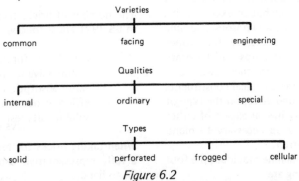

Figure 6.2

Parts of a Brick (figure 6.3)

Bricks are mainly made from clay, calcium silicate (normally known as sand-limes) and concrete. A brief explanation of the manufacture of each type follows.

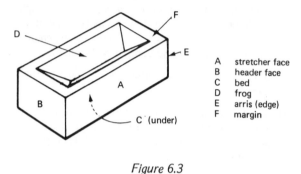

A	stretcher face
B	header face
C	bed
D	frog
E	arris (edge)
F	margin

C (under)

Figure 6.3

MANUFACTURE AND TESTING

Clay Bricks

Clay has provided the basic material for the construction of building units from very early times. It can be defined as earth which forms a sticky coherent mass when mixed with water. Clay is plastic and readily mouldable when damp, but if dried it becomes hard and brittle and will retain its shape. If it is then heated to high temperatures it becomes even harder, is no longer susceptible to the action of water and by no known process can its plasticity be restored. Many classes of clay are suitable for brickmaking and one type or another is found in most parts of the United Kingdom.

The chief constituents of clay are silica (60 per cent) and alumina (20 per cent) (average figures), in addition to which there are smaller proportions of iron oxide, magnesia, lime, etc. Silica (sand) produces hardness, resistance to heat, durability and prevents shrinkage, cracking and warping; but an excess of this constituent makes a brick brittle and porous. Alumina gives the plasticity which is necessary for proper moulding; but this shrinks and warps and becomes extremely hard when burnt. From these two statements it will be obvious that the chemical constituents of the clay will have a profound effect on the type of brick produced. Where a clay has an excess of either of these constituents it may be necessary to blend with a clay from another district.

Bricks may be hand or machine moulded; the four methods of machine-moulding are

(1) the wire-cut process
(2) the stiff plastic process
(3) the semi-dry press process
(4) the slop-moulded process.

The complete wire-cut process is described below.

After removal and storage of the topsoil, any unsuitable subsoil is removed until satisfactory material is located. This is then excavated by means of heavy plant and is taken to the brickworks by railway wagons, conveyor belts, lorries, etc. The clay is preferably left to weather over winter during which time the frost tends to break it up. Blending, if required, is the next process, where the clay is fed into a machine which further breaks it up and feeds it on to a conveyor belt, from which any stones or other obvious impurities can be removed manually. It then goes into a primary grinder which consists of two 10-tonne rollers rotating about an axis, after which secondary grinders such as crushing rolls or ball mills complete the grinding process. The clay, to which water has been added as required, is fed into an extruder which de-airs it and forces it out of a die or mouthpiece in a continuous rectangular column. The clay column is next textured or coloured by spraying, brushing, scoring or similar treatment, and cut to the specified size by a series of wires. Drying is the stage before burning since excess water must be removed before the clay is subjected to high temperatures, otherwise the bricks may twist and warp, or even burst apart. The drying is carried out in long tunnels where a gentle heat is applied with the aid of re-circulating fans. When this stage is complete the bricks are stacked on flat railway wagons and drawn slowly through a tunnel kiln where the temperature is gradually raised to 1000°C at about the centre. From this point the bricks are slowly cooled until emerging at the far end where they are sorted, stacked and sometimes banded (see figure 6.4).

Sampling and Testing Clay Bricks

The number of bricks required for the tests outlined in BS 3921 are as follows

compliance for dimensions	24
compressive strength test	10
water absorption test	10
efflorescence test	10
soluble salts test	10

When taking bricks for testing it is important to take a fully representative sample and not the first 10 or 24 to hand.

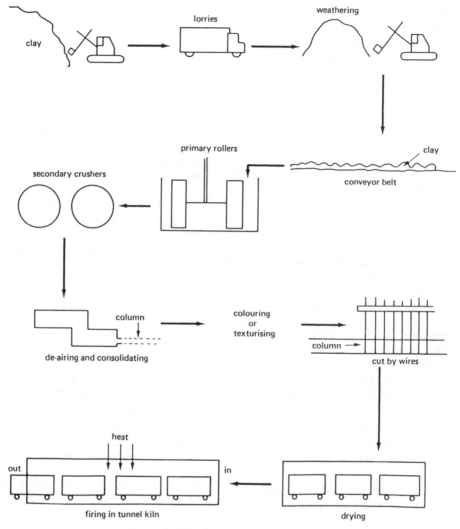

Figure 6.4 Typical wire cut process

Compliance for Dimensions

A load of bricks must comply with the measurements shown below for dimensional tolerance. If the overall measurement falls outside these tolerances the load does not comply. For the test 24 bricks are placed on a flat surface in contact, any loose projections being first removed, and the measurements are checked by steel tape.

Dimension (mm)	Overall measurement of 24 bricks in contact (mm)
65 (thickness)	1605–1515
102.5 (width)	2505–2415
215 (length)	5235–5085

Compressive Strength Test

Each bed face is measured to the nearest millimetre and the smaller area taken as that for calculation purposes.

(1) Solid bricks and bricks with frogs intended to be laid frog downwards and perforated bricks: immerse in water at room temperature for 24 hours or otherwise saturate before testing.

(2) Bricks intended to be laid frog up: saturate as before, drain for 5 min and wipe. Fill the frogs with mortar not weaker than 28 N/mm^2 and not stronger than 42 N/mm^2, cure and store in water until required for testing.

(3) Double-frogged bricks to have both frogs filled similarly, the second from within 4–8 hours after the first.

To test, the sample is placed in a suitable crushing machine between two 3-ply sheets at least as long and as wide as the specimen. Apply the load gradually until failure occurs; the compressive strength of the specimen is calculated by dividing its maximum failing load by its area and expressed in N/mm^2. The arithmetic mean of the ten specimens tested is to be taken.

Water Absorption

Brickwork absorbs rainwater as it runs over the surface and in very cold weather this absorbed water may be subjected to freezing. All bricks used in the United Kingdom except those designated as internal quality will withstand frost attack when used in external walls, except where conditions of prolonged saturation exist. While many bricks will withstand constant soaking and freezing, others will quickly crumble and fall apart. No laboratory test is laid down by BS 3921 that will predict the frost resistance of a clay brick; however, bricks having a low rate of water absorption coupled with a high compressive strength may be considered frost resistant, for example, class A or B engineering bricks.

There are three tests mentioned by BS 3921 for water absorption, including a works control test. One of these tests is as follows

(1) Weigh the brick dry, after oven drying has taken place: this is weight A.
(2) Weigh again, after 24 hours' immersion in water: weight B.
(3) Weigh again after 5 hours' boiling in water, leaving overnight to cool in water: weight C. Then

$$\text{water absorption} = \frac{C-A}{A} \times 100 \,\%$$

[Weight B is required for calculation of the saturation coefficient, $(B-A)/(C-A)$.]

Efflorescence

This is the term given to deposits of soluble salts formed on or near the surface of a porous material as a result of evaporation of the water in which they were dissolved. New brickwork often shows efflorescence when it dries out for the first time after erection, but rarely persists after this. It is mostly quite harmless and usually disappears in a short period of time owing to the effects of wind and rain. The occurrence of efflorescence is normally due to the presence of salts in the bricks or in the mortar in which the bricks have been bedded. The soluble salts

are usually in the form of sulphates of calcium, magnesium, potassium or sodium which occur in clays used for brick manufacture: but these can also occur in subsoils and will dissolve in wet weather, rising up the brickwork because of capillary action. Occasionally bricks become contaminated as a result of being stacked on ground made up of ashes or some other material containing salts. If the unsightly white powder is persistent it may be necessary to brush it off with a stiff brush, but hosing down is not to be recommended since the wetting would dissolve the salts which would re-enter the brickwork and show again at a later date. The test laid down in BS 3921 for efflorescence is as follows.

Each specimen is wrapped in a waterproof cover such as polythene, except for the stretcher face, which is exposed uppermost and stored in a well-ventilated room. A flask of distilled water is inverted on the specimen with its mouth in contact with the exposed face. Further amounts of distilled water are added to the flask as required and after a week or so the specimen is examined for efflorescence and graded as follows (see figure 6.5).

nil | none perceptible
slight | nor more than 10 per cent of the exposed area thinly covered
moderate | heavier than slight, covering up to 50 per cent of the area
heavy | heavy deposit covering more than 50 per cent of the area but no flaking

Note   It is very rarely that bricks are found liable to moderate efflorescence.

Soluble Salts

While soluble salts are the basic cause of efflorescence, they are also the cause of sulphate attack under certain conditions, which is a far more serious problem. For sulphate attack to occur in brickwork, three materials must be present

(1) tri-calcium aluminate (C_3A): this is contained in ordinary Portland cement in varying amounts, normally 8–13 per cent
(2) soluble sulphates: as listed under efflorescence occurring in bricks, subsoils, sands
(3) water: always present in brickwork during construction, but this is insufficient in itself to cause sulphate attack

The tri-calcium aluminate constituent in O.P.C. can react with sulphates in solution to form a compound called calcium sulpho-aluminate or ettringite. When

Figure 6.5

this occurs in brickwork the effect is an overall expansion, followed by progressive disintegration of the mortar joints.

The first signs of sulphate attack are seen when the mortar joints turn a lighter colour and begin to crack at their centres. This is followed by the faces of the joints spalling away and the bedding mortar becoming weak and crumbly. This decomposition may cause brickwork to lean and eventually collapse. Sulphate attack rarely occurs in less then 3 years and can be prevented in any one of three ways.

(1) use frost resistant bricks containing only minute quantities of salts
(2) use sulphate-resisting, super-sulphated or high-alumina cement: these cements contain limited amounts of tri-calcium aluminate
(3) do not allow the brickwork to become saturated.

To test bricks for soluble salts content, the samples must be crushed and analysed in a laboratory. Table 6.1, taken from BS 3921, shows classification of bricks by compressive strength and water absorption.

Table 6.1

Class	Compressive strength in N/mm^2 (av. of 10)	Water absorption % by mass (av. of 10)
Engineering A	n.l.t. 70	n.m.t. 4.5
Engineering B	n.l.t. 50	n.m.t. 7.0
Damp-proof course 1	n.l.t. 5	n.m.t. 4.5
Damp-proof course 2	n.l.t. 5	n.m.t. 7.0
All others	n.l.t. 5	no limits
D.p.c.1 recommended for use in buildings		
D.p.c.2 recommended for use in external works		

Table 6.2

Location	Constructional Weather Conditions	Clay Brick Quality
Backing to external solid walls	Cold	Ordinary
Inner leaves of cavity walls	Warm	Internal
Internal walls and	Cold	Ordinary
partitions	Warm	Internal
Outer leaves of cavity walls above d.p.c.	All	Ordinary
Outer leaves of cavity walls below ground or below d.p.c. but 150 mm above ground	Cold	Special
External facing to solid construction	Warm	Ordinary
Walls within 150 mm of ground level	Cold	Special
	Warm	Ordinary
Parapets	Cold	Special
	Warm	Ordinary
Sills, copings	All	Special
Earth-retaining walls	All	Special
As d.p.c.	All	D.p.c. bricks or engineering

The compressive strength of clay bricks is not always a good guide to their durability and, when selecting bricks for specific situations, attention should be paid to BDA Practical Note 3, 1973, part of which is shown in table 6.2.

Calcium Silicate Bricks

These are made by bonding together a strong aggregate such as sand, crushed stone or a mixture of the two, and hydrated lime. The materials are mixed with water, together with any pigments necessary for colouring, and are accurately pressed to the size and shape required. The bricks are then transferred to an autoclave where they are subjected to high-pressure steam for a period of some hours. This causes the lime to combine with the sand to form calcium silicate. Carbon dioxide from the atmosphere acts slowly on the bricks, converting the lime content eventually to calcium carbonate, and therefore their strength and hardness improves over the years. They vary from an off-white to pink or grey in their natural colour but the addition of pigments allows the manufacture of facings in various pastel shades, some of which may be textured if required. Some of their advantages are as follows

(1) low cost, accuracy and uniformity of shape which makes laying and bonding easy
(2) smooth surface and light colours make them suitable for internal use without plastering
(3) invariably free from efflorescence unless introduced after laying from external sources
(4) unaffected by repeated freezing and thawing, unless exceptionally severe, when a class 4 brick or better should be used (see table 6.3)
(5) resistance to fire and the transfer of heat and sound is similar to that of clay bricks.

On the other hand these bricks may suffer from shrinkage cracking after laying, and movement joints must be provided at 7.5 to 9 m intervals, and more often where changes in wall thickness or height occur or openings are present. The bricks should be laid dry or almost dry to help minimise this drying shrinkage. They are liable to attack by severe frost action when saturated with calcium chloride, common salt or sea water, and will suffer from acid attack if used where they will be exposed to acid fumes or splashing. If they are used for the construction of inspection chambers and may be in contact with sewage they should be rendered internally. BS 187 deals with their properties, sizes and testing; there is no water absorption test on calcium silicate bricks but a drying shrinkage test is included. Table 6.3 gives details of the strengths attained by each class of bricks.

From table 6.3, for brickwork below d.p.c. a class 3 brick or better should be used, and for external facework, class 2 or better. While calcium silicate bricks reach engineering class B strengths they do not have the necessary water absorption figures for this class.

Concrete Bricks

These are manufactured by weigh-batching graded limestone or other aggregates, cement and water, and compressing them to produce bricks with or without frogs. They are normally steam-treated to accelerate weight-gain, and stacked prior to dispatch. As with clay or calcium silicate bricks they may be banded for offloading by crane if required. The natural brick colour varies from pink to grey, depending on the aggregates used in the manufacture, but the addition of pigments enables the supply of a good range of colours. Their texture can be smooth, rustic or split-faced. Concrete bricks have a good resistance to rain penetration, a high degree of fire resistance and, owing to their density, good sound insulation properties. They can be used in almost all situations, except as d.p.c. bricks, depending on their compressive strength category, which is given in BS 6073, part of which is shown in table 6.4.

The positions in which each of the categories of bricks can be used are shown in table 6.5.

From table 6.5, for brickwork below d.p.c., a brick with a compressive strength of 20 N/mm^2 or over is suitable. For concrete bricks used in sulphate-bearing soil, sulphate-resisting cement is the binder used in the bricks. There is no water absorption test detailed in BS 6073 for concrete bricks, but a test for drying shrinkage is included, as for calcium silicate bricks.

Comparison of Drying Shrinkage

All building materials, except materials such as metals, glass and plastics expand and contract on wetting and drying. Comparison of the three types of brick discussed is given in table 6.6.

Table 6.3

Designation	Class	Mean compressive strength of ten bricks n.l.t. (N/mm^2)	Predicted lower limit of compressive strength n.l.t. (N/mm^2)
Loadbearing bricks	7	48.5	40.5
	6	41.5	34.5
or	5	34.5	28.0
	4	27.5	21.5
facing bricks	3	20.5	15.5
Facing brick or common brick	2	14.0	10.0

Table 6.4

	Compressive Strength Category (N/mm^2)					
Minimum compressive strength of average of 10 bricks	7.0	10.0	15.0	20.0	30.0	40.0

Table 5.5

Element of Construction	Details	Minimum Brick Quality (N/mm^2)
Inner leaf of cavity walls and internal walls	Plastered Unplastered	7.0 7.0
Backing to external solid walls		7.0
External walls, including outer leaf of cavity walls and facing to solid construction	Above d.p.c. near ground Below d.p.c. but more than 150 mm above ground level Within 150 mm of, or below ground level	15.0 15.0 20.0
External free-standing walls		15.0
Parapets	Rendered Unrendered	20.0 20.0
Sills and copings		30.0
Earth-retaining walls		30.0

Table 6.6

Material Used	Normal Range of Drying Shrinkage (%)	Movement Joints Spaced at Intervals of (m)
Concrete bricks	0.02–0.06	6
Calcium silicate bricks	0.01–0.035	7.5–9
Clay bricks	0.00–0.015	12

Some other advantages of concrete bricks are their dimensional accuracy (±2 mm), consistency of colour and texture, increasing strength and reasonable costs. They must be dry or almost dry before laying, to avoid excessive drying shrinkage, and movement joints in long walls should be placed at 6 m intervals to accommodate tensile stresses.

Compressive Strength Test

The test differs from that on clay bricks in so far as any frogs are left unfilled, and bed face areas are calculated as shown below. Once again the bricks are saturated before testing.

(1) Using bricks without frogs, the area of the smaller bed face is used for calculations.

(2) For bricks with one frog, the area used is the area of the bed face less the area of the frog.
(3) If the bricks are double frogged, the area to be used is the smaller net bed face, that is, the area of the bed face less the area of the larger frog.
(4) For bricks with holes, the gross area of the smaller bed face is taken.

The test from this point is similar to that for clay bricks.

Drying Shrinkage Test

Brief details of this test are as follows. Four specimens are required and a small steel ball is cemented into the centre of both header faces of each. An accurate measuring apparatus incorporating a suitable dial gauge is necessary, such as that shown in figure 6.6. The gauge has a conical recessed end which locates on to the steel ball in the upper header face as shown. The bottom steel ball locates on to a conical recessed support.

The bricks are immersed in water for 4—7 days and measured, after which they are oven-dried for up to 22 days and measurements are taken at specified times during this period. Thus the drying shrinkage can be determined; for further details refer to BS 187.

Metric Modular Bricks

These sizes were briefly popular between 1970 and 1980, and buildings can be seen constructed from these bricks. Because of their poor appearance and lack of versatility, however, they are no longer available.

Work size (mm)	Format size (mm)
190 x 90 x 90	200 x 100 x 100
190 x 90 x 65	200 x 100 x 75
290 x 90 x 90	300 x 100 x 100
290 x 90 x 65	300 x 100 x 75

Cut Bricks and Specials

In certain bonding situations (see Volume 2) it is necessary to use cut bricks in order to avoid the occurrence of straight joints. These are shown in figure 6.7.

BS 4729 specifies over 40 standard specials; some of the more common of these are shown in figure 6.8.

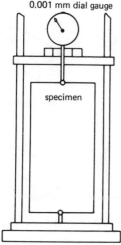

0.001 mm dial gauge

specimen

Figure 6.6

snapped header

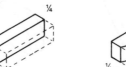
queen closer

king closer

bevelled closer

small bevelled bat

quarter bat

three-quarter bat

large bevelled bat

Figure 6.7 Terms applied to cut bricks

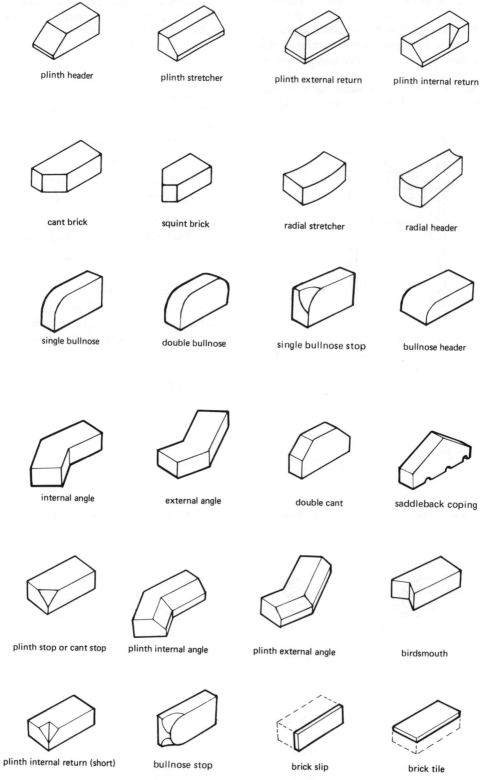

plinth header

plinth stretcher

plinth external return

plinth internal return

cant brick

squint brick

radial stretcher

radial header

single bullnose

double bullnose

single bullnose stop

bullnose header

internal angle

external angle

double cant

saddleback coping

plinth stop or cant stop

plinth internal angle

plinth external angle

birdsmouth

plinth internal return (short)

bullnose stop

brick slip

brick tile

Figure 6.8 Standard specials (BS 4729)

7
MORTARS

Mortar can be thought of as a 'gap-filling adhesive' which evens out any slight irregularities in size and shape between bricks, gives an even bed and ensures uniform distribution of loads. It also influences the compressive strength, durability and resistance to rain penetration through brickwork.

Mortar consists basically of a binder, an aggregate and water, to which plasticisers, colouring pigments, etc., can be added to achieve one objective or another.

MATERIALS

Materials for mortars must comply with the following British Standards

(1) ordinary Portland and rapid-hardening cement: BS 12
(2) sulphate-resisting cement: BS 4027
(3) high-alumina cement: BS 915
(4) sand: BS 1200
(5) water, mains water preferred: refer to BS 3148
(6) lime (non-hydraulic and semi-hydraulic): BS 890
(7) plasticisers: BS 4887
(8) pigments: BS 1014.

Numbers 1 to 5 of this list are covered in the chapter on concrete; details concerning the rest are as follows.

Lime

Pure lime (calcium oxide) is obtained by burning pure chalk or limestone (calcium carbonate) in a kiln. During the burning process carbon dioxide is given off and after burning, quicklime is left. This may be sold as lump quicklime, but more often than not it is steam-treated at the works for the chemical process of slaking and dispatched as a dry white powder in 25 kg moisture-resisting bags. Slaking is accompanied by the release of a considerable amount of heat, coupled with a large expansion.

Other limes contain various amounts of impurities (clays) which confer different properties on pure lime and because of these impurities certain limes are not steam-treated, since this process can lead to the loss of certain properties.

Three types of lime are available for mortars, eminently hydraulic, semi-hydraulic and non-hydraulic.

Eminently Hydraulic Lime

This lime is similar in properties to O.P.C., but neither hardens so quickly nor attains such high strengths. The limestone from which it is produced contains impurities such as alumina, silica and iron oxide which, on burning, combine with a proportion of the lime and are converted into compounds similar to those present in cement. Thus it is quite suitable for use in weak mortars and concretes and will set under water if required. It is not supplied in hydrated form since its hydraulic properties cannot be guaranteed when slaking has occurred.

Semi-hydraulic Lime

This contains smaller amounts of impurities and thus will harden much more slowly. It is normally added to cement—sand mixes to provide workability, but can be used with sand only for the construction of non-loadbearing internal walls.

Non-hydraulic Lime

These limes are known as 'fat' or 'rich' limes and will not harden under wet conditions. Setting takes place by a process called carbonation (taking in air), thus it develops little strength in a long time. It is used in mortar for its other properties, which are listed later.

To test for hydraulicity is simple enough: break the fresh burnt lime into lumps, place in a muslin bag and immerse for 5 seconds in water. Drain off and empty into a beaker or similar. If slaking occurs within 5 min, the lime is non-hydraulic; if slaking takes between 5 min and 2 hours the lime is semi-hydraulic and if the lime is inactive for more than 2 hours it is classed as hydraulic.

Mortars containing non-hydraulic or semi-hydraulic limes have excellent working qualities, good water retention and bonding properties. Less shrinkage occurs and a more attractive colour is produced than when using cement—sand mortars.

It is not necessary to add lime to masonry cement, which has its own plasticiser added, and it must not be added to high-alumina cement or a flash set may occur.

Plasticisers

These may be in liquid or powder form and it is important to keep to the supplier's instructions regarding amounts to add, and mixing times. Plasticised mortars should not be made in a roll-pan mixer since these fail to entrain sufficient air bubbles, also prolonged drum mixing tends to entrain an excessive amount of bubbles, with a subsequent weakening of the mortar. Liquids should be added to the mixing water whereas powders are added straight to the mixer. They have the effect of entraining micro-bubbles of air in the mortar, breaking down surface tension and thus improving workability.

Plasticisers may be added to all cements except masonry cement, and freshly laid mortars containing these additives have improved resistance to frost action, both during and after setting.

Pigments

Coloured cements, also covered by BS 12, are white or O.P.C. to which between 5 and 10 per cent of pigment has been added at the grinding stage. Use of these cements will usually give better distribution of colour throughout the job than will be obtained by adding pigments to the mixer. When using the latter method it is important to keep to the manufacturer's instructions since the addition of too much pigment, especially of carbon black, will affect the final strength of the mortar. To keep to exactly the same proportions too is vital since any variations will show up on the finished brickwork as different shades.

Types of Mortar

Mortars are divided into four main types

(1) cement—sand
(2) lime—sand
(3) cement—lime—sand
(4) cement—sand plus plasticiser.

Cement—Sand

A mix of O.P.C. and sand usually produces a harsh, unworkable mortar unless in the ratio of about 1:3, which is far too strong for most purposes (see table 7.1). Joints are liable to cracking owing to excessive drying shrinkage and unless the correct strength of brick is used with this mortar, vertical cracking may be caused in the brickwork because of lack of flexibility. Therefore masonry cement—sand is preferred,

which will give a workable mix as well as producing the required strength when used in the correct proportions (table 7.1). It is important to follow the manufacturer's instructions when using this type of cement.

Lime—Sand

The strength of mortars made with lime as the binder varies according to the type of lime used. Only the hydraulic limes give any real strength and mixes of 1:2 or 1:3 are comparable to 1:7 or 1:8 sand—cement and their uses therefore are somewhat limited. Semi-hydraulic lime sand mixes can be used for internal non-loadbearing walls, but this and non-hydraulic lime are more often used to provide workability in cement—lime—sand mortar.

Cement—Lime—Sand (Compo)

Used in varying proportions these materials provide a range of mixes which are able to meet most of the requirements of a good mortar. They are more workable than cement—sand and stronger than lime—sand. The cement, lime and sand can be delivered separately to site and added to the mixer as required, but it is preferable to soak the lime overnight in water and add the resulting lime putty to the mixer with the other materials. This will improve the workability by up to 50 per cent.

Ready-mixed lime—sand complying with BS 4721 is an alternative to using separate materials and is delivered in a wet condition to site. This is added to the mixer with the required amount of cement and sufficient extra water to give the necessary workability. Ready-mixed lime—sand is known as 'coarse stuff', and should be stored on a lean concrete base and protected against the elements during winter. If it is allowed to stand for long periods in hot sunshine it may harden and will require crushing before mixing unless a roll-pan mixer is used.

Cement—Sand plus Plasticiser

Plasticisers provide an alternative to lime for improving the working qualities of cement—sand mortars. The effect of the air bubbles that are entrained in the mix is to increase the volume of cement paste and fill the voids in the sand. It is generally considered easier to mix than cement—lime—sand, but produces darker mortars that are liable to stain light coloured bricks especially calcium silicates.

Black Mortar

This is made by grinding down ashes and/or brick

rubble and mixing with hydraulic lime. Cement is not normally used as the binder since it may react with impurities that may be present in the ashes, thus the strength attained by this mortar will be low. The darker joints give a contrasting colour and the mortar is relatively cheap but it is not often seen nowadays because ashes and hydraulic lime are not readily available. Use of cement, sand, a plasticiser and black pigment is usually preferred.

Mortars for Use in Sulphate-bearing Soils

As with the bricks used under these circumstances, the mortar too must be able to resist sulphate attack. It is therefore important to use sulphate-resisting cement as the binder in ground containing moderate amounts of sulphate, and where high concentrations occur the use of super-sulphated cement may be necessary.

REQUIREMENTS FOR MORTAR

Workability

If a mortar is harsh the output of bricklayers will be reduced. Picking up and spreading will be slower, and difficulty will be experienced in placing cross joints. It is therefore important that the specified mix should contain sufficient lime or plasticiser.

Adequate Strength

The final strength of a mortar must not exceed that of the bricks used. The use of too much cement will produce a rigid mortar which may cause vertical cracking in areas of brickwork if any movement were to occur (figure 7.1). Use of the correct mortar in these circumstances will result in any cracking following the joints, which will be much easier to correct (figure 7.2).

Figure 7.1

Figure 7.2

Plasticity Retention

This should be sufficiently long for the bricks to be laid and adjusted, but stiffening must not be over-delayed. Poorly specified mixes may dry out before jointing up has been completed, especially if dry bricks are used. On the other hand, if too many courses are laid on an over-plasticised mortar the bed joints may begin to squeeze out from the lower courses.

Durability

This must be adequate for the situation. A weak mix may be suitable for internal walls but would weather very badly on chimney stacks, for example.

Good Bond with the Bricks Used

Strong mortars are not the best from the point of view of rain penetration owing to their high shrinkage, which tends to leave fine cracks between the joint and the bricks. Also sand lime and concrete bricks have a high drying shrinkage which would further aggravate the problem.

WORKING CHARACTERISTICS

Gauging

It is of little use selecting the appropriate mix for a situation if materials are measured out by the shovelful. As with concrete, use of the appropriate gauge boxes is the only accurate way when batching by volume. Weigh-batching is preferred since use of this method permits the highest level of control.

Mixing

The cementitious properties of the cement may be largely lost if mortar is allowed to stand for long periods and has to be re-mixed. Mortar should be used within 2 hours of adding water and any surplus after this time should be discarded.

When mixing by hand the materials should be carefully measured out on a clean, hard surface and turned continuously until of an even colour throughout. Water can then be added and the process repeated until the mortar is of the required consistency.

Winter Working

Sands should be covered down on delivery and may require heating before use. Mortars used in cold weather should be a little stronger than usual and should be aerated rather than adding lime for workability, since these harden more quickly and have greater frost resistance. Below 4 °C mixing water can be heated up to 40 °C and the temperature of the completed work must not be allowed to fall below 4 °C until the mortar has hardened, which may take up to 7 days, depending on the type of cement used. If completed work is allowed to freeze the water in the mortar will freeze from the face inwards and at best the face of the joint will crumble and fall out. Unless the walling is closely covered down and insulated the joints may freeze completely with a resulting loss of strength, and the work will have to be pulled down and re-built.

Experiment

To study the action of soluble sulphates on mortar cubes made from different cements, make suitably sized mortar cubes using the same proportions of

(1) ordinary Portland cement and sand
(2) sulphate-resisting cement and sand
(3) super-sulphated cement and sand
(4) high-alumina cement and sand

Mark each cube for identification and place in a closed jar containing a sulphate solution. Make regular checks on the cubes and note any deterioration. The mix can be altered if required, including the addition of lime or an air-entraining plasticiser. The strength of the sulphate solution too can be increased if required.

PRE-BAGGED DRY MIXES

These consist of mortar, rendering and concrete mixes which are delivered to site in 50 kg bags. All that is required is the addition of sufficient water for workability. The mixes are prepared from dried aggregates, cement and plasticisers if necessary, and use of this method has several advantages over traditional methods.

(1) All proportioning is carried out by weigh-batching perfectly dry materials, thus mixes never vary and the colour of the joints is constant.
(2) The bags containing the materials are tough, dust-proof and have waterproof linings and if properly stored will not deteriorate for many months.
(3) Stockpiling of loose materials is avoided; twenty bags stacked ten high cover an area of less than 2 m^2.
(4) Where materials have to be delivered in busy roads with restricted access, bagged materials can be quickly and cleanly off-loaded, and transferred where required.
(5) Where work is to be carried out on upper floors, the use of dry-bagged mortars is the cleanest and most convenient method.
(6) For very small jobs this is an economical method, any unused part-bags being returned for use elsewhere.

Two or three different mixes are available and special mixes are obtainable on request from the manufacturers. The bags should be stored on site as for cement and used in rotation, that is, first in, first out.

Table 7.1 Recommended Mortar Mixes

Use	Construction	Built in	Recommended Mixes
Internal walls	Partition walls, etc.	Mild weather	1:3 hydraulic lime—sand 1:3:10 cement—lime—sand 1:8 cement—sand + plasticiser 1:7 masonry cement—sand
		Cold weather	1:2 hydraulic lime—sand 1:2:8 cement—lime—sand 1:7 cement—sand + plasticiser 1:6 masonry cement—sand
External walls	Above d.p.c. in sheltered positions	Mild weather	1:2 hydraulic lime—sand 1:2:8 cement—lime—sand 1:6 cement—sand + plasticiser 1:6 masonry cement—sand
		Cold weather	1:2 hydraulic lime—sand 1:1:6 cement—lime—sand 1:6 cement—sand + plasticiser 1:5 masonry cement—sand
	In exposed positions such as parapet walls, free-standing walls and work below d.p.c.	Mild weather	1:1:5 cement—lime—sand 1:5 cement—sand + plasticiser 1:4½ masonry cement—sand
		Cold weather	1:1:5 cement—lime—sand 1:5 cement—sand + plasticiser 1:4 masonry cement—sand
	Work below ground level in sulphate-bearing soils	All weathers	1:5 sulphate-resisting cement—sand + plasticiser
	Work below ground level in ground containing high concentrations of sulphate	All weathers	1:5 super-sulphated cement—sand + plasticiser 1:5 high-alumina cement—sand + non-alkali plasticiser
Heavy engineering work	Large retaining walls, d.p.c.s, power houses, etc. Clay brickwork of 48.5 mN/m² crushing strength and over	All weathers	1:0—¼:3 cement—lime—sand

8

SUBSTRUCTURES

WALLING UP TO DAMP-PROOF COURSE LEVEL

The setting out of brickwork at foundation level is an operation which requires care and accuracy. The following method is recommended when working in trenches below ground level

(1) Attach the ranging lines to the external wall line positions on all return profiles.
(2) Place a 3 mm mortar screed on the concrete surface at each external angle.
(3) Using the large plumb level, plumb down the position of the ranging line to the mortar screed. (This method of transfer is suitable for trenches up to 1 m in depth; if the depth exceeds 1 m it is advisable to use a large straightedge and a stay board; this is a two-man operation.) (See figures 8.1 and 8.2.)
(4) A builder's square can then be used to mark the completed right-angle on the mortar screed.

(5) The concrete foundations should be checked for level.
(6) Having determined the accuracy of the concrete surface, a gauge staff is used to obtain the number of courses from d.p.c. level to the concrete foundations; this is termed 'gauging down' (figure 8.3). The level is taken from the corner datum or it can be derived from the profiles if the method of fixing and levelling profiles is used (figure 8.4).
(7) The external wall is then built up to d.p.c. level and protective planking placed on the top of this walling. The inner leaf is then built up to ground level and the cavity is then filled with fine concrete up to ground level, chamfering down from the inner leaf to the outer wall (figure 8.5). The planking on top of the external wall prevents any staining of the face brickwork, filling of the cavity being carried out within the area of the building. If any pipes or ducts are required to pass through the walling, sand courses should be inserted as the work proceeds.

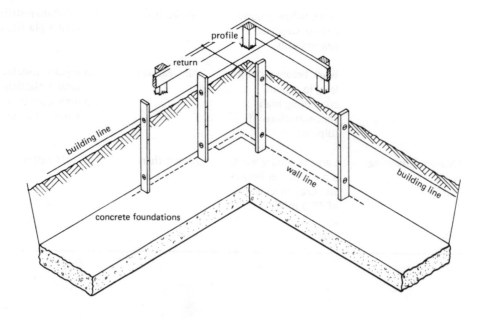

Figure 8.1 Transferring wall lines

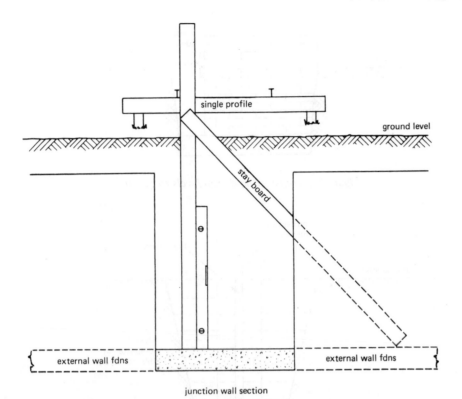

Figure 8.2 Method of transferring wall lines in deep trenches

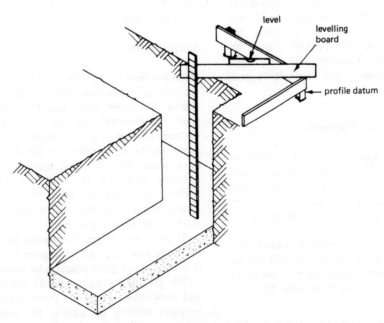

Figure 8.3 Method of gauging down from profile datum

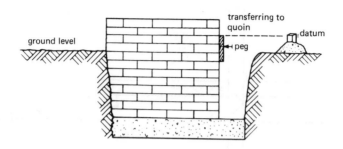

Figure 8.4 Transferring datum levels to quoin

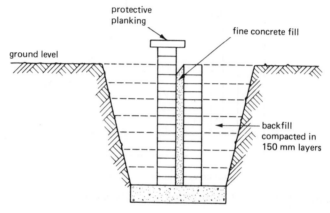

Figure 8.5 Fine concrete cavity fill and trench backfilling

(8) Backfilling of trenches should only be carried out after all walls are built up to d.p.c. level, the mortar completely set and the concrete fill inserted in the cavity. The backfill should be placed in layers not exceeding 150 mm, large stones and rocks rejected and compaction provided for each layer.

Construction of Foundation Brickwork

It is normal practice to leave the face of foundation brickwork from the trowel, but owing to the restricted amount of working space the bricklayer's productivity is considerably reduced; this situation can be substantially improved if the width of the trench is increased.

Since the method of excavation is usually by mechanical plant this part of the cost would not be greatly increased, but there would be an increase in the cost of the concrete; this would be offset by

(1) greater productivity
(2) quality of work below ground level considerably improved

(3) other work and crafts could be programmed to start earlier.

Building of Walls

When building foundation brickwork it is advisable to set out all walls one course high; if the situation allows, the inner walls may be built up to ground level. The external walls are then built to within two or three courses of ground level. It is essential at this point to check all walls for dimensions and squareness; when this operation has been completed the external walls are then set out in facing bricks, making any necessary adjustments to the bonding arrangements; jointing of the face brickwork also starts at this level (figure 8.6).

If the depth of the foundation brickwork, that is, from the concrete base to ground level, exceeds 1 m it is considered good practice to introduce 'bumpers' which greatly increase stability during construction and backfilling. These are three-quarter bats built in in pairs butting up against the external leaf, the spacing usually being six courses vertically and every five or six bricks horizontally staggered. While this incurs a three-course straight joint at the centre of the two headers, very little strength is lost and no cutting on face is involved (see figure 8.7).

Placing of Materials

Before any materials are placed for brickwork below ground level consideration should be given to the following

(1) reducing fatigue for the craftsman
(2) safety and preventing 'fall in' of trench sides
(3) easy access to replenish materials.

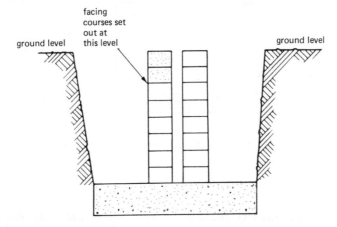

Figure 8.6 Setting out facing brickwork using increased trench width

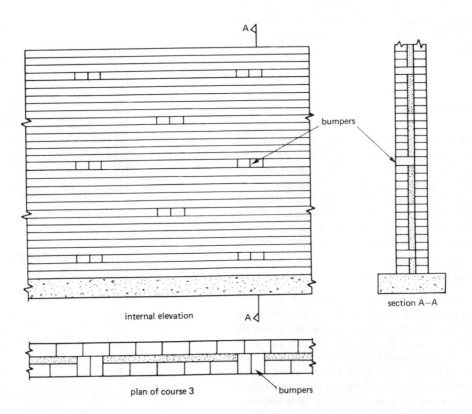

Figure 8.7 Use of 'bumpers' or stabilising headers

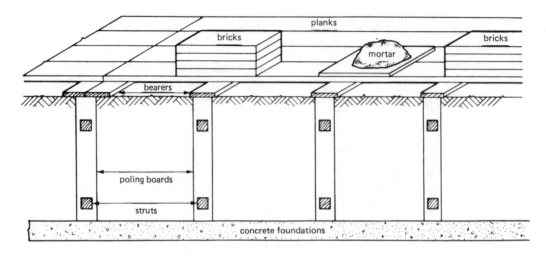

Figure 8.8 Materials placed for work below ground level

To comply with the above considerations it is essential to provide material stacks close to the position where they are required; brick stacks should not exceed 450 mm; platforms should be provided to support materials, prevent 'fall in' and allow access for uninterrupted conveyance. Mechanical plant should not be allowed within 1.5 m of any excavations (figure 8.8).

SLEEPER WALLS

Sleeper walls are dwarf walls that are constructed under suspended timber floors. The purpose of the sleeper wall is to support the floor joists and to prevent any sagging occurring in the length of the joists. As floor joists usually span the shortest horizontal distance of a room the direction of the sleeper walls is across the longest distance.

The spacing of the sleeper walls is determined by

(1) the span of the room
(2) the dimensions of the floor joists to be used.

Figure 8.9 shows the plan of a room with a span of 4.0 m between the sleeper walls; if the joists are spaced at 450 mm, the dimensions required would be 50 x 200 mm. Figure 8.10 is the plan of the same room, but a sleeper wall is built in the centre and the joists have the same spacing of 450 mm; because of the reduction in the span of the supports the dimensions of the joists are now reduced to 50 x 100 mm. Figure 8.11 is the plan of a room with a span of 5.0 m,

with joists spaced at 400 mm. The dimensions required would be 63 x 225 mm, but with two sleeper walls spaced 1.6 m apart the dimensions, with the same spacing of joists, would be 38 x 100 mm.

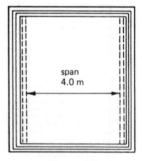

span
4.0 m

Figure 8.9

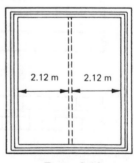

2.12 m   2.12 m

Figure 8.10

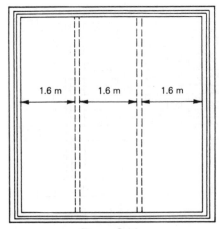

Figure 8.11

Wall Plates

Sleeper walls may be provided with a timber wall plate and a d.p.c. must be placed under the wall plate. The purpose of the wall plate is to distribute the load over the entire length of walling. Figure 8.12 shows a section of a cavity wall with a sleeper wall parallel to the inner leaf and a maximum of 50 mm between the two walls. The Building Regulations require the minimum distance from the upper surface of the oversite concrete to the underside of the floor joists to be not less than 125 mm, and at least 75 mm to the underside of the wall plate; therefore the minimum height of a sleeper wall is one course.

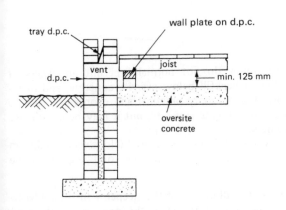

Figure 8.12

Ventilation

It is essential that all floor timbers are kept in a dry state and receive a constant supply of fresh air. When timber becomes damp and the moisture content

exceeds 20 per cent of the weight of the timber, and a source of ventilation is not available, then the conditions for dry rot are provided. To provide ventilation for suspended timber floors, ducts are formed in the cavity wall. This is achieved by 'boxing' in the air vent on the external wall and inducing air beneath the timber floor. Because the duct bridges the cavity a tray d.p.c. must be formed above the air vent, extending at least 150 mm on either side of the vent.

Underfloor ventilation requires air vents to be provided in opposite walls and where possible the vents should be aligned, ensuring constant currents of air for the area below the floor (figure 8.13).

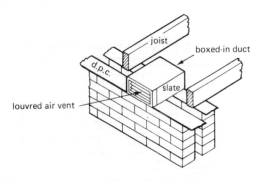

Figure 8.13

Honeycombs

Sleeper walls are built in honeycomb form to facilitate the passage of air; the honeycombs can be termed 'piercings' or 'holes'; the width of the honeycombs is determined by

(1) the length of the wall
(2) the thickness and height of the wall.

The maximum width of a honeycomb should not exceed 110 mm and the minimum is 38 mm but they can vary according to the two factors above.

Construction of Honeycomb Sleeper Walls

To provide honeycomb sleeper walls with stability the following recommendations are made

(1) the sleeper wall should be adequately tied into the inner leaf of the cavity wall
(2) the first and top courses must be of solid construction
(3) sleeper walls of half brick in thickness should not be built above 1.0 m in height
(4) when the depth of the floor exceeds 1.0 m, the wall thickness must be at least one brick (figure 8.14).

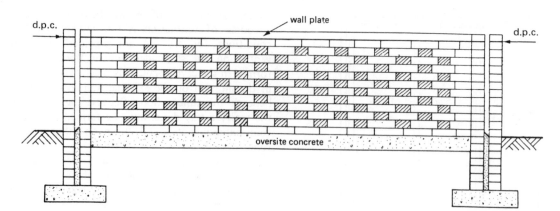

Figure 8.14

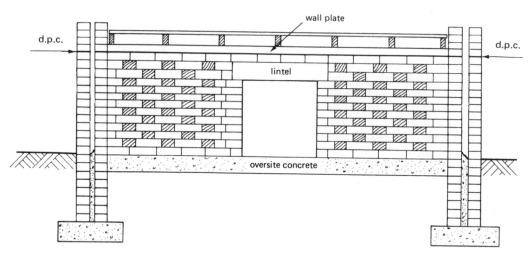

Figure 8.15

Underfloor Access

Underfloor areas often require inspection; work is also necessary when heating installations, electricity supply and other services are designed to occur below floor level; this requires openings to be formed in the sleeper walls. Openings should be at least 600 x 600 mm; the lintel should be reinforced concrete, and the walling below the lintel seating, and openings reveals, should be of solid construction (figure 8.15).

Sleeper Walls on Sloping Sites

As previously explained, stepped foundations are provided on sloping sites, but the area of the building contained within the external walls will also have a similar amount of slope (see figure 8.16). Local

Authorities do not normally permit an excess of 600 mm backfill below oversite concrete, and therefore it is not only practical and economical also to step the oversite concrete, but it may be mandatory to do so.

Where it is necessary to build sleeper walls on stepped oversite concrete, the following should be considered.

(1) If the direction of the sleeper wall is parallel to the line of the step, the position of the wall should be at the overlap of the concrete, the first course of the wall starting with a header course (figure 8.17).

(2) If the direction of the sleeper wall is at right-angles to the line of the steps, the course containing the honeycombs is determined by the step above; this may require two or more courses to be solid at the lower level of the wall (figure 8.18).

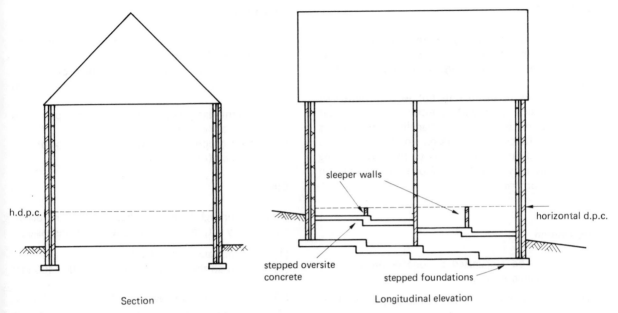

Section

Longitudinal elevation

Figure 8.16

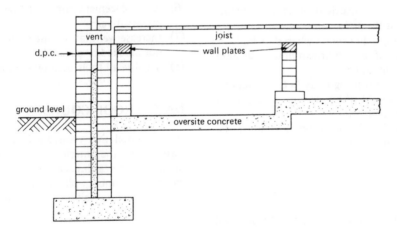

Figure 8.17

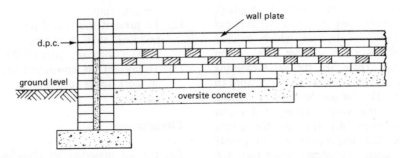

Figure 8.18

DRY ROT

There are two types of dry rot and one type of wet rot, but the most persistent and destructive is the true dry rot fungus known as *Serpula lacrymans* (weeping fungus).

This occurs in damp (but not wet) unventilated situations such as cellars, pantries, below timber ground floors, behind wood panelling, etc. and is caused by a fungus which lives on wood and moisture. The first indication of dry rot to the unsuspecting householder may be when his foot passes through an infected floorboard.

Recognition

Dry rot can be recognised by one or more of the following characteristics.

(1) Irregularities or waviness in the surface of skirtings or panellings.
(2) The presence of a mouldy, musty smell.
(3) Fine, white strands passing along the surface of timber or brickwork.
(4) The surface of floorboards or skirtings splitting along and across the grain.
(5) Woodwork giving a dull sound when struck.
(6) The appearance of reddish-brown powder around joints or cracks in timber.
(7) The presence of a fruiting body (a thick, yellow, pancake-like growth) which denotes that dry rot has existed for some time.
(8) Test with a knife: if the wood is infected the blade will push in and withdraw easily.

Cause

Dry rot is caused either by contact with affected wood or by the germination of microscopic spores on sufficiently damp timber. Fruiting bodies discharge these spores which may be spread by insects, rodents, draughts, etc., over very large distances. Once the spores are deposited on damp timber in stagnant, unventilated atmospheres they germinate, sending out fine, hair-like strands of hyphae which quickly produce sheets of mycelium — rather like cotton wool in appearance. These are the water-carrying tubes which will pass over concrete and brickwork to reach other timber. The fungus develops by feeding on the cellulose in the wood, leaving it dry and friable, hence the term 'dry rot'. As the attack becomes established the fungus grows a fruiting body on which further spores are formed which form the basis for the next attack.

Prevention

(1) Keep areas below ground floors ventilated and free from dampness.
(2) Timber should be impregnated with preservatives.
(3) Use well-seasoned wood and never paint damp wood.
(4) Efficient damp-proof courses without punctures or gaps are essential.
(5) Mortar droppings in cavities should be kept to a minimum (use of laths).

Cure

After a thorough inspection has revealed the occurrence of dry rot, drastic treatment may be necessary.

(1) Find the cause and correct it.
(2) Cut out all the infected timber and 300–400 mm past this.
(3) Carefully carry this away and burn it immediately.
(4) Apply a blowlamp to all brickwork and concrete around the infected area.
(5) Apply three coats of preservative to surrounding timberwork.
(6) All replacement timber to be impregnated with preservative.
(7) Increase ventilation, ascertain that airbricks are not blocked or covered.
(8) Sterilise all tools used on completion.

DAMP-PROOF COURSES

It is essential that moisture is prevented from penetrating into the interior of buildings. Dampness in buildings is a health hazard to the occupants; it causes damage to timber, plaster and paintwork and ultimately structural failure may occur.

Functions of Damp-resistant Materials

(1) To prevent water entering the interior of the building by passage from the ground into the sub-structure.
(2) To prevent the lateral passage of water into the building, that is, where the outer leaf forms contact with any part of the internal structure.
(3) To prevent the downward passage of water from all work above roof level into the interior of the building.

Classification

Damp-proof materials are classified as flexible, semi-rigid or rigid.

Flexible Materials

Lead BS 1173. Weight 19.5 kg/300 mm². When used to prevent upward movement of water it should have a lapped joint of at least 100 mm; if used to prevent the downward passage of water a welted joint should be formed.

Lead does not extrude under pressure; it corrodes if in contact with cement and limes but it can be treated to prevent corrosion by applying a coat of bitumen. Although it is expensive it is very suitable for loadbearing structures.

Copper BS 2670. Weight 2.2 kg/m². The use of copper in preventing the movement of water upwards and downwards is the same as for lead.

Copper will not extrude under pressure, but it should receive bituminous treatment when in contact with soluble salts. When exposed to weather it often causes staining of the brickwork face.

Bitumen BS 743. The following are forms of bituminous materials

> hessian base
> fibre base
> asbestos base
> hessian base with lead
> fibre base with lead
> asbestos base with lead
> sheeting with hessian base

All the above should have a lapped joint of at least 100 mm to prevent the upward passage of water; to prevent downward movement of water the joint should be lapped and sealed with bitumen. All bituminous materials may extrude under pressure; when used in loadbearing structures of long-life requirements the bitumen should contain lead.

Polythene BS 743. Black polythene should have a minimum thickness of 0.5 mm. When it is used to prevent the movement of water downwards it should have a lapped joint of a distance at least equal to the d.p.c.; to prevent the downward passage of water the joint should be welted. Polythene will not extrude under pressure; if adhesion is required the polythene will require treatment.

Black Pitch/Polymer To prevent water moving upwards the joint should be lapped at least 100 mm. The joint required when preventing the downward movement of water is a lapped/sealed joint. This material does not extrude under pressure; it will accommodate lateral movement.

Semi-rigid Materials

Mastic Asphalt BS 1097 or BS 1418. There are no joint problems with this material; it may extrude if the pressure exceeds 65 kN/m². To provide adhesion for mortar the asphalt should have sand beaten into its surface while still in a heated condition.

Rigid Materials

Bricks BS 3921. Bricks should have a maximum water absorption not exceeding 4.5 per cent. Bricks are not suitable as d.p.c. when the downward movement of water is required to be prevented. Joints should be completely filled with mortar and adequate lap provided. The bedding mortar should be 1:3 cement/sand.

Slates BS 3798. All slates should be at least 230 mm in length; they should be able to pass the wetting and drying test, also the acid immersion test. Slates are suitable for resisting upward and lateral water movement; the minimum thickness of slate should not be less than 4 mm. Joints should be formed with adequate lap and the bedding mortar should be 1:3 cement/sand.

Epoxy/Resin/Sand The resin content of this material should be at least 15 per cent, a hardening additive should be used and the minimum effective thickness should be 6 mm, This material does not extrude and there are no joint problems.

Treatment of Damp-proof Course

All bituminous materials should be laid on flat surfaces; when used at the base of walls they should be placed on a flat mortar bed at least 6 mm in thickness. When external walls are required to be rendered the d.p.c. should project beyond the rendering. Walls should not be rendered below the d.p.c. Lead and copper should also be placed on a mortar bed if there is a possibility of undulations occurring in the bedding surface. Joints should be avoided in d.p.c.s used over door and window openings.

Selection of Damp-proof Course

For a damp-proof course to be effective it is essential to select the material that will comply with the following considerations

(1) the position in the building where the d.p.c. is required
(2) design of the structure
(3) the total loading on the d.p.c.

(4) other forces that may be applied to the d.p.c.
(5) will the d.p.c. extrude or suffer from 'creep'?
(6) the ability of the material to accommodate change of direction and level
(7) cost considerations
(8) other building materials.

Positions for Damp-proof Courses

External Walls

A damp-proof course must be inserted in all external walls, that is, cavity or solid construction, at a height of not less than 150 mm above the finished external ground level. Its function is to prevent the upward passage of water (figure 8.19).

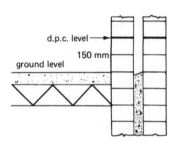

Figure 8.19

Sloping Sites

The damp-proof course must be inserted at least 150 mm above the finished external ground level at the lowest point, following the line of the slope but maintaining at least 150 mm in all places. The damp-proof course can therefore be stepped within the vertical joints of the wall. Pitch/polymer is very suitable because of its ability to accept adjustments laterally and vertically (figure 8.20).

Floor Membranes and Wall d.p.c.

When the surface of the oversite concrete is level with the wall d.p.c. the floor membrane must be lapped by the internal wall d.p.c. a distance of at least 100 mm and a sealed joint formed (figure 8.21).

When the oversite concrete is formed below the level of the internal wall d.p.c., the membrane must be turned up the face of the wall and lapped over the d.p.c. at least 75 mm. A sealed joint is not necessary in this position (figure 8.22). Further details are given in Volume 2, chapter 1.

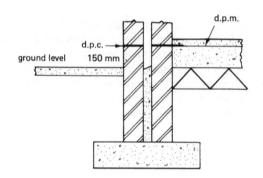

Figure 8.21

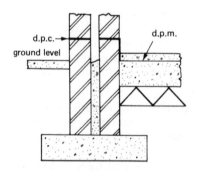

Figure 8.22

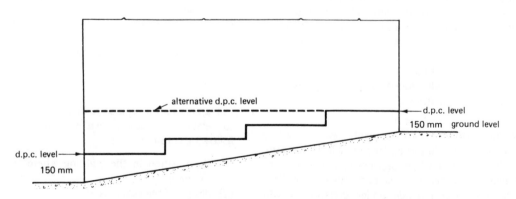

Figure 8.20

Suspended Timber Floors

It is essential to provide a d.p.c. where structural timber comes into contact with concrete or brickwork, therefore the d.p.c. is inserted below the timber wall plate and on top of the sleeper wall (figure 8.23). When floor joists are built into the wall it is necessary to treat the ends of the joists with a preservative and to place the joists on the internal wall d.p.c.

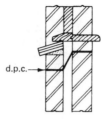

Figure 8.25

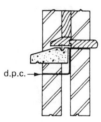

Figure 8.26

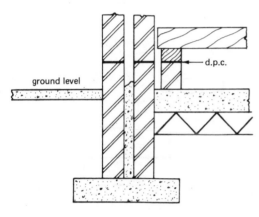

Figure 8.23

Window Sills

When bricks, concrete or other porous materials are used to form sills, a d.p.c. should be inserted under the sill or the brickwork immediately below, extending to the back of the sill and turned up within the cavity to lap over the brickwork of the internal wall. The height of tray should be 150 mm and it should extend at least 100 mm on each side of the sill. This treatment should also be used for impervious materials if they contain any mortar joints in the length of the sill. The d.p.c. should be flexible and able to accommodate change of level (figures 8.24 to 8.26).

Door and Window Reveals

In cavity wall construction it is usual for the inner leaf of brickwork to meet the outer leaf at the formation of reveals. To prevent the lateral passage of water a vertical d.p.c. should be inserted. The d.p.c. should be flexible and formed without joints (figure 8.27).

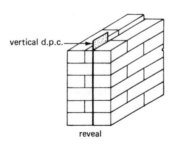

Figure 8.27

Over Door and Window Openings

A flexible d.p.c. should be placed over the top of all external frames or lintels, extending at least 150 mm on each side of the frame or lintel; it should be turned up within the cavity and lap over the inner leaf. The height of the tray should be at least 150 mm. A flexible d.p.c. is required for this position. The function of the d.p.c. is to protect the inner leaf of the brickwork. This treatment should be carried out for all types of lintels (figures 8.28 to 8.30).

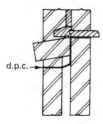

Figure 8.24

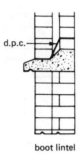

d.p.c.

boot lintel

Figure 8.28

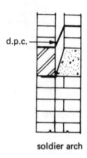

d.p.c.

soldier arch

Figure 8.29

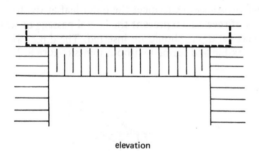

elevation

Figure 8.30

Parapet Walls

Parapets can take the form of solid or cavity construction, occurring above eaves level. They are exposed to all forms of weather conditions, therefore damp-proofing treatment should be carried out to prevent water passing downwards into the building below. Cavity parapet walls should be provided with tray d.p.c.s which are stepped down at least 150 mm to the wall on the roof side; the water barrier on the flat roof surface is taken up the wall and lapped by the wall d.p.c.; or a cover flashing can be used for this position. All copings should be placed on a d.p.c., but for low parapet walls it is not necessary to provide a separate wall d.p.c. The minimum height for the wall d.p.c. is 150 mm above roof surface level. It is essential to use flexible d.p.c. for parapet walls (figures 8.31 and 8.33).

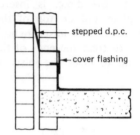

stepped d.p.c.

cover flashing

Figure 8.31

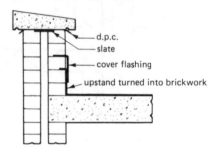

d.p.c.
slate
cover flashing
upstand turned into brickwork

Figure 8.32

Chimney Stacks

This is also an exposed feature occurring above roof level, therefore treatment is required to prevent water passing downwards into the walling below. The damp-proof course used for this position is a copper or lead tray with an attached apron. The tray is placed at least 150 mm above the lowest point of intersection between roof and stack; it covers the entire brickwork of the chimney; complete protection is afforded by using soakers and flashings around the walls of the stack (figures 8.33 and 8.34).

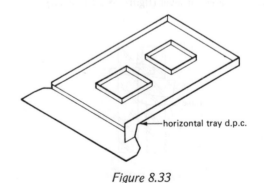

horizontal tray d.p.c.

Figure 8.33

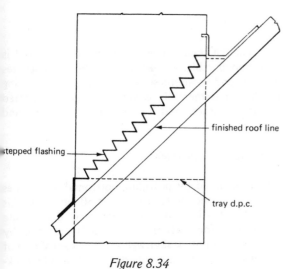

stepped flashing

finished roof line

tray d.p.c.

Figure 8.34

Test for d.p.c. materials

The following experiment can be used to test different damp-proof materials for efficiency and defects if incorrectly positioned. Requirements are one large plastic tray, a number of bricks of the same type, cement/lime sand mortar, bituminous d.p.c. slate, and a solution of sodium sulphate.

(1) Build the required number of brick piers in the tray, ensuring all piers are a distance apart. All piers should be built four courses in height. Insertion and treatment of the d.p.c. to be made according to requirements (figure 8.35).

(2) After a period of 1 week all piers should be completely dried out. Mark a line 25 mm above the bottom of the tray and fill to this level with the solution of sodium sulphate.

(3) Inspect the level of the solution every 3 days,

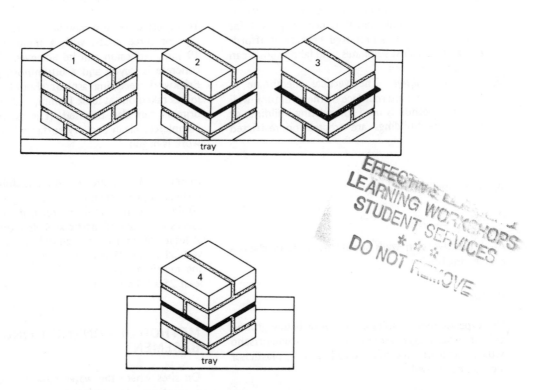

Figure 8.35 Pier 1, built without any d.p.c.; pier 2, built with a bituminous d.p.c. set in 20 mm from the face; pier 3, built with a bituminous d.p.c. projecting 20 mm from the face; pier 4, built with a d.p.c. consisting of two pieces of slate side by side with a 20 mm gap in between

topping up when required. Every 7 days inspect and record the rise of the solution.

(4) This experiment must be continued for a period of at least 28 days.

WATERPROOFING BASEMENTS

When basements are required to have a dry interior they must receive a waterproofing treatment which is termed tanking. This involves the lining of the basement structure with an impervious material and thus a tank is constructed using water-resistant material. A simpler explanation is that the basement has been placed in a waterproof envelope.

Tanking a basement requires the provision of a continuous waterproof membrane applied horizontally over the base concrete and in complete continuity with the vertical membrane used for the external walls of the basement. There are two methods of tanking basements

(1) *External tanking:* This requires the vertical membrane to be applied on the external face of the structural walls and it is normally used for new buildings; it is preferred because it affords protection against sulphates, and labour costs can usually be reduced.

(2) *Internal tanking:* The vertical membrane is formed on the internal face of the structural wall and this method is used for existing buildings or when new buildings are erected on restricted sites.

Materials Used for Tanking

Bricks

All protective skin walls should be built in class A engineering bricks.

Mortars

The type of mortar suitable for work below ground level in waterlogged ground should be cement/sand with a ratio of mix not exceeding 1:4; plasticiser should not be used.

Concrete

This should be O.P.C., sand and aggregate, the ratio not exceeding 1:2:4; sulphate-resisting cement, or Aquacrete can be used if additional protection is required.

Membranes

The material used is mastic asphalt; this material is extremely durable, semi-rigid, unaffected by sulphates in soils and is jointless and capable of accepting building movement. It is applied while in a heated plastic condition and can be used vertically and horizontally.

Protection of Membranes

This is achieved by providing horizontal membranes with a fine concrete screed at least 50 mm thick; this prevents abrasion of the membrane during site operations. The vertical membrane is protected by building a half-brick skin wall on the external face of the membrane; this prevents damage caused by backfilling, and sulphates. The skin wall should be built clear of the asphalt face, the space between wall and asphalt being grouted up in cement mortar as the work proceeds.

Application of Mastic Asphalt

To be effective as a membrane, asphalt is applied in two or three coats, with a minimum thickness of 19 mm. Adhesion can be achieved if the surface of concrete is keyed and a bituminous primer is used. Brickwork receiving asphalt should have all mortar joints raked out at least 12 mm, and also be dust free; primers can also be used. It is essential to provide completely clean and dry surfaces before any asphalt work is begun.

Joints   Where the horizontal membrane meets the vertical asphalt membrane an angle fillet at least 50 mm in width should be formed. Joints formed in successive coats of mastic asphalt should be staggered at least 150 mm for horizontal joints and 75 mm for vertical joints. When terminating vertical membranes the mastic asphalt should be turned into a chase at least 25 x 25 mm in the face of the brickwork.

METHOD OF CONSTRUCTING A NEW BASEMENT

On sites where the water table is above the level of excavations, dewatering operations should be begun before the excavation work is started, and should continue until the basement construction is completed.

When the water table is reduced to below the foundation level the basement area is excavated and also the foundations. Clean hardcore is then placed in position, compacted and it can receive a blinding layer of coarse aggregate (figure 8.36). A polythene membrane can be placed on the hardcore to allow construction operations to begin. The site concrete and foundation concrete are then poured and the surface is left rough to provide a key for the mastic asphalt. When the concrete surface is dry, it should be primed and the two- or three-coat asphalt applied (figures 8.37 and 8.38).

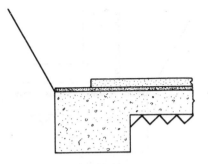

Figure 8.39

A loading coat of concrete is then placed over the fine concrete screed; the thickness of the loading coat is determined by

(1) the water pressure on the basement
(2) loading of the structural walls
(3) imposed loads on walls and floor (figure 8.40).

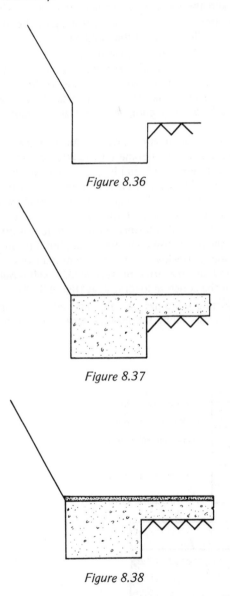

Figure 8.36

Figure 8.37

Figure 8.38

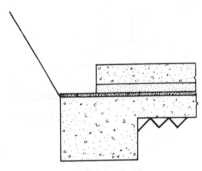

Figure 8.40

The internal structural wall is next erected to 150 mm above ground level; all mortar joints on the external face should be raked out to a depth of 12 mm (figure 8.41). As soon as possible the vertical

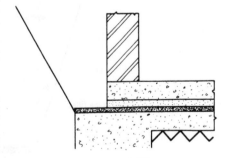

Figure 8.41

A protective screed of fine concrete at least 50 mm thick is next placed over the horizontal membrane; this protects the membrane from abrasive damage and allows operatives to work on the surface (figure 8.39).

mastic asphalt membrane should then be applied to the prepared face of the wall, forming an angle fillet where the two planes meet; the membrane is continued over the top of the structural wall and provides a damp-proof course for the ground floor (figures 8.42 and 8.45).

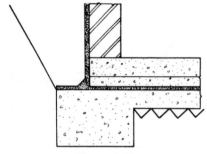

Figure 8.42

The half-brick protective skin can now be set out, allowing at least 25 mm from the vertical membrane, and built up to ground level, grouting up with cement mortar as the work proceeds (figure 8.43).

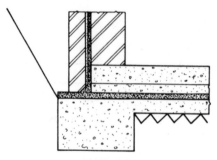

Figure 8.43

When all construction work has been completed the backfilling of the earth can begin. This should be carried out in layers of 150 mm, each layer being well consolidated. During this operation land drains can be laid parallel to the external wall, allowing water to be taken away from the building (figure 8.44).

Figure 8.45 shows the completed section incorporating the tanking; the external face brickwork would be started two to four courses below ground level and the damp-proof course inserted at 150 mm above ground level. The cavity walling can be formed at ground-floor level of the building.

Figure 8.47 shows external application with the earth face supported by timbering; the struts are fixed between the walings and the walling. As the work proceeds the struts are constantly withdrawn and refixed, until complete removal of timbering is achieved.

Internal application involves forming the vertical mastic asphalt membrane on the internal face of the structural wall, and building a protective skin wall internally. The horizontal membrane ends at the face of the internal wall, requiring an internal two-coat fillet at the junction of the two planes. If the protective skin wall is formed with concrete it is usual to omit the protective concrete screed, but if the internal skin wall is brickwork, a concrete protective screed is inserted and the brick protective skin wall is built on top of the concrete loading coat (figure 8.48).

Existing buildings can also be tanked by forming block chases along the internal face of the wall. This

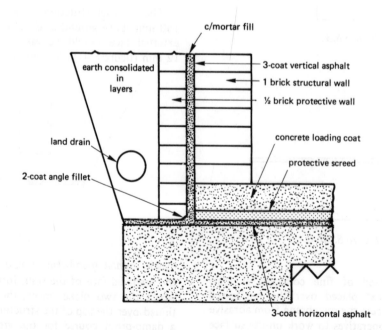

Figure 8.44

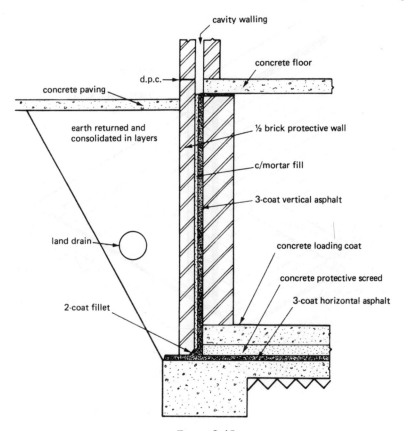

Figure 8.45

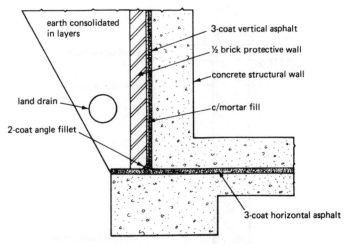

Figure 8.46 Basement with structural walls formed in concrete

method can only be used successfully if the wall thickness is at least 1½ bricks. The vertical membrane is formed up the entire wall face, angle fillets being formed within each chase; the external angles should be chamfered off to form rounded corners for the asphalt. The concrete loading coat supports the internal skin wall. This method is expensive in labour and materials and should only be used if it is not possible to use any other method of tanking (figure 8.49).

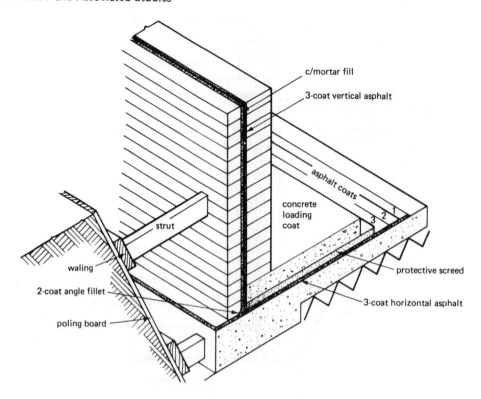

Figure 8.47

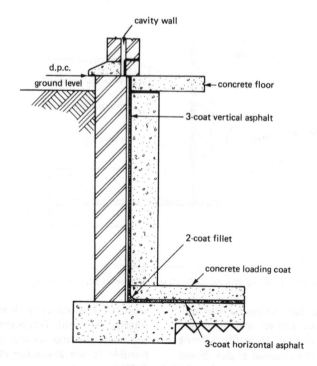

Figure 8.48

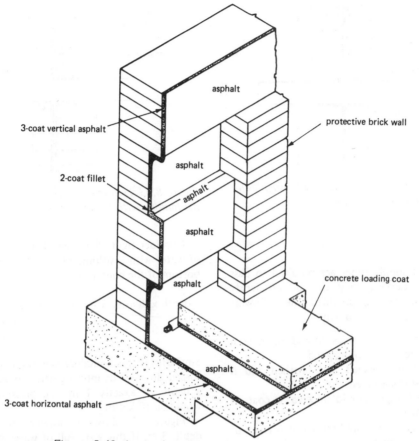

Figure 8.49 *Internal tanking of existing basement using block bonding method*

When columns, piers or stanchions are required to be incorporated in basements that are subject to water pressure the horizontal asphalt membrane should be formed around the base of the column or pier; this requires four angle fillets at each junction. The concrete loading coat should be placed to cover the membrane extending to the base of the column (figure 8.50).

Figure 8.51 shows the treatment provided for the base of a pier and above when the maximum head of water is expected to be above the basement floor level. A chase must be formed in the column or pier to accommodate the vertical asphalt.

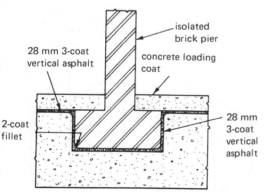

Figure 8.50

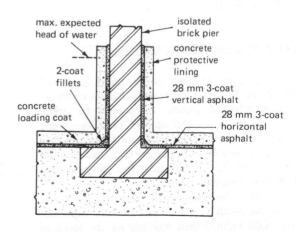

Figure 8.51

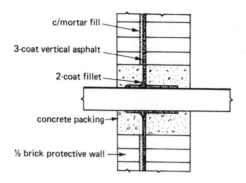

Figure 8.52 Treatment of pipe passing through external tanking with concrete packing

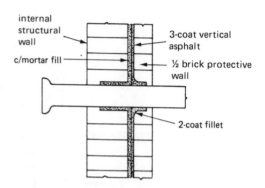

Figure 8.53 Treatment of pipe passing through external tanking with brickwork around pipe

Services and Pipes

When service and supply pipes are required to pass through vertical tanking they can be efficiently married into the tanking provided the following recommendations are complied with.

(1) The distance between pipes should be at least 300 mm to allow for making good.
(2) Pipes should be cleaned and brushed and if possible the surface keyed.
(3) The prepared pipe should receive a coat of bitumen primer.
(4) The primed portion should receive two coats of mastic asphalt sleeving.
(5) The sleeved portion of the pipe is built into the wall with the mastic asphalt sleeve projecting at least 75 mm before any tanking is applied.
(6) The vertical mastic asphalt membrane is then applied and incorporates the mastic asphalt on the sleeve.
(7) A two-coat fillet should be formed at the junction of the sleeve and vertical membrane (figures 8.52 and 8.53).

WATER PRESSURE ON BASEMENTS

Since the pressure of water at any depth is exerted equally in all directions, basements that are constructed in waterlogged ground are subject to lateral pressure on the walls and upward pressure on the floor. In the case of a large basement these pressures may be considerable so it is important that they are calculated, and the wall and floor thicknesses designed with these figures in mind.

To calculate the pressure of water at any depth, multiply the depth by the density of water (approximately 1000 kg/m^3)· then multiply by the force of gravity (approximately 10 m/s^2).

Consider figure 8.54, which shows a container full of water with a volume of 1 m^3. The pressure on the bottom of the container is

$$\text{depth} \times \text{density} \times \text{force of gravity} = 1 \times 1000 \times 10$$
$$= 10\ 000 \text{ N/m}^2$$
$$= 10 \text{ kN/m}^2$$

Similarly, if this container was emptied out and then forced downwards into a greater volume of water to a depth of exactly 1 m, the pressure on the base would be 10 kN/m^2.

Figure 8.55 shows a container of area 1 m^2 and depth 3 m. If this were full of water the pressure on the base would be

$$\text{depth} \times \text{density} \times \text{force of gravity} = 3 \times 1000 \times 10$$
$$= 30\ 000 \text{ N/m}^2$$
$$= 30 \text{ kN/m}^2$$

Figure 8.56 shows a basement subject to water pressure to a depth of 2.5 m. Calculate

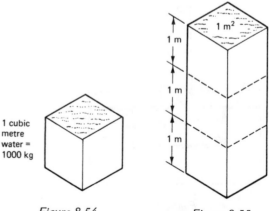

Figure 8.54   *Figure 8.55*

(a) the maximum pressure on the base
(b) the average pressure on the sides
(c) the total pressure on the underside of the floor.

(a) Maximum pressure is given by

depth x density x force of gravity = 2.5 x 1000 x 10
$$= 25\ 000\ \text{N/m}^2$$
$$= 25\ \text{kN/m}^2$$

(b) Average pressure on the sides is

$$\frac{\text{maximum pressure}}{2} = \frac{25}{2} = 12.5\ \text{kN/m}^2$$

(c) Total pressure on underside of floor is

area of floor x maximum pressure = 6 x 5 x 25
$$= 750\ \text{kN}$$

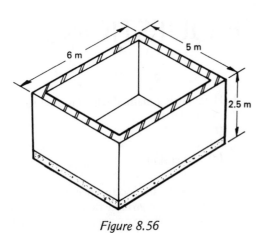

Figure 8.56

MULTIPLE CHOICE QUESTIONS

Select your options from the questions below, underline your selection, for example (b), and check your answers with those on page 117.

1. In open timbering, the poling boards are retained in position by using:
   (a) puncheons
   (b) struts
   (c) lacings
   (d) walings

2. Setting-out lines for walls which divide semi-detached houses are derived from:
   (a) return profiles
   (b) gauge staffs
   (c) single profiles
   (d) building squares

3. Curing of concrete is achieved by providing:
   (a) cold surface conditions
   (b) constant hot air
   (c) intermittent surface heat
   (d) damp conditions

4. Stabilising a bricklayer's line in windy conditions is carried out by using:
   (a) line pins
   (b) corner blocks
   (c) heavy lines
   (d) tingle plates

5. The area of a brick which surrounds the frog is termed the:
   (a) margin
   (b) stretcher face
   (c) arris
   (d) header face

6. In tanking construction the cement mortar fill is used to:
   (a) increase the wall strength
   (b) prevent water penetration
   (c) strengthen the protective wall
   (d) protect the vertical asphalt

7. The type of aggregate used for the manufacture of lightweight concrete blocks is:
   (a) crushed sandstone
   (b) pulverised fuel ash
   (c) granite chippings
   (d) broken old bricks

8. Hatching is a method which is used to indicate:
   (a) materials
   (b) volume
   (c) density
   (d) dimensions

9. To prevent the occurrence of dry rot below hollow timber ground floors, it is necessary to:
   (a) ensure all d.p.c.s are correctly placed
   (b) use only well-seasoned timber
   (c) put a damp-proof membrane below the over-site concrete
   (d) ensure a through draught of fresh air

10. Plasticisers used in mortars increase:
    (a) strength
    (b) density
    (c) adhesion
    (d) workability

11. A disadvantage with clayey subsoils is that in long periods of dry weather they tend to:
    (a) lift
    (b) shrink
    (c) explode
    (d) collapse

12. The type of material suitable for a flexible d.p.c. is:
    (a) blue slate
    (b) bituminous felt
    (c) engineering bricks
    (d) mastic asphalt

13. Honeycombs in sleeper walls:
    (a) improve speed of construction
    (b) increase underfloor warmth
    (c) provide underfloor ventilation
    (d) reduce the brickwork cost

14. The use of wall plates on the top of sleeper walls is to:
    (a) provide equal distribution of load
    (b) increase the floor strength
    (c) reduce the number of floor joists
    (d) form a level bedding surface

15. The maximum thickness for hardcore when used below oversite concrete is normally:
    (a) 300 mm
    (b) 500 mm
    (c) 400 mm
    (d) 600 mm

16. Backfilling and compaction of trenches should be carried out in layers of:
    (a) 150 mm
    (b) 200 mm
    (c) 300 mm
    (d) 450 mm

17. The first reading taken from an instrument position is termed the:
    (a) foresight
    (b) datum sight
    (c) intermediate sight
    (d) backsight

18. A trench 4.000 m in length, 1.000 m in depth and 1.000 m wide is to be excavated; the sub-soil is clay. The amount to be carted away will be:
    (a) 6.500 m^3
    (b) 4.500 m^3
    (c) 5.00 m^3
    (d) 5.500 m^3

19. A d.p.c. placed under a stone sill should extend on each side of the sill at least:
    (a) 75 mm
    (b) 100 mm
    (c) 150 mm
    (d) 200 mm

20. Sand-lime bricks are hardened by:
    (a) steaming in an autoclave
    (b) firing in a continuous kiln
    (c) the addition of Portland cement
    (d) continuous spraying with water

21. Before a foundation may be concreted, the natural foundation must be inspected by the:
    (a) clerk of works
    (b) factory inspector
    (c) safety officer
    (d) building control officer

22. In order to qualify for the weekly tool allowance, a bricklayer may be expected to provide:
    (a) bevel, square, dividers, carborundum stone
    (b) dividers, square, boat level, jointer
    (c) bevel, square, corner blocks, line level
    (d) corner blocks, jointer, carborundum stone, square

23. To keep the bed joints to the required thickness, it is advisable to:
    (a) level the work frequently
    (b) lay the bricks frog-uppermost
    (c) use the gauge staff regularly
    (d) take great care with bed joints

24. Coarse aggregate is mainly retained on a sieve of:
    (a) 3 mm
    (b) 4 mm
    (c) 5 mm
    (d) 6 mm

25. A horizontal d.p.c. is placed to stop moisture:
    (a) rising up from the ground
    (b) running down into the foundations
    (c) passing from the outer leaf to the inner leaf
    (d) flooding the cavity

26. The number of layers in which the slump cone is filled and the number of times each layer is rodded is:
    (a) 2; 25
    (b) 3; 25
    (c) 3; 35
    (d) 4; 35

27. The minimum lap for a flexible d.p.c. is:
    (a) 50 mm
    (b) 75 mm
    (c) 100 mm
    (d) 150 mm

28. A horizontal member used in trench timbering is a:
    (a) waling board
    (b) poling board
    (c) foot prop
    (d) puncheon

29. Dry rot in timber is the result of:
    (a) chemical decay
    (b) insect attack
    (c) old age
    (d) fungal attack

30. A d.p.c.:
    (a) provides a level bed
    (b) provides a bed for floor joists
    (c) holds back moisture
    (d) binds the lower courses together

31. Intermediate level points between two previously established levels can be determined by using:
    (a) boning rods
    (b) builder's squares
    (c) trammel rods
    (d) gauge staffs

32. The horizontal d.p.c. is placed not less than 150 mm above the:
- (a) finished floor level
- (b) foundations
- (c) external ground level
- (d) oversite concrete

33. Bricks that are subjected to severe weather conditions should be:
- (a) permeable
- (b) porous
- (c) dense
- (d) sulphate resisting

34. Having set out a rectangular building, the accurate check for squareness of the corners would be to:
- (a) use an instrument called a sitesquare
- (b) measure each side to confirm the correct dimensions
- (c) measure the diagonals
- (d) place a builder's square against each corner

35. The application of asphalt to basements in order to prevent penetration of water is described as:
- (a) flaunching
- (b) tanking
- (c) benching
- (d) haunching

36. A building line is:
- (a) the line in front of which no building is allowed
- (b) part of a bricklayer's tool kit
- (c) the datum line of a building
- (d) a line seen through the eyepiece of a surveyor's level

37. The usual test for the workability of concrete is the:
- (a) compression test
- (b) sieve test
- (c) silt test
- (d) slump test

38. A brick scutch should be used to:
- (a) adjust quoin bricks
- (b) cut small bevelled bats
- (c) cut perforated bricks
- (d) trim rough-cut bricks

39. Care should be taken when using a plumb rule during periods of:
- (a) strong winds
- (b) severe frost
- (c) patchy fog
- (d) prolonged rain

40. If a working drawing is made to a scale of 1:50, a length of 5.5 metres would be shown on the drawing by a length of:
- (a) 55 mm
- (b) 110 mm
- (c) 275 mm
- (d) 550 mm

ANSWERS TO MULTIPLE CHOICE QUESTIONS

1. (b)	**14.** (a)	**27.** (c)
2. (c)	**15.** (d)	**28.** (a)
3. (d)	**16.** (a)	**29.** (d)
4. (d)	**17.** (d)	**30.** (c)
5. (a)	**18.** (c)	**31.** (a)
6. (d)	**19.** (c)	**32.** (c)
7. (b)	**20.** (a)	**33.** (c)
8. (a)	**21.** (d)	**34.** (c)
9. (d)	**22.** (a)	**35.** (b)
10. (d)	**23.** (c)	**36.** (a)
11. (b)	**24.** (c)	**37.** (d)
12. (b)	**25.** (a)	**38.** (d)
13. (c)	**26.** (b)	**39.** (a)
		40. (b)

INDEX

118

SHOP THEORY

PREPARED BY THE HENRY FORD TRADE SCHOOL, 1942

REPRINTED BY LINDSAY PUBLICATIONS INC

SHOP THEORY

Revised Edition

PREPARED BY

THE SHOP THEORY DEPARTMENT
HENRY FORD TRADE SCHOOL
DEARBORN, MICHIGAN

SHOP THEORY
Revised Edition
prepared by
> The Shop Theory Department
> Henry Ford Trade School
> Dearborn, Michigan

Original Copyright 1934, 1941, 1942 by Henry Ford Trade School

Originally published by McGraw-Hill Book Co, New York

Reprinted by
> Lindsay Publications Inc
> Bradley IL 60915

ISBN 1-55918-006-4

   4 5 6 7 8 9 0

PREFACE

From the beginning Shop Theory has played an important part in the educational program of Henry Ford Trade School. Because there was little material in print suited to our use, mimeographed sheets were prepared by our instructors. These were distributed in class and in time the student accumulated much information on many subjects.

Requests for these sheets were received from other schools and individuals in such numbers that we finally bound them in paper covers and sold them to those interested. More than 150,000 copies of previous editions have been furnished to high schools, colleges, industrial and vocational schools, United States Army and Navy schools, and many individuals in the United States and foreign countries.

During this process of developing the material now incorporated in this book many instructors contributed to make it a usable tool for student and teacher. Although it is impossible to give proper credit to all the instructors who helped bring the text to its present form, the following have contributed much and are entitled to recognition: Edward H. Bailey; Vincent C. Gourley, M. A.; Albert M. Wagener, M. E.; Roy E. Wipert; and John P. Heinz, who also did valuable work in rearranging and proofreading this edition.

We hope that because of the improvement in form and printing, this revised edition of Shop Theory may better meet the need of a larger number of students who are trying to fit themselves for the present emergency.

Frederick E. Searle, Superintendent
Ford Industrial Schools

ACKNOWLEDGMENTS

The Henry Ford Trade School gratefully acknowledges the courtesy of the following companies in permitting the use of various cuts and descriptive material from their textbooks and catalogues: Abrasive Machine Tool Company; Acme Machine Tool Company; Allen Manufacturing Company; American Swiss File and Tool Company; American Tool Works Company; Anderson Brothers Manufacturing Company; Armstrong Brothers Tool Company; Avey Drilling Machine Company; Baker Brothers, Inc.; Blanchard Machine Company; Bonney Forge and Tool Works; Boyar-Schultz Corporation; Brown Instrument Company; Brown & Sharpe Manufacturing Company; Bullard Company; Charles Bond Company; Carborundum Company; S. W. Card Manufacturing Company; Cincinnati Bickford Tool Company; Cincinnati Grinders, Inc.; Cincinnati Milling Machine Company; Cincinnati Planer Company; Cleveland Twist Drill Company; Desmond-Stephan Manufacturing Company; Henry Disston and Sons, Inc.; Detroit Torch and Manufacturing Company; Eugene Dietzgen Company; Ex-Cell-O Corporation; Farrel Birmingham Company, Inc.; Fellows Gear Shaper Company; Foster Machine Company; Frederick Post Company; Gleason Works; Goodell-Pratt Company; Gould and Eberhardt; Greenfield Tap and Die Corporation; Heald Machine Company; John Bath and Company, Inc.; Jones & Lamson Machine Company; Kearney and Trecker Corporation; Landis Tool Company; Leeds and Northrup Company; Lucas Machine Tool Company; Lodge and Shipley Machine Tool Company; Lufkin Rule Company; Millers Falls Company; Moline Tool Company; Monarch Machine Tool Company; Morse Twist Drill and Machine Company; National Twist Drill and Tool Company; Nicholson File Company; Norton Company; Oliver Instrument Company; Pratt and Whitney Aircraft; Ransom Grinding Machine Company; Reed Manufacturing Company; Reed-Prentice Corporation; Rivett Lathe and Grinder, Inc.; Simonds Saw and Steel Company; Skinner Chuck Company; South Bend Lathe Works; Standard Fuel Engineering Company; Stanley Rule and Level Division of the Stanley Works; Swann Chemical Company; Charles A. Strelinger Company; J. T. Slocomb Company; L. A. Sayre Company; L. S. Starrett Company; Taft-Peirce Manufacturing Company; Taylor and Fenn Company; Union Tool Company; Walton Company; Warnock Manufacturing Company; Warner and Swasey Company; O. S. Walker Company, Inc.; J. H. Williams Company; Western Tool and Manufacturing Company; Whitman and Barnes; Wilson Mechanical Instrument Company, Inc.

TABLE OF CONTENTS

Page

Chapter 1

DECIMAL EQUIVALENTS

The measurements made in a machine shop are usually taken in inches or fractional parts of an inch. Most of the precision tools used in the shop read in thousandths of an inch. The usual graduations on a rule are in 64ths, 32nds, 16ths, and 8ths of an inch.

Before a student can read a rule or the precision measuring tools efficiently, he must be thoroughly familiar with fractions and decimal fractions. Since he is often called upon to change decimals to fractions and fractions to decimals in making measurements, and in reading and checking blue prints and sketches, he should understand this operation thoroughly. Precision measuring tools, such as micrometers and vernier tools, are read in thousandths or fractional parts of a thousandth of an inch. For example, 1/16 is read sixty-two and one-half thousandths, 1/32 is read thirty-one and one-quarter thousandths, 1/64 is read fifteen and five-eighths thousandths, etc. It will be noted that these readings give the full decimal values for the corresponding fractions. However, since the precision tools commonly used in the shop cannot be read closer than

one-tenth of one thousandth, it is customary for a mechanic to use only those figures up to and including the tenth-thousandth figure, or to four decimal places. For example, the complete decimal value of 1/64 is .015625, which in the shop is commonly read as fifteen and six-tenths thousandths, fifteen being the whole number of thousandths while six is six-tenths of one one-thousandth, or a fractional part of a thousandth.

To change a fraction to a decimal, divide the numerator by the denominator. For example, in changing 3/16 to a decimal, $3.0000 \div 16 = .1875$.

When reading a rule it is sometimes convenient to read either way from some large dimension line. That is, in measuring 47/64 of an inch it is easier to find 3/4 and subtract 1/64 from it than to count the divisions from the end of the rule. The following table contains the decimal values and fractional values in halves, fourths, eighths, sixteenths, thirty-seconds, and sixty-fourths, for the fractions most commonly used in the shop. These values should be memorized.

Read the decimal equivalent to the

1/64	–	–	–	–	–	.015625	7/16	28	14	7	–	–	.4375
1/32	2	–	–	–	–	.03125	1/2	32	16	8	4	2	.500
1/16	4	2	–	–	–	.0625	9/16	36	18	9	–	–	.5625
3/32	6	3	–	–	–	.09375	5/8	40	20	10	5	–	.625
1/8	8	4	2	1	–	.125	11/16	44	22	11	–	–	.6875
3/16	12	6	3	–	–	.1875	3/4	48	24	12	6	3	.750
1/4	16	8	4	2	1	.250	13/16	52	26	13	–	–	.8125
5/16	20	10	5	–	–	.3125	7/8	56	28	14	7	–	.875
3/8	24	12	6	3	–	.375	15/16	60	30	15	–	–	.9375
	64	32	16	8	4		1	64	32	16	8	4	1.0000

fourth decimal place and express in thousandths. For example, 23/64 is read three hundred fifty-nine and four tenths, thousandths.

Memorize the quarters, eighths, sixteenths, one thirty-second, three thirty-seconds, and one sixty-fourth.

Fractions	64ths	32nds	16ths	8ths	4ths	Decimal Equivalents	Fractions	64ths	32nds	16ths	8ths	4ths	Decimal Equivalents
1/64	-	-	-	-	-	.015625	33/64	-	-	-	-	-	.515625
1/32	2	-	-	-	-	.03125	17/32	34	-	-	-	-	.53125
3/64	-	-	-	-	-	.046875	35/64	-	-	-	-	-	.546875
1/16	4	2	-	-	-	.0625	9/16	36	18	-	-	-	.5625
5/64	-	-	-	-	-	.078125	37/64	-	-	-	-	-	.578125
3/32	6	-	-	-	-	.09375	19/32	38	-	-	-	-	.59375
7/64	-	-	-	-	-	.109375	39/64	-	-	-	-	-	.609375
1/8	8	4	2	-	-	.125	5/8	40	20	10	-	-	.625
9/64	-	-	-	-	-	.140625	41/64	-	-	-	-	-	.640625
5/32	10	-	-	-	-	.15625	21/32	42	-	-	-	-	.65625
11/64	-	-	-	-	-	.171875	43/64	-	-	-	-	-	.671875
3/16	12	6	-	-	-	.1875	11/16	44	22	-	-	-	.6875
13/64	-	-	-	-	-	.203125	45/64	-	-	-	-	-	.703125
7/32	14	-	-	-	-	.21875	23/32	46	-	-	-	-	.71875
15/64	-	-	-	-	-	.234375	47/64	-	-	-	-	-	.734375
1/4	16	8	4	2	-	.250	3/4	48	24	12	6	-	.750
17/64	-	-	-	-	-	.265625	49/64	-	-	-	-	-	.765625
9/32	18	-	-	-	-	.28125	25/32	50	-	-	-	-	.78125
19/64	-	-	-	-	-	.296875	51/64	-	-	-	-	-	.796875
5/16	20	10	-	-	-	.3125	13/16	52	26	-	-	-	.8125
21/64	-	-	-	-	-	.328125	53/64	-	-	-	-	-	.828125
11/32	22	-	-	-	-	.34375	27/32	54	-	-	-	-	.84375
23/64	-	-	-	-	-	.359375	55/64	-	-	-	-	-	.859375
3/8	24	12	6	-	-	.375	7/8	56	28	14	-	-	.875
25/64	-	-	-	-	-	.390625	57/64	-	-	-	-	-	.890625
13/32	26	-	-	-	-	.40625	29/32	58	-	-	-	-	.90625
27/64	-	-	-	-	-	.421875	59/64	-	-	-	-	-	.921875
7/16	28	14	-	-	-	.4375	15/16	60	30	-	-	-	.9375
29/64	-	-	-	-	-	.453125	61/64	-	-	-	-	-	.953125
15/32	30	-	-	-	-	.46875	31/32	62	-	-	-	-	.96875
31/64	-	-	-	-	-	.484375	63/64	-	-	-	-	-	.984375
1/2	32	16	8	4	2	.500	1 inch	64	32	16	8	4	1.000

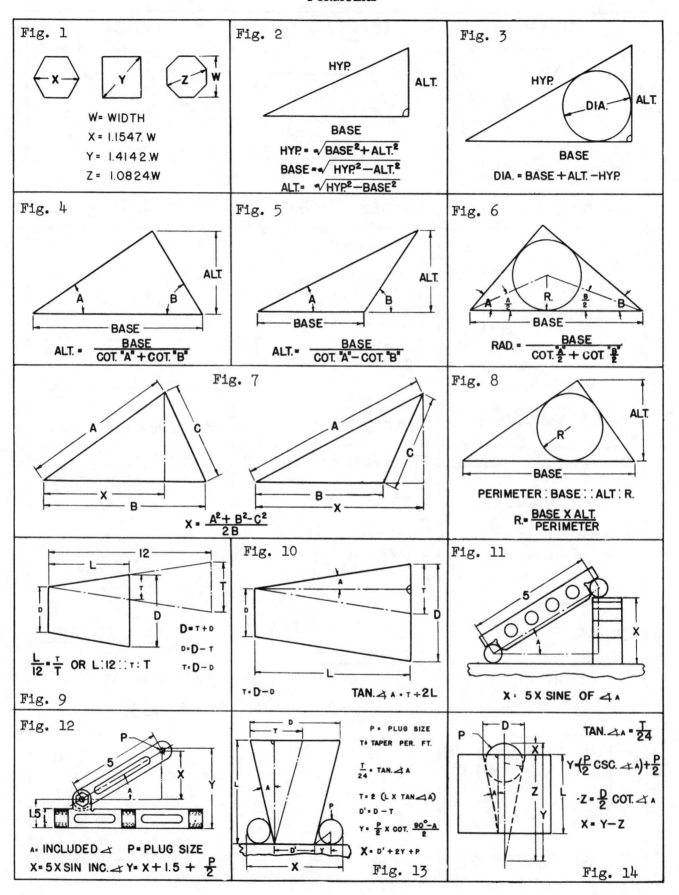

Fig. 1

W = WIDTH
X = 1.1547. W
Y = 1.4142.W
Z = 1.0824.W

Fig. 2

$HYP. = \sqrt{BASE^2 + ALT.^2}$
$BASE = \sqrt{HYP.^2 - ALT.^2}$
$ALT. = \sqrt{HYP.^2 - BASE^2}$

Fig. 3

DIA. = BASE + ALT. − HYP.

Fig. 4

$ALT. = \dfrac{BASE}{COT. "A" + COT. "B"}$

Fig. 5

$ALT. = \dfrac{BASE}{COT. "A" - COT. "B"}$

Fig. 6

$RAD. = \dfrac{BASE}{COT. \frac{"A"}{2} + COT. \frac{"B"}{2}}$

Fig. 7

$X = \dfrac{A^2 + B^2 - C^2}{2B}$

Fig. 8

PERIMETER : BASE :: ALT. : R.

$R. = \dfrac{BASE \times ALT.}{PERIMETER}$

Fig. 9

$\dfrac{L}{12} = \dfrac{T}{T}$ OR L:12 :: T:T

D = T + d
d = D − T
T = D − d

Fig. 10

T = D − d   TAN. ∠ A = T ÷ 2L

Fig. 11

X = 5 X SINE OF ∠ A

Fig. 12

A = INCLUDED ∠   P = PLUG SIZE
X = 5 X SIN INC. ∠ Y = X + 1.5 + $\dfrac{P}{2}$

Fig. 13

P = PLUG SIZE
T = TAPER. PER. FT.

$\dfrac{T}{24}$ = TAN. ∠ A

T = 2 (L X TAN ∠ A)
D' = D − T
Y = $\dfrac{P}{2}$ X COT. $\dfrac{90°-A}{2}$
X = D' + 2Y + P

Fig. 14

TAN. ∠ A = $\dfrac{T}{24}$

Y = ($\dfrac{P}{2}$ CSC. ∠ A) + $\dfrac{P}{2}$

Z = $\dfrac{D}{2}$ COT. ∠ A

X = Y − Z

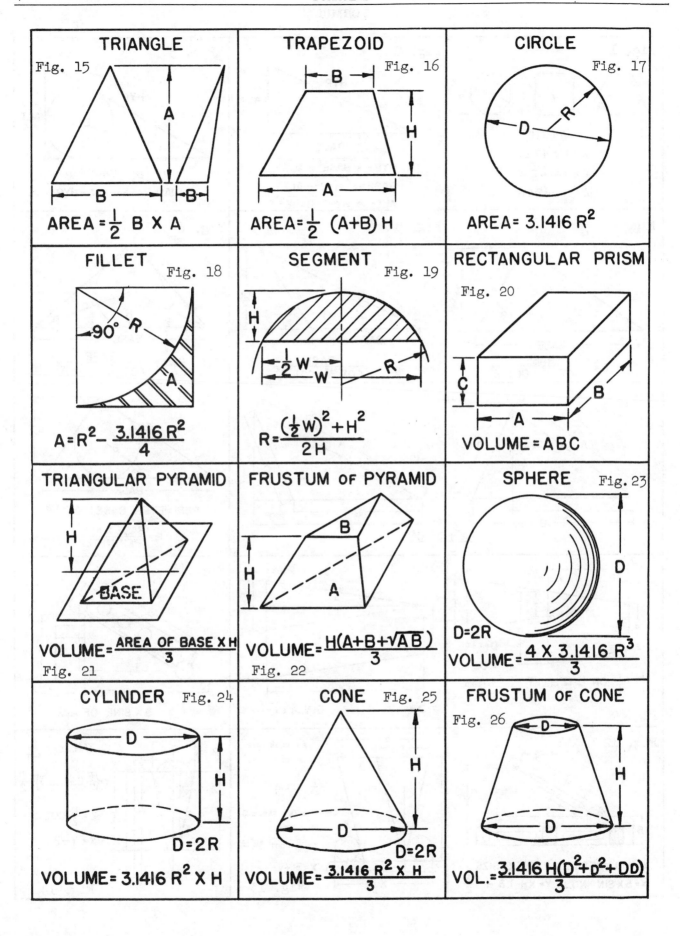

TRIANGLE — Fig. 15

$$\text{AREA} = \frac{1}{2} \text{ B} \times \text{A}$$

TRAPEZOID — Fig. 16

$$\text{AREA} = \frac{1}{2}(A+B)H$$

CIRCLE — Fig. 17

$$\text{AREA} = 3.1416 R^2$$

FILLET — Fig. 18

$$A = R^2 - \frac{3.1416 R^2}{4}$$

SEGMENT — Fig. 19

$$R = \frac{(\frac{1}{2}W)^2 + H^2}{2H}$$

RECTANGULAR PRISM — Fig. 20

$$\text{VOLUME} = ABC$$

TRIANGULAR PYRAMID

$$\text{VOLUME} = \frac{\text{AREA OF BASE} \times H}{3}$$

Fig. 21

FRUSTUM OF PYRAMID

$$\text{VOLUME} = \frac{H(A+B+\sqrt{AB})}{3}$$

Fig. 22

SPHERE — Fig. 23

$$D = 2R$$

$$\text{VOLUME} = \frac{4 \times 3.1416 R^3}{3}$$

CYLINDER — Fig. 24

$$D = 2R$$

$$\text{VOLUME} = 3.1416 R^2 \times H$$

CONE — Fig. 25

$$D = 2R$$

$$\text{VOLUME} = \frac{3.1416 R^2 \times H}{3}$$

FRUSTUM OF CONE — Fig. 26

$$\text{VOL.} = \frac{3.1416 H(D^2 + D^2 + Dd)}{3}$$

Chapter 3

SMALL TOOLS

The condition in which a mechanic keeps the various tools he uses determines his efficiency as well as the judgment that others pass upon him in his daily work. A workman is always judged by the way he handles his tools. For instance, an ordinary steel rule which is battered or otherwise damaged so that it is difficult to read will often cause a workman to scrap a piece of work, thereby losing considerable time and money. A caliper with points bent out of shape will inevitably cause trouble if used in that condition. A divider with points dull or loose in the joint cannot be made to do satisfactory work. A micrometer that has been strained out of shape will not give an accurate measurement.

Every mechanic should have a tool box of his own where he keeps his tools when he is not using them. There should be a place for every tool and each tool should be kept in its place.

All tools should be wiped clean before they are placed in the tool box, and if not to be used again for some time they should be oiled to prevent rusting.

Tools that are being used on the machine or bench should be kept within easy reach of the operator and placed so that they cannot fall on the floor. They should never be placed on the finished parts of a machine.

1. Describe the most common hammers and tell the purpose for which each is used.

A. The most common hammers are the ball peen (Fig. 27), straight peen (Fig. 28), cross peen (Fig. 29), and claw hammer (Fig. 30). Hammers are made with wooden handles and steel heads and vary in size from 6 ounces to 2½ pounds. The principal parts of a hammer are the peen, eye, face, and post. The upper part of the machinists' hammer,

Fig. 27

Fig. 28

called the peen, is made in three common shapes--the ball peen for riveting and the straight and cross peen types for swaging. The eye is the hole that receives the handle, the face is the lower part of the head, and the post is the portion between the face and the eye. The claw hammer is used by the carpenter for driving or pulling nails.

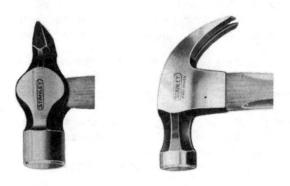

Fig. 29   Fig. 30

2. What are soft hammers and why are they used?

A. Hammers with heads made of lead, copper, babbitt (Fig. 31) or rawhide (Fig. 32) are known as soft hammers. They are used to seat work in a machine vise, to drive a mandrel, or in any similar operation where the steel hammer might injure the work.

Fig. 31

Fig. 32

3. Why should a hammer handle be gripped near the end?

A. A hammer handle should be gripped near

the end so that full leverage may be obtained. A solid blow cannot be delivered when it is held too close to the head.

4. Why does the eye in a hammer head taper from each end toward the middle?

A. The hammer handle is formed to fit one end of the tapered eye and prevent the head from slipping up the handle, and in the other end of the tapered eye the handle is expanded by wedges to fit the eye and hold the head securely.

5. Why must a hammer handle be set square with the head?

A. The hammer handle must be set square with the head to insure the proper balance.

6. What is meant by "peening"?

A. Peening is the stretching of metal by hammering, such as hammering the end of a rivet or the end of a bolt so the nut will not jar loose. Peening is used to stretch babbitt to fit tightly in a bearing, to straighten bars by stretching the short side, and in numerous other operations.

Goggles must be worn when chipping, grinding, working near cyanide pots, etc. (Fig. 33). The chipping guard (Fig. 34) should be placed so as to protect fellow workmen as well as machinery.

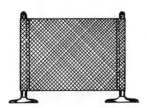

Fig. 33. Goggles

packing block of wood or metal under it to keep it from working down in the vise. Protect finished work with brass or copper jaw caps (Fig. 36). Place the work so that it can be chiseled toward the solid jaw. Refer to Chisels and Chipping, page 47, for an illustration of the proper use of these tools.

7. Describe the hack saw.

A. The hack saw (Fig. 37) consists of a frame and a thin steel blade with teeth formed on one edge. It is six to twelve inches long, approximately one-half inch wide, and usually about .027" thick.

Fig. 37. Hack Saw

8. What is a chisel?

A. A chisel is a tool or instrument made from octagon-shaped steel, having a cutting edge at one end of the blade, and used in dressing, shaping, or cutting. It is driven by a hammer or mallet. Some of the different types are shown below and on the following page.

Fig. 38. Flat Cold Chisel

The flat cold chisel is used for chipping flat surfaces, and often for cutting thin sheet metal.

Fig. 34. Chipping Guard   Fig. 35. Vise

Fig. 36. Vise Jaw Cap

A vise (Fig. 35) is used for holding work while it is being finished, assembled, etc., at the bench. Do not hammer a vise handle. When holding work in a vise, put a

The cape chisel (Fig. 39) is used for chipping grooves, keyways, and often for slots where a sturdy narrow chisel is needed.

Fig. 39. Cape Chisel

The round nose chisel is used for roughing out small concave surfaces of filleted corners, etc.

Fig. 40. Round Nose Chisel

The oil groove chisel is used for chiseling oil grooves, etc.

Fig. 41. Oil Groove Chisel

The diamond point chisel is used for cutting V-shaped grooves or for chipping in corners.

Fig. 42. Diamond Point Chisel

The star drill is really a multiple-pointed chisel and is used for drilling stone and concrete.

Fig. 43. Star Drill

The drift punch is used for aligning holes for bolts or rivets.

Fig. 44. Drift Punch

The pin punch is used to drive taper pins, cotter pins, etc.

Fig. 45. Pin Punch

The prick punch is used to mark scribed or layout lines with small indentations.

Fig. 46. Prick Punch

The center punch is used to mark the location for the drill and assist in starting it properly.

Fig. 47. Center Punch

Fig. 48. Scriber

The scriber is used to mark lines in measuring or on layout work.

Fig. 49

9. What is a wrench?
   A. A tool with jaws or openings for turning or twisting bolts, nuts, etc.

10. How are wrenches named?
    A. (1) From their shape, as "S" wrench, angle wrench, etc.
       (2) From the object on which they are used, as tap wrench, pipe wrench, etc.
       (3) From their construction, as spanner wrench, ratchet wrench, etc.
    Some of the most common wrenches are shown on the following page.

11. Explain the use of each of the following wrenches: single-end, double-end, closed-end, adjustable hook spanner, adjustable pin face, T-socket, off-set socket, ratchet, pipe, strap, chuck key, hollow set-screw, and monkey wrench.

The single-end wrench shown in Fig. 50 is used on jobs requiring a wrench with one size only.

Fig. 50.   Single-end Wrench

The double-end wrench shown in Fig. 51 is used on jobs that require a wrench which has openings of different sizes.

Fig. 51.   Double-end Wrench

The use of the closed-end wrench shown in Fig. 52 is much the same as that of the single-end wrench. The closed end of the wrench eliminates the danger of the jaws spreading.

Fig. 52.   Closed-end Wrench

The adjustable hook spanner wrench (Fig. 53) is used on nuts having notches cut in the periphery to receive the hook located at the end of the wrench.

Fig. 53.   Adjustable Hook Spanner Wrench

The adjustable pin face wrench is used to adjust nuts having holes in their face to accommodate pins in the end of the adjustable legs (see Fig. 54).

Fig. 54.   Adjustable Pin Face Wrench

The T-socket wrench (Fig. 55) is made with different types of sockets, such as square, hexagon, and octagon. It is generally used on jobs where the nuts are almost inaccessible, as on engines, motors, bodies, wire wheels, etc.

Fig. 55.   T-Socket Wrench

The off-set socket wrench (Fig. 56) is used on nuts requiring a greater leverage than that which can be obtained with the T-socket wrench, or on jobs where the T-socket wrench cannot be used.

Fig. 56.   Off-set Socket Wrench

The ratchet wrench (Fig. 57) is similar to the socket wrench except that the handle works as a ratchet. This is especially useful when only a short swing of the handle is permissible. Another advantage of this wrench is that it is not necessary to remove it until the bolt or nut is tight.

Fig. 57.   Ratchet Wrench

The Stillson pipe wrench (Fig. 58) and the chain pipe tongs (Fig. 59) are used to turn pipes. Be careful in using these wrenches that the finish on plated pipes is not scratched or otherwise damaged.

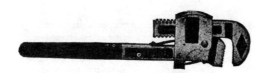

Fig. 58. Stillson Pipe Wrench

Fig. 59. Chain Pipe Tongs

The strap wrench (Fig. 60) is used for turning plated pipes, removing bezels, etc., or for revolving any job on which the surface finish must be preserved.

Fig. 60. Strap Wrench

The chuck key (Fig. 61) is used for adjusting chuck jaws.

Fig. 61. Chuck Key

The hollow set-screw wrench (Fig. 62) is used to adjust hollow (safety) set-screws.

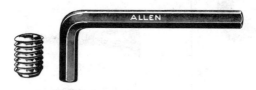

Fig. 62. Hollow Set-Screw Key

The monkey wrench is a heavy adjustable wrench for use on heavy work (see Fig. 63). Its use on small tools is not encouraged. The three following rules for the use of the monkey wrench are very important:

1. Never hammer with the solid jaw of the wrench.

2. Always adjust the movable jaw so that it is tight against a nut.

3. Point the jaws in the direction of the force applied. This will prevent the jaws from springing and the wrench will be less likely to slip off a nut.

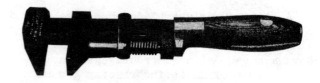

Fig. 63. Monkey Wrench

Do not use a pair of pliers instead of a wrench (see Fig. 64).

Fig. 64. Pliers

12. What is a screw driver?
A. A screw driver is a metal blade with a handle on one end with the other end flattened to fit screw slots.

13. What kind of steel is generally used to make screw drivers?
A. Tool steel is generally used. Screw drivers under one-half inch in size are usually made of round stock (see Fig. 65).

Fig. 65

14. Why are the larger sizes of screw drivers made with square blades?
A. The blade of the screw driver shown in Fig. 66 is made square so that a wrench may be applied to assist in turning a screw.

Fig. 66

15. What is the purpose of the double-end
    off-set screw driver shown in Fig. 67?

    A. This type of screw driver is used for
turning screws that cannot be reached with a
straight screw driver.

16. How should the blade of any screw driver
    be ground?

    A. A screw driver blade should be ground
so that the faces will be almost parallel
with the sides of the screw slot. Fig. 68
shows a screw driver correctly ground.

    If a blade is incorrectly ground as
shown in Fig. 69, it has a tendency to slip
out of the slot and leave a burr.

    Excessive heat due to grinding, shown
by a blue color, will draw the temper and
cause the driver to become soft.

17. Describe the helical ratchet screw driver
    shown in Fig. 70.

    A. The helical ratchet screw driver drives
or draws screws by pushing on the handle. It
also has a ratchet movement and can be locked
rigid. It may be changed from right-hand to
left-hand or locked by moving the shifter.

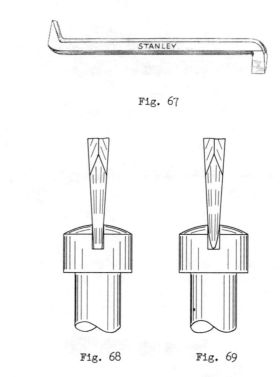

Fig. 67

Fig. 68   Fig. 69

Fig. 70

18. What are the three most important classes
    of screws and what are the names of the
    principal types of screw heads in each
    class?

A. Machine screws, set screws, and wood
screws are the most important classes. Fig.
71 names and illustrates the principal screw
heads in these classes.

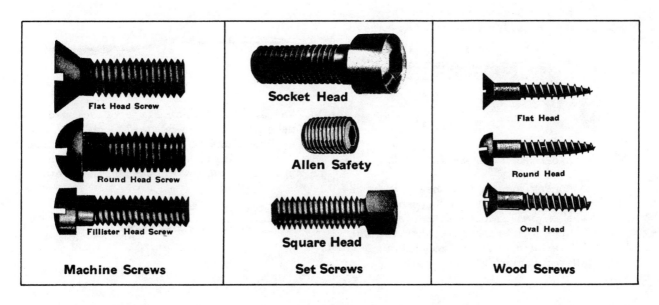

Fig. 71. Different Types of Screws and Heads

19. What do flat head screws and fillister or socket head screws have in common?

A. These screws have heads that must be recessed into the work. The 82° countersink is used to form the recess for a flat head screw. A counterbore, the diameter of which is slightly larger than the screw head, and a pilot, the diameter of which is equal to the screw body size, is used to form the recess for a socket head screw.

20. What is a drill? Describe some of the various types of drills and drill equipment.

A. A drill is a round steel shaft with grooves formed on its periphery, one end, known as the point, is ground to form suitable cutting edges (see Figs. 72 and 73). The shank end of a drill may be either tapered or straight. Most drills have a tapered shank, which can be fitted into a spindle, socket, or sleeve (see Figs. 74 and 75). All drills with a diameter less than 1/8 have straight shanks, however, straight shanks in other sizes are available. Drills with straight shanks are held in a drill chuck (Figs. 76 and 77).

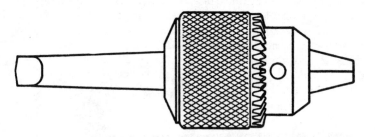

Fig. 76. Jacobs Drill Chuck

Fig. 77. Skinner Drill Chuck

Figs. 78 and 79 show two drills for drilling wood, steel, or other material, one driven by hand and the other by an electric

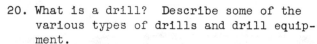

Fig. 72. Taper Shank Drill

Fig. 73. Straight Shank Drill

Fig. 74. Socket

Fig. 75. Sleeve

Fig. 78. Hand Drill

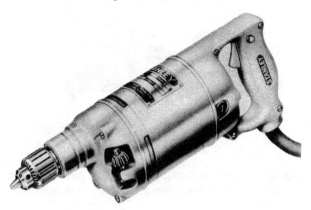

Fig. 79. Portable Electric Drill

motor. The drill holder in Fig. 80 is used
to hold a drill while drilling in a lathe.
A drill drift is used to remove drills from
sleeves and sockets. A plain drift is shown
in Fig. 81 and a safety drift in Fig. 82. The
drill vise in Fig. 83 is used to hold work
while it is being drilled.

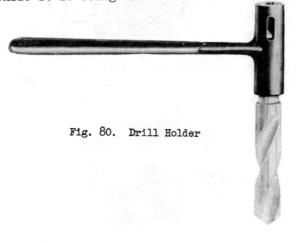

Fig. 80. **Drill Holder**

Fig. 81. Plain Drill Drift

Fig. 82. Safety Drill Drift

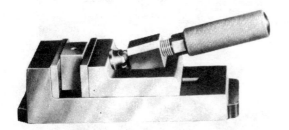

Fig. 83. Drill Vise

21. How are the countersink, the counterbore,
    and the combined drill and countersink
    used?
    A. The countersink, shown in Fig. 84, is
used for countersinking the tops of holes
which are to receive flathead screws or bolts.

Fig. 84. Countersink

The counterbore shown in Fig. 85 is
used for enlarging a hole on the axis of one
already drilled, so that the head of the bolt
or cap screw will come flush with the surface
of the work.

Fig. 85

Fig. 86 shows a combined drill and
countersink which is used for drilling the
center holes in work which is to be mounted
on centers. It is also used to spot the center in a piece that is to be drilled.

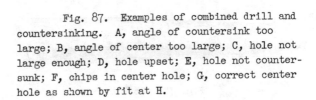

Fig. 86

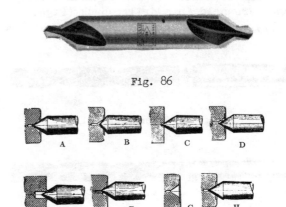

Fig. 87. Examples of combined drill and
countersinking. A, angle of countersink too
large; B, angle of center too large; C, hole not
large enough; D, hole upset; E, hole not countersunk; F, chips in center hole; G, correct center
hole as shown by fit at H.

22. Explain what a reamer is used for and
    name 3 types of reamers.
    A. A reamer is a tool for enlarging a
hole, round, straight, smooth, and up to
size. Fig. 88 shows a hand reamer which is
a finishing tool used to remove from .002" to
.003" stock. The machine reamer shown in
Fig. 89 is used for reaming in the drill press

Fig. 88. Hand Reamer

Fig. 89. Machine Reamer

or lathe. It removes approximately 1/64"
stock. Fig. 90 shows an expansion reamer.
This reamer is adjustable for reaming holes
of special size. Use a solid reamer whenever
possible.

Fig. 90. Expansion Reamer

Chart for the expansion of reamers

Size	Limit	Size	Limit
1/4" to 15/32"	.005"	1" to 1 23/32"	.010"
1/2" to 31/32"	.008"	1 3/4" to 2 1/2"	.012"

23. Name some of the common types of taps and
    the tools for revolving them.

(a)

(b)

(c)

Fig. 91. Hand taps for internal threading: (a) taper; (b) plug; (c) bottoming.

A T tap wrench used for revolving small taps
is shown in Fig. 92. Fig. 93 shows a hand
tap wrench for revolving larger taps. Fig.
94 shows a tapping chuck which is used for
holding taps in machine tapping operations.

24. How can a broken tap be removed?
    A. The tap extractor shown in Fig. 97 is
used to remove a broken tap. It must be used
carefully, as it is a fragile tool.

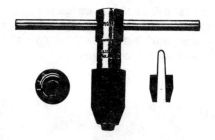

Fig. 92. T Tap Wrench

Fig. 93. Hand Tap Wrench

Fig. 94. Tapping Chuck

Fig. 95. Nut or Extension Tap

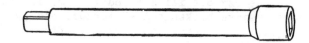

Fig. 96. Tap Extension

Fig. 97. Tap Extractor

25. What is a drill and tap gage?

A. Fig. 98 shows this gage which is used to select and check tap drills for the most commonly used machine screws.

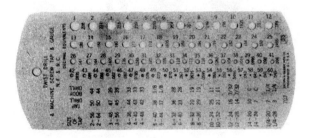

Fig. 98.   Drill and Tap Gage

26. What tools are used in cutting an external thread by hand?

A. External threading is done with a die, such as the one shown in Fig. 99. This die is inserted in the die stock of Fig. 100 and turned by hand.

Fig. 99.   Round Split Die

Fig. 100.   Die Stock

27. What is an "ezy-out"?

A. The ezy-out illustrated in Fig. 101 is used to remove broken bolts, screws, etc., from holes. In using it, a hole is first drilled in a broken bolt or screw, the size being a little smaller than the minor diameter of the thread. An ezy-out of the proper size is then inserted and revolved counter-clockwise (see Fig. 103).

Fig. 101.   Ezy-out

Fig. 102.   Fig. 103.
Broken Screw   Using the Ezy-out

28. What are the different types of rules generally used in the shop?

A. Figs. 104 to 107 illustrate several rules which are commonly used in the shop.

Fig. 104.   Six Inch Steel Rule

Fig. 105.   Hook Rule

Fig. 104 shows the common six inch steel rule, which is used in many ways. Fig. 105 shows the hook rule, used for measuring inside dimensions where the hook is an advantage. The rule depth gage of Fig. 106 is used to find the depth of holes, slots, etc. The tape-rule in Fig. 107 is used where a flexible measuring instrument is desired. The boxwood folding rule, the zig-zag rule, and the carpenters' steel square, which are commonly used by carpenters, are illustrated on pages 27 and 28.

Fig. 106.
Rule Depth Gage

29. Describe the combination square and some of the equipment used with it.

A. Fig. 108 shows the

Fig. 107.   Steel Tape-Rule

Fig. 108.   Combination Set

combination set, consisting of the combination square, a center head, and a protractor
(the center head and the protractor being
frequently used with the combination square).
This set has a multitude of uses, especially
in laying out centers and angles.   Fig. 109
shows several common applications of the combination square and one application of the
center head.

Fig. 109.   Combination Square and Center
Head Applications.

30. What devices are commonly used to hold
work?

    A. In addition to the drill vise in Fig.
83, several types of clamps are used to hold
work.   Fig. 110 shows a small toolmakers'
steel clamp, which has two interchangeable
blocks for varying the capacity of the clamp
(the small block being shown on the left).
Fig. 111 shows toolmakers' parallel clamps
and Fig. 112 shows a C clamp.   The V blocks

and clamps shown in Fig. 113 are used frequently in holding round work.

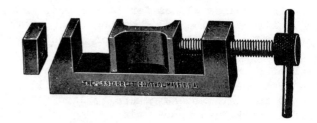

Fig. 110.   Toolmakers' Steel Clamps

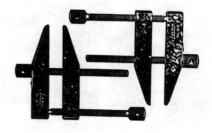

Fig. 111.   Toolmakers' Parallel Clamps

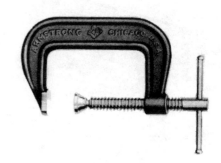

Fig. 112.   C Clamp

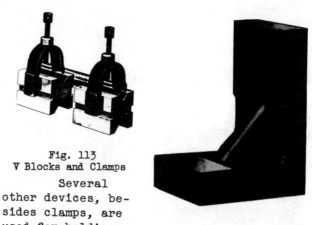

Fig. 113
V Blocks and Clamps

    Several
other devices, be-
sides clamps, are
used for holding
work in various ways.
The angle plate in Fig. 114 is used to hold
work for laying out or machining.   The box

Fig. 114.   Angle Plate

parallel or cube shown in Fig. 115 is used
mostly in the grinding department for holding
work while 90° angles are being ground, and
the parallels in Fig. 116 are used to support
work in parallel planes for laying out, shaping, milling, grinding, checking, etc.   The
cat head in Fig. 117 is used for supporting
irregular shaped revolving work in a steady
rest of a lathe, the work being adjusted centrally by set screws. The hold downs illustrated in Fig. 118 are generally used for
holding work in a shaper vise. The step
block in Fig. 119 is used for clamping work,
as shown in Fig. 120.

Fig. 115. Box Parallel or Cube

Fig. 116. Parallels

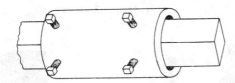

Fig. 117. Cat Head

Fig. 118. Hold Downs   Fig. 119. Step Block

31. Explain how a surface plate and a
    straight edge are used.
    A. The surface plate shown in Fig. 121 is

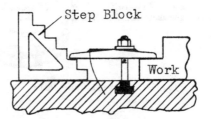

Fig. 120. Application of Step Block

used to check flat surfaces. The straight
edge in Fig. 122 is used to check the ways of
lathes, shapers, planers, etc., for being flat
as well as straight. The flat scraper (see
Fig. 158) is used to remove the high spots of
the ways.

Fig. 121. Surface Plate

Fig. 122. Straight Edge

    When the surface plate or straight
edge is being used, it is coated lightly with
a marking material such as Venetian red or
Prussian blue, and then rubbed over the surface that is to be trued. The marking material is left on the high places, thus showing
the high spots and are to be removed with a
scraper. This operation is repeated until the
surface shows a good bearing at all points.
Small work is rubbed on the plate. Every part
of the surface plate should be used as evenly
as possible, for rubbing in one place will
wear the surface unevenly.

32. Name several tools and pieces of equipment
    commonly used on lathe work.
    A. Fig. 123
shows the straight
tool holder, used
for threading and
general turning.
Figs. 124 and 125
show left-hand and
right-hand tool

Fig. 123. Straight Tool Holder

Fig. 124.   Left-hand Tool Holder

Fig. 125.   Right-hand Tool
Holder

Fig. 126.   Straight Cut-
ting-off Tool Holder

Fig. 127.   Right-hand
Off-set Cutting-off Tool Holder

Fig. 128.   Threading Tool
Holder with Spring Head

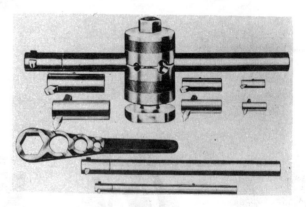

Fig. 129.   Forged Boring
Tool Holder

holders.   The left-hand tool holder is used for turning material with a large diameter and for facing toward dead center (do not take a heavy cut toward dead center).   The right-hand tool holder is used for facing toward live center and for turning close to the chuck.

Figs. 126 and 127 show holders used for cutting-off work.   The holder in Fig. 128 is used for producing threads of a fine finish with close limits.

Fig. 129 shows a holder for a forged boring tool.   The knurling tool in Fig. 130 is used to produce a knurled finish, as shown in Fig. 131.

The boring bar holder in Fig. 132 is used to hold large heavy duty boring bars. The lathe center (Fig. 133) is used to support revolving work and the lathe mandrel (Fig. 134) is used to support work with standard holes while it is being machined.

Fig. 131.   Knurling Tool in Use

Fig. 132.   3-Bar Boring Tool Set

Fig. 130.   Knurling Tool

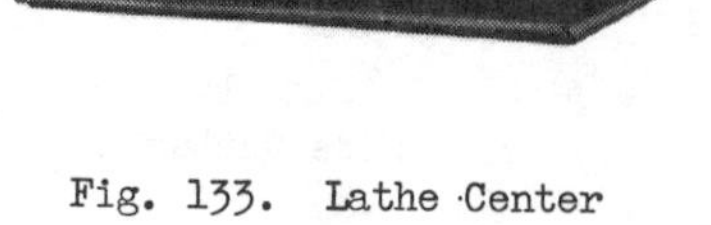

Fig. 133.   Lathe Center

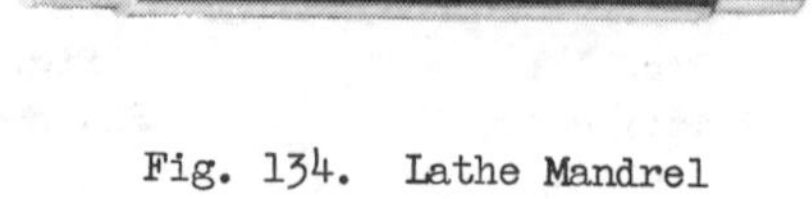

Fig. 134.   Lathe Mandrel

The expanding mandrel in Fig. 135 is used to support work with special size holes while machining it. The bent tail lathe dog (Fig. 136) is used for driving round or regular work (square, hexagon, etc.) and the clamp dog (Fig. 137) for rectangular work.

33. Name some of the equipment and cutters used on the milling machine.

A. The milling machine arbor in Fig. 138 is used for revolving and driving milling machine cutters. The micrometer offset boring head shown in Fig. 139 is adjustable and is used to drive a boring tool. The Woodruff keyway cutter and holder in Fig. 140 are used in cutting Woodruff keyways.

The slitting saw in Fig. 141 is used for milling narrow slots, the plain cutter for milling slots the same width as the cutter, the side cutter for milling vertical and horizontal surfaces, and the angular cutter for milling angles. The convex cutter is used for cutting concave surfaces and the concave cutter for cutting convex surfaces. The involute form cutter is used for cutting gear teeth. The inserted tooth side cutter is the most economical of the large cutters, as broken teeth can be replaced easily. The slab cutter is used for milling flat surfaces where a large cutting area is desired. The gear hob is used on gear hobbers for cutting teeth in gears. The shell end mill is used

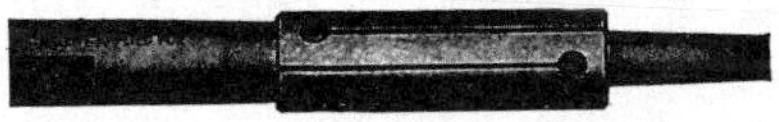

Fig. 135. Expanding Mandrel

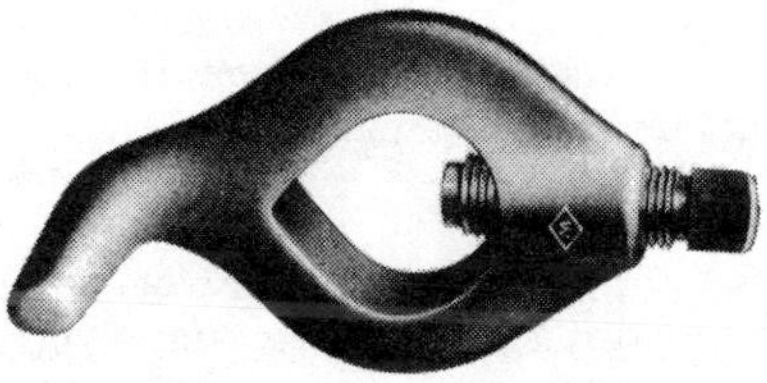

Fig. 136. Bent Tail Lathe Dog

Fig. 137. Clamp Dog

Fig. 138
Milling Machine Arbor

Fig. 139
Micrometer Offset Boring Head

Fig. 140
Woodruff Keyway Cutter
and Holder

Fig. 141
Slitting Saw

Fig. 142
Plain Cutter

Fig. 143
Side Cutter

Fig. 144
Angular Cutter

Fig. 145
Convex Cutter

in end milling when a large cutting area is wanted.

Fig. 146
Concave Cutter

Fig. 147
Involute Form Cutter

Fig. 148
Inserted Tooth Side
Cutter

Fig. 149. Slab Cutter

Fig. 150. Gear Hob

Fig. 151
Shell End Mill

The two lip end mill in Fig. 152 is used for milling slots in solid metal and the helical end mill in Fig. 153 is used for milling slots, fillets, keyways, etc.

34. What equipment is used for dressing and truing grinding wheels?

A. The metallic abrasive wheel dresser (Fig. 154) and the magazine wheel dresser (Fig. 155) are used for dressing grinding wheels. The diamond nib in Fig. 156 is used for both dressing and truing.

Fig. 152. Two Lip End Mill

Fig. 153. Helical End Mill

Fig. 154. Metallic Abrasive Wheel Dresser

Fig. 155. Magazine Wheel Dresser    Fig. 156
Diamond Nib

35. Name the various types of bearing scrapers.

A. The bearing scraper shown in Fig. 157 is used, as its name implies, for scraping bearings. Fig. 157A, the three-corner scraper is used mostly for removing the burrs or sharp internal edges from soft bushings, etc. Do not use a file for this operation. Flat scrapers are used to remove high spots from flat bearing surfaces.

Fig. 157. Bearing Scraper

Fig. 157A. Three-Corner Scraper

Fig. 158. Flat Scraper

The oil stone shown in Fig. 159 is used to sharpen scrapers, cutting tools, etc. Keep the surface of a stone well oiled.

Fig. 159. Oil Stone

36. What gages, measuring instruments, and equipment are commonly used in measuring and setting up precision work in the toolroom?

A. Fig. 160 shows a micrometer, the most commonly used precision tool in the toolroom. This tool was invented in 1848 by Jean Palmer, a Frenchman, and has been improved from time to time until the modern micrometer has resulted. There are other types of micrometers besides the outside micrometer shown here, which is the type generally used. Fig. 199 shows the thread micrometer (which is an outside micrometer) and Fig. 170 shows the depth micrometer. Still another type is the inside micrometer. All of these are to be studied later in the course.

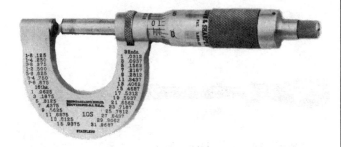

Fig. 160. Common Outside Micrometer

Fig. 161 shows a sine bar, which is used for checking angles, tapers, etc., and Fig. 162 shows an individual set of Johansson gage blocks, used for precision measuring. These blocks are rectangular pieces of tool

Fig. 161. Sine Bar

steel, hardened, ground, and finished to an accuracy within a few millionths of an inch. A full set consists of eighty-one blocks, and these will make 120,000 different size gages, in steps of .0001", from a minimum size of

.200" to a maximum size of 12.000". Fig. 163 shows an application of a sine bar, using Johansson gage blocks.

Fig. 162.   Johansson Gage Blocks

Fig. 163. Application of a Sine Bar, Using Johansson Gage Blocks.

The surface gage shown in Fig. 164 is used for layout work, for leveling and lining

Fig. 164. Surface Gage

up work, and for locating centers on rough work. The surface gage may be set to a combination square in the following way, using the straight end of the scriber (see Fig. 165). First, set the square on the layout table, as shown in the illustration, being

Fig. 165. Setting a Surface Gage

sure that the blade of the square is resting on the table and is clamped in place in the head. Then set the standard of the surface gage, clamp in place, adjust the scriber to the approximate height desired, and clamp in place on the standard. Finally, set the scriber against the desired index mark on the blade of the square, and, by means of the adjusting screw on top of the base of the gage, set the scriber to the desired location on the blade of the square.

Two of the many applications of the surface gage are shown in Fig. 166.

Fig. 166
Surface Gage Applications

Fig. 167 shows a planer gage, which is used for setting the cutting tool on a planer (it may also be used on bench work). This gage can be set to a micrometer (Fig. 168), surface gage, or caliper.

Fig. 167. Planer Gage

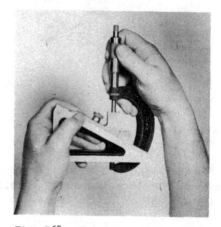

Fig. 168. Setting a Planer Gage

Figure 169 shows a solid square, which is used for checking square work.

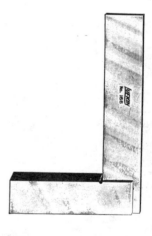

Fig. 169. Solid Square

Figs. 170 and 171 show depth gages, one with a micrometer scale and one with a vernier scale. These are used for measuring the depth of holes and grooves.

Fig. 170. Micrometer Depth Gage

Fig. 172 shows a vernier height gage, which is used for locating center distances, finding the height of projections, etc.

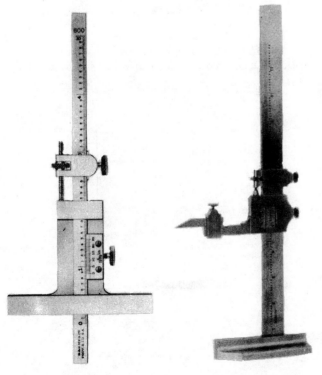

Fig. 171
Vernier Depth Gage

Fig. 172
Vernier Height Gage

Fig. 173 shows a vernier caliper, used to determine inside and outside dimensions to the thousandth of an inch. The caliper is graduated to read on one side for outside measurements, and on the other side for inside measurements.

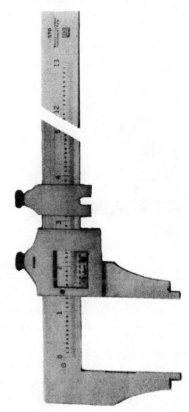

Fig. 173. Vernier Caliper

The gear tooth vernier caliper in Fig. 174 is used for finding the chordal thickness and the corrected addendum of gear teeth, worms, etc.

The principle of the Vernier, with which these last four tools are equipped, will be explained later.

Fig. 175 shows the Test Indicator, which is generally used in connection with a height gage. The indicator is mounted on a hardened steel tool post holder (only a portion of the holder being shown). The scale is graduated in thousandths, and an indicator with either a single or double scale can be obtained (the double scale can of course be read from both front and rear). The contact point is spherical in shape and can be turned 90° on either side of the indicator.

Fig. 176 shows a dial indicator, which is commonly used for general shop work. The dial alone is shown in Fig. 177. It is graduated in half-thousandths from 0 to 25 to 0, so that plus or minus measurements are indicated, and readings of less than a half-

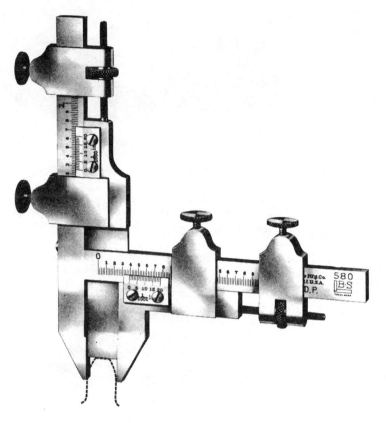

Fig. 174. Gear Tooth Vernier Caliper

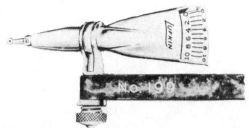

Fig. 175. Test Indicator

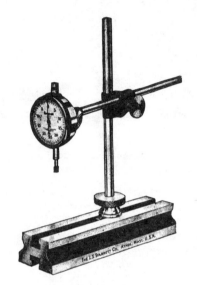

Fig. 176. Dial Indicator

Fig. 177. Dial

Fig. 178. Vernier Bevel Protractor

thousandth can be estimated. The dial can
be turned to bring the zero in any position.

Fig. 178 shows a vernier bevel
protractor, which is used for checking angles. Toolmakers' buttons, which are used
in locating holes for boring, are shown in
Fig. 179, with a sectional view showing
the method of attaching a button to the

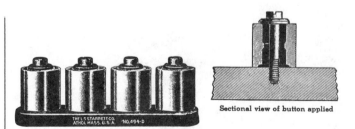

Sectional view of button applied

Fig. 179. Toolmakers' Buttons

work. Fig. 180 illustrates the application
of these buttons.

Fig. 180. Application of Toolmakers' Buttons

The ring gages in Figs. 181 and 183
are used for external measuring and the plug
gages in Figs. 182 and 184 are used for internal measuring. Fig. 185 shows outside and
inside caliper gages.

Fig. 186. Bench Centers

The bench centers in Fig. 186 are used
for checking work mounted on centers. Fig.
187 shows a set of thickness gages. These are
used for measuring the space between two surfaces.

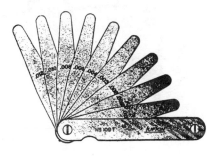

Fig. 187. Thickness Gage

Fig. 181. Ring Gage   Fig. 182. Plug Gage   Fig. 183. Ring Thread Gage

Fig. 184. Plug Thread Gage

Fig. 188 shows a telescoping gage,
which is used for internal measuring, as

Fig. 185. Outside and Inside Caliper Gages

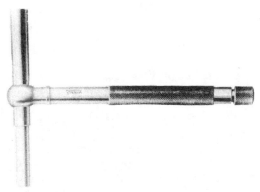

Fig. 188. Telescoping Gage

illustrated in Fig. 189 (this illustration shows the gage being used on a blanking die). Fig. 190 shows the method of taking the measurement from a telescoping gage with a micrometer.

Fig. 195 is used for checking and setting 60° threading tools, while the gages in Figs. 196 and 197 are used for grinding and setting threading tools for cutting Acme threads and

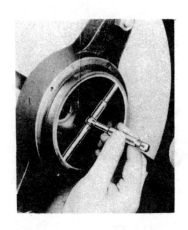
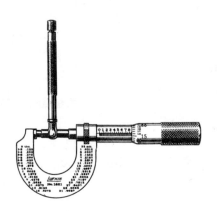
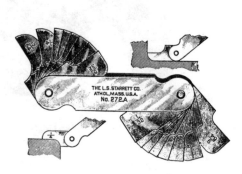

Fig. 189. Telescoping Gage in Use on a Die

Fig. 190. Taking the Measurement of a Telescoping Gage

Fig. 191. Radius Gage

The radius and fillet gage in Fig. 191 consists of a number of steel blades marked in fractions of an inch, each blade having the corresponding radius accurately formed in it. The gage is used to check male and female radii, a female radius being called a fillet (see Figs. 192 and 193).

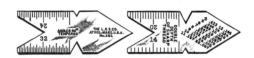

Fig. 195. Center Gage

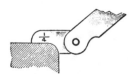
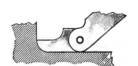

Fig. 192
Checking a Radius

Fig. 193
Checking a Fillet

Fig. 196. Acme Thread Tool Gage

Figs. 194 to 199 illustrate several gages used in checking threads and threading tools. The screw pitch gage in Fig. 194 has a series of blades which are accurately notched and numbered according to the measure of the thread pitches in common use. It is used for checking the number of threads per inch of screws and nuts. The center gage in

Fig. 194. Screw Pitch Gage

Fig. 197. Worm Thread Tool Gage

worm threads. Fig. 198 shows the method used
in checking threads with an ordinary micrometer, while Fig. 199 shows the screw thread
micrometer for measuring threads.

Fig. 198. Checking threads with
three wires and common micrometer

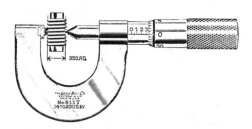

Fig. 199.   Screw Thread Micrometer

37. What are the principal tools used in sol-
    dering operations?

    A. Figs. 200 to 203 show soldering cop-
pers, the square pointed copper being the one
used for general work. The electric soldering
copper is heated by electricity. The flat file

Fig. 200. Square Point Copper

Fig. 201. Bottoming Copper

in Fig. 204 is used for abraiding or smoothing metal, and is fitted into the file handle
shown in Fig. 205. The tinners' snips in Fig.
206 are used for cutting soft metals.   The
ladle (Fig. 207) is used for melting solder or
other materials. The blow torch in Fig. 208
is used for heating soldering coppers, melting materials in ladles (though a different

Fig. 202. Hatchet Copper

Fig. 203. Electric Copper

Fig. 204. Flat File

Fig. 205. File Handle

Fig. 206. Tinners' Snips

Fig. 207. Ladle

type of torch, known as a fire pot, is preferable for this), or for other needs in which
heat must be applied locally.

38. What tools are commonly used in woodwork-
    ing and carpenter work?
    A. Figs. 209 and 210 show two folding

Fig. 208. Blow Torch

wood rules, the boxwood rule and the well-known zig-zag rule, both used for general measuring by the person working with wood.

The hand saw in Fig. 211 is the type generally used by carpenters. The compass-saw (Fig. 212) is used for making circular holes. The level shown in Fig. 213 is used in testing horizontal and vertical surfaces.

Fig. 209. Boxwood Folding Rule

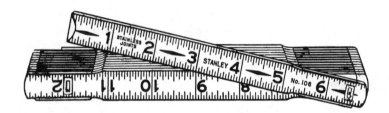

Fig. 210. Zig-zag Rule

Fig. 211. Hand Saw

Fig. 212. Compass Saw

Fig. 213. Wood Level

The marking gage in Fig. 214 is used for scribing lines parallel to an edge. The iron plane (Fig. 215) is used for producing a

Fig. 214. Marking Gage

Fig. 215. Iron Plane

Fig. 216. Drawing Knife

smooth finish on wood, and the drawing knife (Fig. 216) is used for peeling bark and roughing out narrow surfaces. The spoke shave (Fig. 217) is used for roughing out cylindrical wood work, and

Fig. 217. Spoke Shave

the flat wood chisel (Fig. 218) is used for cutting slots, etc. The wood turning tools in Fig. 219 are used when a piece of wood is turned in a lathe.

Fig. 218. Flat Wood Chisel

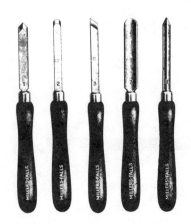

Fig. 219. Wood Turning Tools

Fig. 220 shows a bit brace, used for revolving bits when drilling in wood. Two types of bits are shown in Figs. 221 and 222. Fig. 223 shows a plumb bob, which is used to establish vertical lines. The steel square in Fig. 224 is used by carpenters for measuring and squaring work.

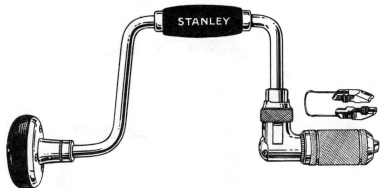

Fig. 220.   Bit Brace

Fig. 221.   Auger Bit

Fig. 222.   Expansion Bit

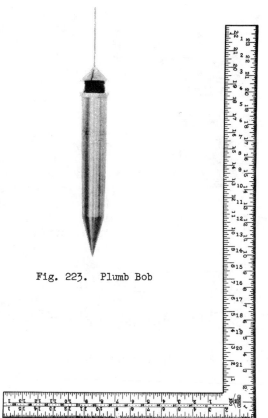

Fig. 223.   Plumb Bob

Fig. 224.   Carpenters' Steel Square

The cabinet clamp in Fig. 225 is used for clamping glued joints while they are drying. The corrugated steel fastener in Fig. 226 is used as a permanent joint reenforcement, while the pinch dog of Fig. 227 is used simply for drawing glued joints together. Glass is cut with the cutter shown in Fig. 228. The fillet sticker in Fig. 229 is used for setting wax fillets, and is heated by the alcohol lamp shown in Fig. 230. The putty knife (Fig. 231) is used for filling holes and joints with putty.

Fig. 228.   Glass Cutter

Fig. 229.   Fillet Sticker

Fig. 225
Cabinet Clamp

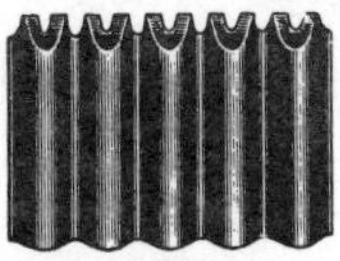

Fig. 226
Steel Fastener

Fig. 227
Pinch Dog

Fig. 230
Alcohol Lamp

Fig. 231. Putty Knife

39. What is a hack saw? Why is it necessary to have different kinds of hack saw blades?

A. Fig. 232 shows a hand hack saw frame, used to hold hack saw blades while cutting metal. Blades having teeth of various sizes are used, according to the kind of metal being cut. Figure 233 shows three blades, with 10, 18, and 32 teeth. The 10-tooth blade is a heavy blade with coarse teeth. It can be used only in a power saw, as it is too heavy for use in a hand frame. The 18-tooth blade is an all-around saw for general work (aluminum, babbitt, tool steel, high speed steel, cast iron, etc.). The 32-tooth blade is recommended for fine stock, tubing, sheet metal, etc.

Fig. 232. Hand Hack Saw Frame

10-Tooth Blade

18-Tooth Blade

32-Tooth Blade

Fig. 233. Hack Saw Blades

Different kinds of steel are used in making hack saw blades for different classes of work, tool steel, high speed steel, or tungsten alloy steel being generally used. A blade should be placed in the frame with the teeth pointing forward, so that the forward stroke will be the cutting stroke, and the blade should be drawn tight enough so that it will not bend.

40. Describe a divider and explain how it is used.

A. A divider (Fig. 234) is a tool for measuring the distance between points, for transferring a distance directly from a rule, and for scribing circles and arcs (Fig. 235). It consists of two legs hardened on the ends and usually has a spring adjustment. The size is determined by the length of a leg from the pivot to the point.

Fig. 234

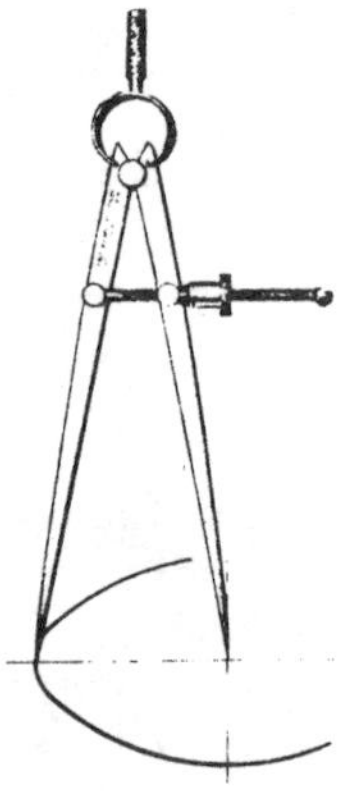

Fig. 235

In scribing a circle, first set the divider to the radius of the circle desired, as shown in Fig. 236. Then set one leg of the divider in the center punched hole and scribe short arcs on opposite sides of the center. If the distance between arcs is not equal to the required diameter, make the necessary adjustment and scribe the circle.

Fig. 236

NOTE: When using a divider on finished work, its surface should first be covered with copper sulfate (blue vitriol) or layout ink (a solution of gentian violet, orange shellac, and alcohol), so that the layout lines will be clearer. Unfinished work should be covered with chalk or with a layout paint made of calcimine.

41. Describe a trammel and explain how it is
    used.

    A. A trammel is a tool used to measure
the distance between points too great to be
reached with an ordinary divider. It consists of two adjustable points on a bar. In
setting the trammel (shown in Fig. 237) to a
given size, the points are first set approximately and the final adjustment made with the
thumb screw. A rule is laid flat on the work
bench, one point of the trammel is placed on
a selected mark on the rule, and the other
point is adjusted until it is set at the required dimension.

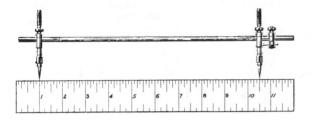

Fig. 237

    The method of scribing a circle is
shown in Fig. 238. Before scribing a complete circle, one should scribe short arcs on
the center line on opposite sides of the center. If the distance between these arcs is
the same as the diameter of the required circle, the setting is correct and no further
adjustment is necessary.

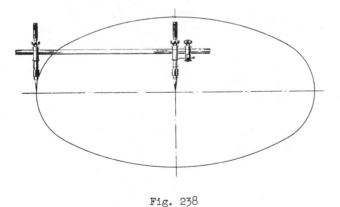

Fig. 238

42. Describe outside calipers and explain how
    they are used.

    A. Calipers are used for measuring dis-
tances between or over surfaces, or for comparing distances or sizes with standards. Calipers have two adjustable legs which work on
a hinged joint. Outside calipers have curved
legs for measuring outside dimensions (Fig.
239).

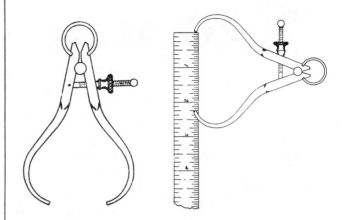

Fig. 239   Fig. 240

    In setting outside calipers with a
rule, first set the calipers at approximately
the given size. Then, with one point resting
on the end of the rule, adjust the calipers
until the other point is opposite the proper
mark on the rule (Fig. 240). To insure an accurate setting the center line of the points
must be parallel to the edge of the rule.

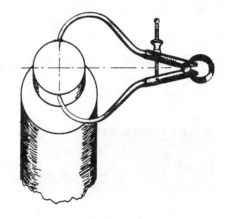

Fig. 241

    Fig. 241 shows how outside calipers
are used to measure the diameter of round
stock. The calipers are first set to the approximate diameter of the stock. Then they
are held at right angles to the center line
of the work and moved back and forth across
the center line while they are adjusted until
the points bear lightly on the work. This is
called "getting the feel." When the tool has
been adjusted properly, the diameter may be
read as indicated in Fig. 240.

43. Describe inside calipers and explain how
    they are used.

    A. Inside calipers are calipers with legs
curved for measuring the diameter of holes,
the distance between two surfaces, the width

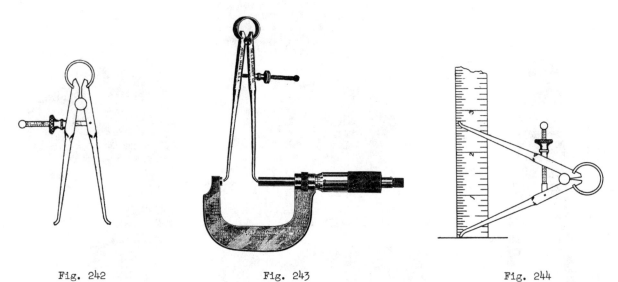

Fig. 242   Fig. 243   Fig. 244

of slots, and other similar jobs (Fig. 242).

    In setting inside calipers, first set a micrometer to the required dimension. Then hold the micrometer and calipers in the position shown in Fig. 243 and adjust until the proper "feel" is secured. Another method is to hold a rule at right angles to a flat surface, set one leg of the inside calipers on the flat surface, and adjust until the other leg is opposite the proper mark on the rule (Fig. 244 shows this method of setting the calipers).

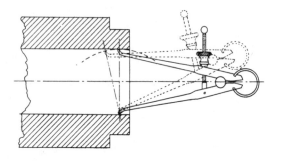

Fig. 245

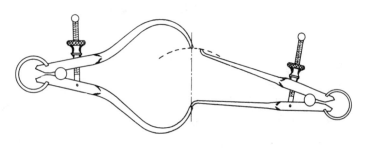

Fig. 246

    To measure the diameter of a hole with inside calipers, first set the calipers approximately to the size of the hole. Then hold one leg against the wall of the hole and adjust the other leg until it just touches a point diametrically opposite, as illustrated in Fig. 245. The next step is to transfer the measurement from the inside calipers to a micrometer. Great care must be taken in adjusting the calipers and transferring measurements. With a little practice a mechanic should "caliper" a hole within one-half a thousandth of an inch.

44. Describe the transferring of a measurement from outside to inside calipers.

    A. When a measurement has to be transferred from outside to inside calipers, both calipers are held so that they are in the position shown in Fig. 246. With one extreme point of a leg of the inside calipers placed on the extreme point of a leg of the outside calipers, adjust the inside calipers until the two extreme points touch lightly. Care must be taken not to "force" either pair of calipers or a true reading will not be obtained.

45. Describe hermaphrodite calipers and explain how they are used.

    A. Hermaphrodite calipers have two legs which work on a hinge joint (Fig. 247). One leg is similar to a leg on a pair of dividers and the other is similar to a leg on a pair of outside calipers. Hermaphrodite calipers may be used to scribe arcs, as shown in Fig. 248, or as a marking gage in layout work, as shown in Fig. 249.

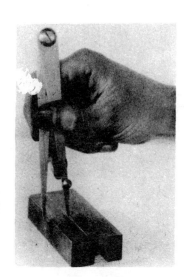

Fig. 247   Fig. 248   Fig. 249

To set hermaphrodite calipers to a rule, adjust the scriber leg until it is slightly shorter than the curved leg. Then, with the curved leg set on the end of a rule, adjust the scriber leg to a point opposite the required line on the rule, as illustrated in Fig. 250.

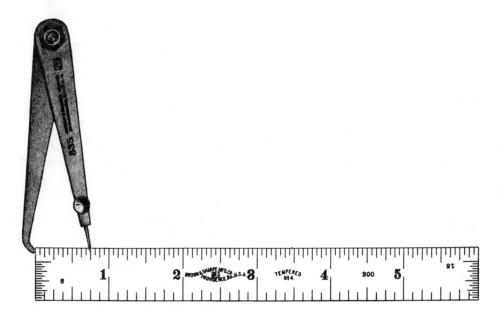

Fig. 250

Chapter 4

RULES

1. What is the difference between a rule and a scale?

A. A rule is a graduated measuring instrument, made of wood, metal, or other suitable material (see Fig. 251). It is usually graduated to indicate inches and fractions of an inch (or centimeters and millimeters). A scale is similar in appearance to a rule, as its surface is graduated into regular spaces, but these spaces differ decidedly from those on a rule, because they are larger or smaller than the actual measurements indicated. A scale therefore gives proportional measurements instead of the accurate measurements obtained with a rule. Figs. 252 and 253 illustrate two scales regularly used in the drawing room for measuring and laying out distances and dimensions.

2. Name some rules commonly used in the shop.

A. There are several different kinds of rules, each being especially adapted to certain classes of work. By choosing the proper rule it is easier for a mechanic to obtain accurate measurements. In addition to the well-known flexible rule shown in Fig. 251, which is used in many ways in the shop, two other rules commonly used are shown in Figs. 254 and 255. The narrow rule is used for work on which other rules are too wide and the hook rule is used where the hook is an advantage in measuring.

3. Briefly describe and tell the purpose of a shrink rule.

A. When molten metal is poured into a mould and allowed to cool and solidify it will con-

Fig. 251. Six-Inch Flexible Rule

Fig. 252. Flat Boxwood Scale

tract. To make allowance for this contraction or shrink and to maintain exact relationship for all dimensions, the pattern maker uses a shrink rule. Since the rate of contraction changes with different materials each shrink rule has the shrink allowance stamped on it. The shrink rule, Fig. 256, is actually 12-1/8 inches long, the additional length being taken up proportionally throughout its length.

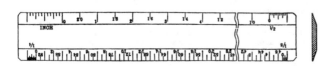

Fig. 253. Triangular Boxwood Scale

Fig. 254. Narrow Rule

Fig. 255. Hook Rule

33

Fig. 256. Shrink Rule

4. How is a six-inch rule usually graduated?

A. As every mechanic uses a rule frequently (Fig. 257 shows one application) it is important to understand this type thoroughly.

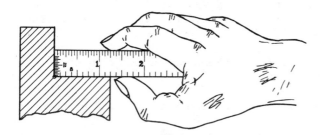

Fig. 257. Using a Rule

One of the most frequent ways of graduating this rule is to divide the inches into halves, quarters, eighths, sixteenths, thirty-seconds, and sixty-fourths. One side of the rule will then usually have coarse graduations, 8ths on one edge and 16ths on the other (see Fig. 258), while the other side

Fig. 258. Rule Graduated in 8ths and 16ths

will have fine graduations, 32nds on one edge and 64ths on the other (see Fig. 259). Since the Ford Motor Company has adopted the decimal system for specifying dimensions on all new designs, the rule in general use is graduated in 32nds and 64ths on one side and in 10ths and 50ths on the other side (see Fig.

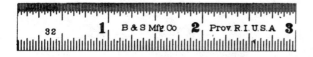

Fig. 259. Rule Graduated in 32nds and 64ths

263). Various other graduations and combinations are of course frequently used, and any desired marking will be furnished by a rule manufacturer on a special order.

5. What is the best method of learning to read a rule?

A. The following procedure is recommended:
   (1) Learn the 8ths and 16ths shown in Fig. 258 (or the 10ths and 50ths if a rule is so graduated).
   (2) Become familiar with the 32nds and 64ths (Fig. 259).
   (3) Thoroughly understand the readings shown in Fig. 260.

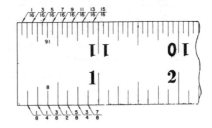

Fig. 260

   (4) Practice until measurements such as those in Figs. 261 and 262 can be read quickly.

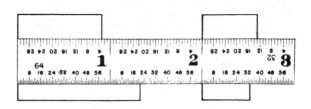

Fig. 261

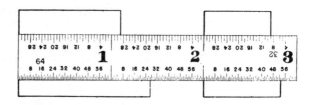

Fig. 262

6. Explain the decimal system used by the Ford Motor Company for specifying dimensions.

A. The Ford Motor Company has adopted the decimal system for specifying divisions of an inch, this system replacing the old one which used common fractions.

It is obvious that rules graduated for the new system must be used in working on jobs dimensioned in decimals. Fig. 263 shows a six-inch flexible rule graduated in 10ths and 50ths.

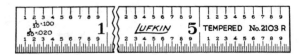

Fig. 263. Six-Inch Rule Graduated in 10ths and 50ths

The system is effective on all new drawings but is not used for changes on drawings showing the old style dimensions. The following outline explains the new system, which saves time and makes calculating easier.

Instead of dimensioning drawings 1/2, 1/4, 1/8, 1/16, 1/32, and 1/64 of an inch, dimensions on drawings of new parts are specified in tenths or hundredths of an inch. Thus, instead of such dimensions as 1-5/16,

2-7/32, 5-33/64 and 6-1/2 on drawings, the mixed decimals 1.32, 2.22, 5.52, and 6.5 are used (see Fig. 264).

It is understood that one or two place decimals carry the same degree of accuracy that the common fractions carried previously, namely, + or - .010 on a finished drawing. When greater accuracy than this is demanded, a three or four place decimal is used and a definite tolerance is specified as .400 - .402.

The use of an odd number for the second place in a dimension of a diameter is to be avoided, because "halving" the diameter to obtain the radius will give a three place number. An effort should therefore be made to keep the second figure even. This of course does not apply to ordinary dimensions which will not be "halved."

1. List and find the combined length of the lettered divisions in Figs. 265(a) and (b).

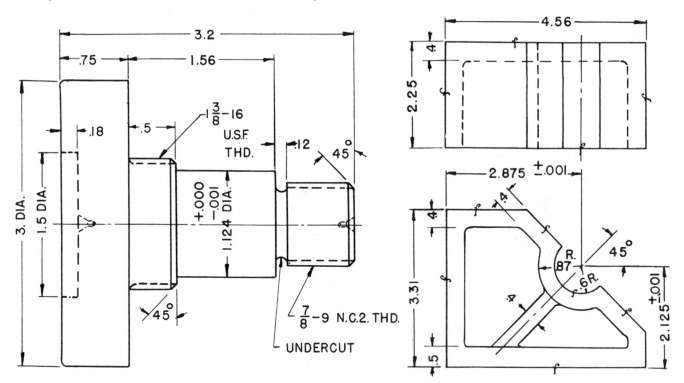

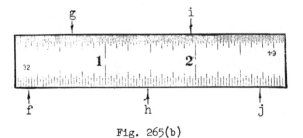

Fig. 264. Decimal System of Dimensioning

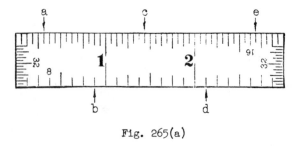

Fig. 265(a)   Fig. 265(b)

2. Measure the lines below and list their
   lengths. Find their sum,
   (a)_____(b)_____ (c)_____
   (d)_____ (e)_____
   (f)__ (g)_____ (h)_____
   (1)_____
   (j)_____

3. Draw lines as near the following lengths
   as possible.
   (a) 3-3/4   (e) 2-7/16   (h) 1.3750
   (b) 1-1/8   (f) .0312   (1) .453125
   (c) 3/16   (g) .1875   (j) 3.765625
   (d) 1/2

4. Measure the lettered dimensions in the following sketches of Fig. 266 and write them
   in the proper order.

5. Draw a sketch of a rectangular bar
   1-3/8 x 2-5/16 x 4-11/32.
   Draw a sketch of a rectangular bar
   1-1/4 x 1-7/16 x 1-1/2.
   Draw a sketch of a bar 1.5 x 3.87.

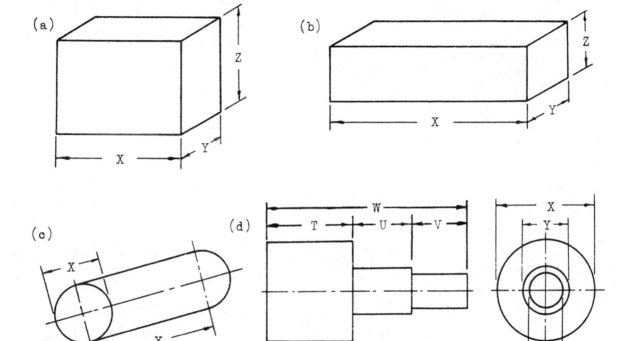

Fig. 266 (a, b, c, and d)

Chapter 5

MICROMETERS

The micrometer is the most commonly used precision tool in the toolroom, used for accurately measuring dimensions. In order to use such a tool effectively and appreciate some of the mechanical principles on which it works, its origin, history, construction, use, and care should be understood.

Fig. 267. Micrometer Caliper of 1848

The micrometer (see Fig. 267) was invented by Jean Palmer, a Frenchman, in 1848. It was not introduced into this country until 1867, when J. R. Brown and L. Sharpe brought back a Palmer micrometer with them after a visit to the Paris Exposition. From this in-

micrometer shows that it is a comparatively new tool. It has been improved from time to time by different persons, until the modern tool in Fig. 271 is the result.

Three different types of micrometers are used in the shop. These are the outside micrometer of Fig. 271 (including the thread

Fig. 268. Micrometer Caliper of 1877

micrometer in Fig. 275), the depth micrometer of Fig. 270, and the inside micrometer shown in Fig. 272.

Fig. 269. Improved
Micrometer Caliper
of 1885

strument they developed a micrometer which was the beginning of our modern micrometer. This micrometer was introduced in 1877 (see Fig. 268), and was the first one of this type to be sold in the United States. Fig. 269 shows an improved micrometer introduced in 1885 by the same concern. This brief history of the

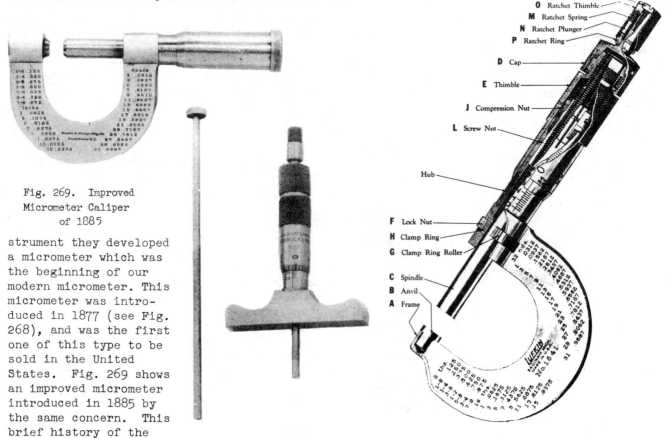

Fig. 270

Fig. 271

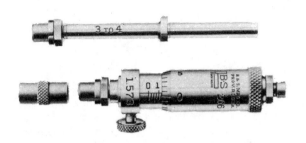

Fig. 272

1. What are the five principal parts of a micrometer?

A. The frame, anvil, spindle, sleeve or
barrel, and thimble (see Fig. 271).

2. Explain how a micrometer is read.

A. A micrometer is very easily read, but
of course, like many other things, rapid work
is obtained only after some practice. When
once learned it can be read at a glance.

The micrometer divides the inch into
1000 parts. As usually made, it has a 40
pitch screw (a screw with 40 threads per inch)
which advances, through a nut, .025 of an
inch per revolution. It is evident that if
only measurements of .025" or less were to be
made, all the graduating could be on the end
of the revolving thimble, and all that would
be necessary besides this would be an indicating line on the stationary part. As a micrometer must have a greater range, however,
it is necessary to have some means of counting and adding together the additional revolutions of the screw. This is done in a very
simple manner by the graduations and numbers
used, which is plainly illustrated in Fig.
273 (the sketches are larger than actual size).
The cross lines on the sleeve are spaced .025"
apart to equal the pitch of the screw. A
revolution line is cut lengthwise on the

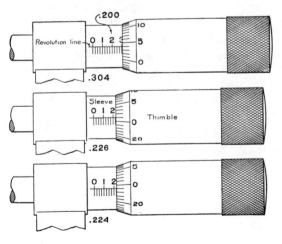

Fig. 273

sleeve, and in connection with the zero line
on the thimble, counts whole revolutions of
the screw. When the end of the thimble matches any of the cross lines and the zero line
matches with the revolution line, the number
of spaces exposed denotes the number of revolutions made. Every fourth cross line is numbered from 0 to 10.

In the first illustration of Fig. 273,
the reading is .304", showing .300" on the
sleeve and .004" on the thimble. In the second illustration the reading is .226" showing
.225" on the sleeve and .001" on the thimble.
The reading in the third illustration is .224",
showing .200" on the sleeve and .024" on the
spindle.

The figures on the sleeve should be
taken as hundreds, that is, as 100, 200, 300,
etc. In Fig. 273 the thimble is purposely
shown close to the lines, as these are the
points where a mistake is most likely to occur. In the .226" reading, while the end of
the thimble may appear to match the cross
line, it is evident that it does not, because
the zero line on the thimble does not coincide
with the revolution line on the sleeve. The
zero line is one space advanced, which is of
course added to the .225", making the reading
.226".

A very good way to learn to read this
instrument is to take a one inch micrometer,
turn it down to zero (or against the anvil),
and then turn the screw back while you count
the graduations on the thimble for four revolutions (which means counting these graduations to 100). The method of counting the
cross graduating will then be easily understood.

READING THE MICROMETER IN TEN-THOUSANDTHS

While the thousandths part of an inch
is the most convenient unit and is fine enough
for the general run of machine work, there are
times when this must be further divided. With
the ordinary micrometer one-half and one-
quarter thousandths are easily estimated. Micrometers graduated with a vernier arranged
so that ten-thousandths can be read on the
thimble are used where finer adjustments are
required.

The vernier used consists of ten divisions which equal, in over-all space, nine divisions on the thimble. Thus, one division
on the vernier equals .1 of .009" or .0009".
Each graduation on the thimble equals .001".
The difference in space between a division on
the thimble and a division on the barrel is
equal to .0010" - .0009" or .0001". The two

zero lines on the vernier coincide with lines
on the thimble when the reading is exact with
respect to the number of thousandths. The
difference between the lines on the thimble
and lines on the vernier at number 1 is
.0001", at number 2 is .0002", at number 3 is
.0003", etc. For example, when the 1st, 2nd,
or 3rd lines coincide, the thimble has moved
past the zero setting .0001", .0002", or
.0003" to bring these lines together.

dle. This anvil may be either fixed or
swiveled. In measuring screw threads, the angle of the point and the sides of the V come
in contact with the cut surface of the thread
so that the reading of the micrometer indicates the pitch diameter or the full size of
the thread less the depth of one thread.

The pitch line may be represented by
a line drawn through the plane AB (Fig. 276)
when the spindle and the anvil are in contact

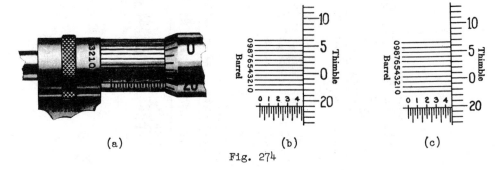

(a)   (b)   (c)

Fig. 274

To Read: First obtain the reading
for thousandths as described in the preceding
section and then add the ten-thousandths
which are indicated by the line on the vernier that coincides with a line on the thimble.

Example: In Fig. 274(b) the reading
is .4690". There are no ten-thousandths to
be added because the two zeros on the vernier
coincide with lines on the thimble. In Fig.
274(c) the seventh graduation on the vernier
coincides with a line on the thimble, indicating that .0007 should be added to the
thousandths reading. The correct reading is
.4690" + .0007" = .4697".

3. Briefly describe the screw thread micrometer and explain its purpose.

A. The screw thread micrometer (Fig. 275)
is a tool for measuring the pitch diameter of
threads. The spindle is ground conical to an
included angle of 60° and is slightly flattened on the end. The anvil has a 60° groove
and is set to align perfectly with the spin-

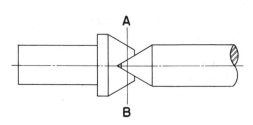

Fig. 276

with each other and the graduations on the
thimble are set at zero. When the spindle is
turned to open the micrometer the reading represents the distance between the pitch line
on the anvil and the pitch line on the spindle, or the pitch diameter. Since the thread
is measured from the angular surfaces, the
actual outside diameter need not be considered.
The thread micrometer is read the same as the
ordinary outside micrometer.

There are two methods of measuring the
pitch diameter of threads in common use, the
thread micrometer, and the three-wire system.
The thread micrometers are fast and convenient
to use, but when set as described above, the
reading is slightly distorted, the amount of
distortion depending upon the helix angle of
the thread being measured. While not as fast
and convenient as the thread micrometer, the
three-wire system is theoretically correct.
The wires must be held very close to size because any error in the wire is multiplied
when the measurement over the wires is calculated.

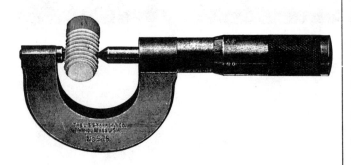

Fig. 275

Pitch Diameter = Major (Outside) Diameter—Depth of Thread

COARSE THREAD SERIES				FINE THREAD SERIES			
Size		Threads Per Inch	Cal. Reading or Pitch Diam.	Size		Threads Per Inch	Cal. Reading or Pitch Diam.
Number or Fraction	Decimal	N	$D - \dfrac{.649519}{N}$	Number or Fraction	Decimal	N	$D - \dfrac{.649519}{N}$
1	.0730	64	.0629	0	.0600	80	.0519
2	.0860	56	.0744	1	.0730	72	.0640
3	.0990	48	.0855	2	.0860	64	.0759
4	.1120	40	.0958	3	.0990	56	.0874
5	.1250	40	.1088	4	.1120	48	.0985
6	.1380	32	.1177	5	.1250	44	.1102
8	.1640	32	.1437	6	.1380	40	.1218
10	.1900	24	.1629	8	.1640	36	.1460
12	.2160	24	.1889	10	.1900	32	.1697
1-4	.2500	20	.2175	12	.2160	28	.1928
5-16	.3125	18	.2764	1-4	.2500	28	.2268
3-8	.3750	16	.3344	5-16	.3125	24	.2854
7-16	.4375	14	.3911	3-8	.3750	24	.3479
1-2	.5000	13	.4500	7-16	.4375	20	.4050
9-16	.5625	12	.5084	1-2	.5000	20	.4675
5-8	.6250	11	.5660	9-16	.5625	18	.5264
3-4	.7500	10	.6850	5-8	.6250	18	.5889
7-8	.8750	9	.8028	3-4	.7500	16	.7094
1	1.0000	8	.9188	7-8	.8750	14	.8286
1 1-8	1.1250	7	1.0322	1	1.0000	14	.9536
1 1-4	1.2500	7	1.1572	1 1-8	1.1250	12	1.0709
1 1-2	1.5000	6	1.3917	1 1-4	1.2500	12	1.1959
1 3-4	1.7500	5	1.6201	1 1-2	1.5000	12	1.4459
2	2.0000	4 1-2	1.8557	1 3-4	1.7500	12	†1.6959
2 1-4	2.2500	4 1-2	2.1057	2	2.0000	12	†1.9459
2 1-2	2.5000	4	2.3376	2 1-4	2.2500	12	†2.1959
2 3-4	2.7500	4	2.5876	2 1-2	2.5000	12	†2.4459
3	3.0000	4	2.8376	2 3-4	2.7500	12	†2.6959
3	3.0000	3 1-2	2.8144	3	3.0000	10	†2.9350

The table of measurements shows the pitch diameters of American National threads for the coarse and fine thread series.

4. What are some of the applications of the common types of micrometers?

A. The 1" micrometer in Fig. 277 is used to measure the outside diameter of a small plug. In Fig. 278 a 2" micrometer is used to measure the diameter of stock in a lathe.

The methods of using a depth micrometer (Fig. 279) and an inside micrometer (Fig. 280) are shown below.

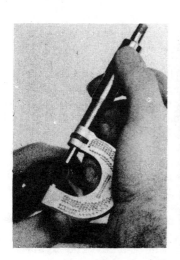
Fig. 277

Fig. 278

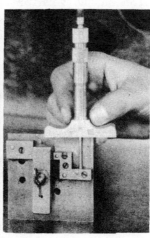
Fig. 279

Fig. 280

VERNIER GAGES

1. Describe the height gage.

A. The height gage (shown in Fig. 281) is a steel upright bar, usually 10" to 18" in height, and is graduated to read thousandths of an inch by means of a vernier scale on the movable jaw. The fixed jaw forms a base. The sharp point of the scriber is used to scribe lines on steel that has had the oxidized surface or scale removed.

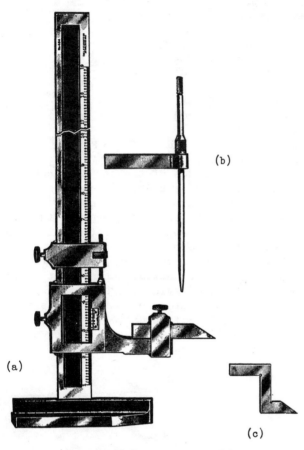

Fig. 281. (a) Height Gage, (b) H Attachment, (c) Offset Scriber.

Fig. 282 illustrates the vernier used, with a scale graduated into 40ths or .025ths of an inch (scale shown on height gage in Fig. 281). The vernier has 25 divisions which are numbered every fifth division, equaling, in total length, 24 divisions on the scale, or for length in inches,

$$24 \times \frac{1}{40} = 24 \times .025 = .600"$$

Thus one division on the vernier equals 1/25 of .600" = .024". Therefore the difference between a division on the vernier and a division on the scale on the bar is .025" - .024" = .001".

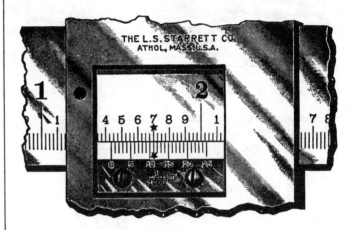

Fig. 282

To read the gage, note how many inches, how many tenths of an inch (or .100"), and how many fortieths of an inch (or .025") the zero mark on the vernier is from the zero mark on the bar. Then note the number of divisions on the vernier from its zero line to the line that exactly coincides with a line on the bar. For example, in Fig. 282 the vernier has been moved to the right one and four-tenths and one-fortieth inches (1.425"), as shown on the bar, and the eleventh line on the vernier coincides with a line on the bar (as indicated by the stars). Eleven thousandths of an inch is to be added to the reading on the bar, so that the total reading is one and four hundred thirty-six thousandths inches (1.436").

Care must be taken in reading the vernier to insure correct readings. Face the light with the vernier held in a horizontal position and tipped slightly so you can look directly down the lines on the vernier plate. In this way it can be more readily determined when a line on the vernier plate coincides with a line on the bar. A magnifying glass should be used so that the reading will be more accurate and cause less strain on the eyes.

NOTE. Be sure that the reading is taken on the same side of the bar each time.

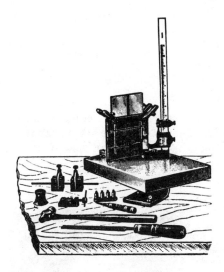

Fig. 283. Using Scriber

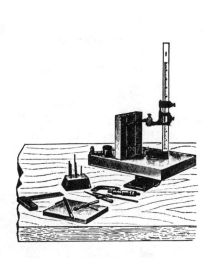

Fig. 284. Measuring Height

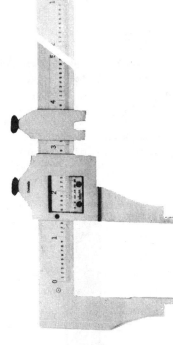

Fig. 286

Remember that one side is graduated for inside and the other for outside measurements.

Figs. 283, 284, and 285 show applications of the height gage.

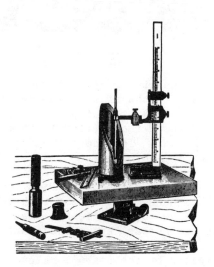

Fig. 285. Measuring Depth

2. Briefly describe the vernier caliper and explain its use.

A. The vernier caliper (Fig. 286) is a tool for checking inside and outside measurements. Both sides of the bar are usually graduated. The jaws are hardened, ground, and lapped parallel with each other. With the jaws in contact, the vernier plate is set on "0" on one side and at a point equal to the thickness of the measuring points on the other side. This makes it possible to check either inside or outside measurements without making any calculations. Points are placed on the bar and slide so that dividers may be set to transfer distances.

Figs. 287, 288, and 289 show the vernier caliper in use.

Fig. 287
Checking External Dimension

Fig. 288
Checking Diameter

Fig. 289
Checking Inside Diameter

3. Describe the gear tooth vernier caliper.

A. The gear tooth caliper shown in Fig. 290 is designed to measure the thickness of a gear tooth on the pitch circle (chordal thickness) and the distance from the top of the tooth to the chord (corrected addendum), at the same time. The vernier on the gear tooth caliper is read in the same way as the vernier on the height gage, except that the distance between the graduations on the bar of the gear tooth caliper is .020" instead of .025", as on the bar of the height gage. The vernier of the gear tooth caliper has only 20 divisions, while the height gage vernier has 25 divisions.

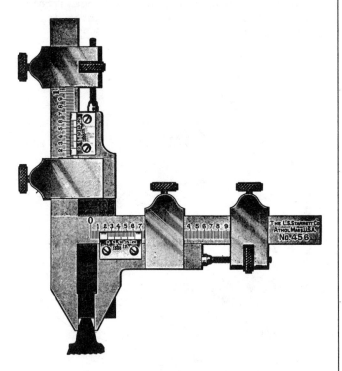

Fig. 290

4. Describe the vernier bevel protractor.

A. The vernier bevel protractor (shown in Fig. 291) is a steel tool having a dial graduated in degrees and a sliding blade which is usually about one-sixteenth of an inch thick. One side of the tool is flat, permitting it to be laid flat upon the work.

The disk of the bevel protractor is graduated in degrees throughout the entire circle. The vernier is graduated so that 12 divisions on the vernier occupy the same space as 23 degrees on the disk. The difference between the width of one of the 12 spaces on the vernier and two of the 23 spaces on the disk is therefore one-twelfth of a degree. The only difference between a vernier on a bevel protractor and a vernier

on a height gage is that the vernier on the bevel protractor is graduated in degrees instead of inches.

Fig. 291

First of all, observe that the vernier covers a space of 23 degrees or 1380 minutes. As it is divided into 12 equal parts (see Fig. 292), each part is equal to one-twelfth of 1380 minutes or 115 minutes, as indicated at B. Since there are 60 minutes in one degree, the space of two degrees indicated by C is equal to 120 minutes. Solving for A, C - B = A or 120 - 115 = 5 minutes, which is the curved distance the vernier moves from one line on the vernier to the next line on the protractor disk.

Fig. 292

To read the protractor, note on the disk the number of whole degrees between 0 on the disk and 0 on the vernier. Then count in the same direction the number of spaces from 0 on the vernier to a line coinciding with a

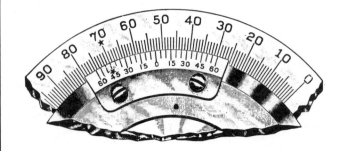

Fig. 293

line on the disk. Multiply this by five and
the product will be the number of minutes to
be added to the number of whole degrees. For
example, in Fig. 293 the number of degrees
between 0 on the disk and 0 on the vernier is
52.   The line 45 on the vernier coincides
with line 70 on the disk, as the stars indicate.   The number of spaces on the vernier
from 0 is nine. Multiplying by five gives 45,
the number of minutes to be added to the number of degrees. The reading of the protractor is therefore 52 degrees and 45 minutes
(52°45').

As the divisions, both on the disk
and the vernier, are numbered both to the
right and the left from zero, any size angle
can be measured, and the readings on the disk
and the vernier are taken either to the right
or the left, according to the direction in
which the zero on the vernier is moved.

Fig. 294 shows some of the applications of the vernier bevel protractor.

The draftsmen's protractor in Fig.
295 is graduated in degrees to read either to
the right or the left, with vernier reading
in spaces of five minutes. The three straight
edges of the protractor are graduated in
inches and sixteenths.

Fig. 294

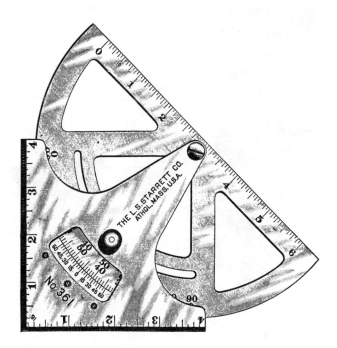

Fig. 295

Chapter 7

CHISELS AND CHIPPING

1. What is a chisel?

A. A chisel is a metal tool or instrument with a cutting edge at the end of the blade, used in dressing, shaping, or cutting. It is driven by a hammer or mallet.

2. What is chipping?

A. Chipping is the process of removing stock with a hammer and chisel.

3. Name two methods of chipping.

A. Hand and pneumatic.

4. Describe the method of chipping by hand.

A. A hammer, weighing from one to one and three-quarter pounds, and a variety of chisels are used for ordinary chipping. The chisels commonly used are flat, but the cape chisel and various forms of side and grooving chisels are also used. The hammer should be held at the extreme end when chipping, grasped by the thumb and second and third fingers, with the first and fourth fingers closed loosely around the handle. The hammer handle may thus be swung more steadily and more freely without tiring the hand as much as if the handle were grasped rigidly by all four fingers.

The chisel should be grasped with the head close to the thumb and first finger, and held firmly with the second and third fingers. The first finger and thumb should be slack, as the muscles are then relaxed and the fingers and hand are less likely to be injured if struck with the hammer. The edge of the chisel should be held on the point where the cut is desired, at an angle that will cause the cutting edge to follow the desired finished surface. After each blow the chisel must be set to the proper position for the next cut. The depth of the cut depends on the angle at which the chisel is held.

5. From what kind of material are chisels made?

A. Chisels are made from a good grade of octagon-shaped steel, generally known as chisel steel. They are forged, usually annealed, and then hardened and tempered.

6. How are chisels classified for size?

A. By the size of the cross section.

7. Why are chisels annealed?

A. Annealing makes a chisel tougher and stronger.

8. What part of a chisel is hardened?

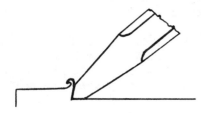

Fig. 296. Correct angle at cutting edge.

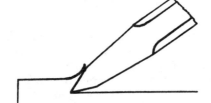

Fig. 297. Angle at cutting edge is too small for general use.

Fig. 298. Point rounded from being used after it became dull.

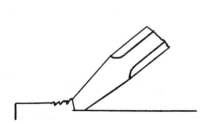

Fig. 299. Dull and incorrectly sharpened. Cutting angle should be smaller.

Fig. 300. Cutting edge slightly rounded to give better cutting action.

A. The cutting edge is hardened, from the end back to approximately one inch from it.

9. What angle should be ground on the cutting edge of a chisel?

A. The correct cutting angle of a chisel depends on the strength of the stock to be chiseled. The softer the metal, the sharper should be the cutting angle of the chisel. An angle of about 70° is suitable for most work. Correct cutting edges are illustrated in Figs. 296 and 300. Avoid tool edges as shown in Figs. 297, 298 and 299.

10. Draw sketches of the flat, cape, diamond point, and round point chisels, showing the "facet" on the flat chisel.

A. The appearance of each of these chisels is illustrated in Fig. 301.

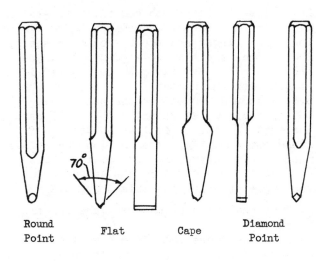

Round Point   Flat   Cape   Diamond Point

Fig. 301

11. How should a cold chisel be sharpened?

A. The chisel should be held at the correct angle and moved across the face of the wheel. The pressure of the chisel against the wheel must not be strong enough to draw the temper from the chisel. The facet should be straight back of the cutting edge, but across the width of the chisel it should be curved a little to give a slightly convex cutting edge.

12. When should the round nose chisel be used?

A. The round nose chisel is used for chipping filleted corners and concave surfaces.

13. When should the flat cold chisel be used?

A. The flat cold chisel is the best chisel for chipping flat surfaces or cutting through thin metal.

14. When should the cape chisels be used?

A. The cape chisel is a narrow chisel for chipping grooves, slots, etc. The round nose cape chisel is used on drill press work, in drawing the drill back concentric with the layout.

15. When should the diamond point chisel be used?

A. The diamond point chisel should be used in cutting V-shape grooves.

16. Give some hints on chipping.

A. Place a chipping guard (a piece of canvas about two feet square attached to two wooden pedestals) in front of the work you are chipping so that flying chips will not injure the fellow in front of you. Wear goggles to protect your eyes.

Use a packing block when chipping in a vise. Never hammer the handle of a vise. The vise jaws should have guards made of some soft material, such as brass or copper, to protect the finish on the work.

17. When chipping, should the cutting edge or the head of the chisel be watched?

A. Watch the cutting edge of the tool. With practice, the ability to hit the head of the chisel without watching it is soon acquired.

18. What is meant by a "Mushroom Head" on a chisel?

A. A Mushroom Head on a tool is a head that has been hammered until the end spreads out so as to resemble a mushroom.

19. Why are "Mushroom Head" chisels dangerous?

A. The mushroomed part of the head may break off and injure some one. The chisel cut cannot be controlled so easily when the head is mushroomed.

20. Write a 100 word story about chisels and chipping, paying particular attention to the points shown in Fig. 302.

Fig. 302

Chapter 8

HACK SAWS AND SAWING

1. Describe the hack saw blade.

A. The hand hack saw blade (Fig. 303), one of the most useful tools in the shop, is a thin blade with teeth formed on one edge, and is 6 to 12 inches long, approximately one-half inch wide, and usually about .027" thick. Held in a frame (Fig. 304), it is used for cutting metal.

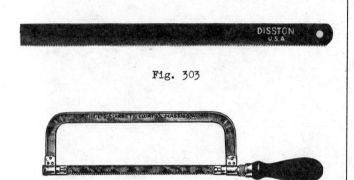

Fig. 303

Fig. 304

2. What kind of steel is used in making hack saw blades?

A. Different kinds of steel are used in making blades for different classes of work. Tool steel, high speed steel, or tungsten alloy steel are generally used.

3. What is meant by "flexible back" and "all hard" blades?

A. A "flexible back" blade is one in which only the teeth are hardened, leaving the back soft. When the entire blade is hardened it is called an "all hard" blade.

4. How should a hack saw blade be placed in a frame?

A. Have the teeth of the blade pointing forward so that the forward stroke will be the cutting stroke (Fig. 304). The blade should be drawn tight enough so that it will not bend. In using a flexible blade the tension should be increased while cutting, as the blade will stretch because of the heat produced by friction.

5. Does a saw blade cut on the return stroke?

A. No; because the teeth are formed so as to cut only on the forward stroke.

6. How should a blade be held in a frame to cut a strip 18 inches long and 2 inches wide from a thin sheet of metal?

A. The blade should be set at right angles to the frame by giving the clips a quarter-turn. A strip of any width up to the capacity of the frame may then be cut (see Fig. 305).

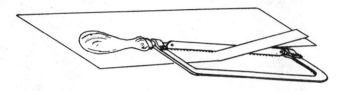

Fig. 305

7. What is the correct number of strokes per minute for a hand hack saw blade?

A. Under ordinary conditions, 35 to 40 strokes per minute is satisfactory. About 50 strokes per minute should be the maximum. Hard material should not be sawed too fast, as this will unnecessarily dull the blade. In cutting drill rod, for example, it is very effective to saw slowly and use greater pressure than is ordinarily used.

8. What is meant by the "set" of a saw?

A. The standard set indicated by general practice is regular alternate; that is, one tooth is turned slightly to the right and the next to the left (see Fig. 306). The teeth should be turned just enough to insure a free, smooth, and rapid cut in a slot a little wider than the blade itself, removing no more stock than is necessary. In certain fine-toothed saws every pair of teeth is set alternately right and left, a style of setting known as double alternate (see Fig. 307).

Fig. 306

(a)

(b)

Fig. 307

9. When should an all hard saw blade be used? A flexible or soft back blade? Why?

A. Use an all hard blade on brass, tool steel, cast iron, rails, etc., and a flexible blade on channel iron, tubing, tin, copper, aluminum, babbitt, etc. An all hard blade is used on the former group of metals because it does not have the tendency to buckle or run out of line, when pressure is applied to it, that the flexible blade has. The flexible blade is used on the latter group because it does not break as easily as the all hard blade on a material with a thin cross section.

10. Why should a blade with the correct number of teeth per inch be used on a particular job?

A. The greatest economy is obtained by using blades of the correct pitch (distance from a point on one tooth to a corresponding point on the next tooth). Fig. 308 shows the importance of correct pitch.

11. On what kind of a job should a blade with 14 teeth per inch be used? Why?

A. Use a blade with 14 teeth per inch on machine steel, cold rolled steel, and structural steel. The coarse pitch makes the saw free and fast cutting (see Fig. 308).

12. When should a blade with 18 teeth per inch be used? Why?

A. Use this blade on solid stock, aluminum, babbitt, tool steel, high speed steel, cast iron, etc. This pitch is recommended for general use, as a blade with 14 teeth per inch is too coarse and leaves the cut too ragged (see Fig. 308).

13. On what kind of a job should a blade with 24 teeth per inch be used? Why?

A. Use this blade on tubing, tin, brass, copper, channel iron, and sheet metal over 18 gage. If a coarser pitch is used, the thin stock will tend to strip the teeth out of the saw blade. Two or more teeth should be in contact with the work (see Fig. 308).

14. When should a blade with 32 teeth per inch be used? Why?

A. Use this blade on small tubing, conduit, and sheet metal less than 18 gage, because these thin materials require a very fine pitch in the saw blade (see Fig. 308).

15. How much pressure is required for sawing with a hand hack saw?

A. A pressure of 20 to 30 pounds per inch of contact area of the teeth has been found to give the most satisfactory results. Enough pressure must be applied so that the saw will not slip over the work, but not enough to cause the blade to buckle or break. More pressure must be applied as the saw becomes dull.

16. What is the effect of using a coolant on a power hack saw blade?

A. By using a coolant the strokes per minute may be increased and the saw will last much longer. The coolant absorbs the heat due to friction and prevents the blade from overheating and losing its temper.

17. How should thin steel stock be supported while being cut with a hack saw?

A. Clamp thin stock between two pieces of wood or soft steel and saw through all three pieces.

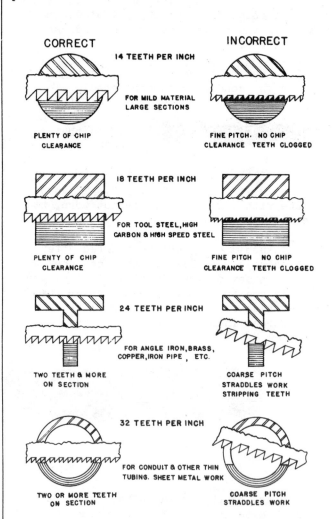

Fig. 308. Chart for selecting correct number of teeth per inch for hack saw blades for various jobs. In hand frames, a blade with 18 teeth per inch is recommended for general work.

18. Name three common causes for the breaking
    of hack saw blades.
    A. (1) Using a coarse blade on thin work;
       (2) Drawing the blade too tightly and
           then canting (tilting) it over;
       (3) Using too much pressure.

19. Describe slotting hack saw blades.
    A. Slotting hack saw blades (illustrated
                            in Fig. 309) are usually
                            eight inches long by
                            one-half inch wide, and
                            of four different thick-
                            nesses (about .049",
       .065", .083", and .109").
                            They are very handy when
                            slotting a few screws
                            for a special job which
    Fig. 309   is needed at once. When
                            slotting saw blades are
not available, two ordinary saw blades may be
placed in the frame side by side and used as
a substitute.

20. If a saw blade does not cut straight, how
    can the condition be corrected?
                            A. First ex-
                         amine the blade
                         to see that the
                         teeth are not
       worn on one side,
                         as shown in Fig.
    Fig. 310   310. If worn,
replace with a new blade. If not worn, start
a new cut.

21. Why should a new cut be started after re-
    placing a worn blade with a new one?
    A. The set of the old blade will be worn
slightly and the cut will be narrower than
the new blade. The new blade will break if
it is forced into the old cut.

24. Explain the importance of holding work in
    the vise correctly.
    A. It is important that the work be placed
in the vise correctly if the greatest efficiency is to be secured from the saw blade.
Expose as much of the surface of the work as
possible, so that a corner may be taken gradually and the maximum number of teeth engaged
throughout the cutting. This prevents strip-
ping, the cut is made quicker and straighter,
and the blade lasts longer. Be sure that the
work is held rigid, and always start the cut
with the least possible angle facing the
thrust of the saw teeth. It is especially
important to consider these conditions in cutting angle iron or material with an odd shape.

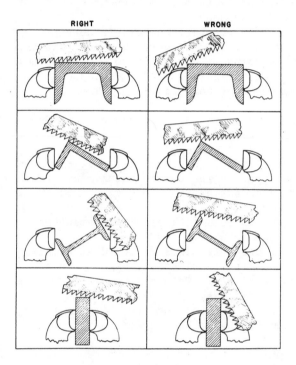

Fig. 312

Fig. 312 shows the right and wrong positions
of shapes commonly cut with a hack saw.

Fig. 311

22. How are teeth formed on hack saw blades?
    A. The teeth are formed by using gang
milling cutters (illustrated in Fig. 311).

23. How should the length of a hack saw
    blade be measured?
    A. The length is the distance between the
centers of the holes in the blade.

25. Name two methods, other than the hand hack
    saw method, that are used in sawing
    metal.
    A. The power hack saw and the band saw
are also used for sawing metal.

26. Give some of the rules to be followed in sawing with a hand hack saw.
    A. (1) Use a blade with the correct pitch.
       (2) Saw as close as possible to the point where the work is clamped, to prevent chattering.
       (3) Do not cut too fast.
       (4) Relieve pressure on the back stroke.
       (5) Do not use excessive pressure.
       (6) Saw carefully when the blade is almost through the cut.

Chapter 9

FILES AND FILING

1. What is a file?

A. A file is a hardened steel instrument having cutting edges or teeth upon its surface, made by rows of straight chisel cuts running diagonally across it (see Fig. 313). It differs from the chisel in having a large number of cutting points instead of one cutting edge and in being driven directly by hand instead of by a hammer. The file is used for abrading or smoothing metals.

4. What is meant by the "safe edge"?

A. The safe edge of the file is the edge on which no teeth are cut. The mill file and the flat file have single teeth cut on the edges, but the hand file usually has teeth on only one edge, the other being termed the safe edge. The pillar file has two safe edges.

5. What is meant by a "blunt" file?

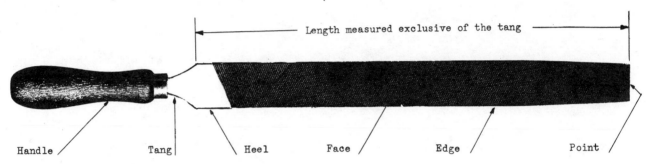

Length measured exclusive of the tang

Handle   Tang   Heel   Face   Edge   Point

Fig. 313

2. What is a rasp?

A. The rasp, shown in Fig. 314, is similar to a file but has coarse teeth raised by a pointed triangular punch.

A. A blunt file is a file that has the same sectional shape throughout its length, from point to tang.

Fig. 314

3. What is the tang of a file?

A. The tang is the pointed portion of the file which is inserted in the handle. CAUTION: Be sure that the handle is firmly attached to the file.

6. Why are files made with convex surfaces?

A. Files are generally made with convex surfaces, that is, they are thicker in the middle than at the ends (see Fig. 315). This is done to prevent all the teeth from cutting

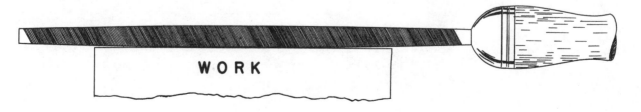

WORK

Fig. 315

at the same time, as that would require too much pressure on a file and make it hard to control. A flat surface could not be obtained if the face of the file were straight, as there is a tendency to "rock" the file. The convex surface helps to overcome the results of rocking.

This convexity of files also serves another purpose. The pressure applied to a file to make it bite into the work also bends the file more or less, and if the file in its natural state were perfectly flat, it would be concave during the cutting operation. This would prevent the production of a flat surface, as the file would cut away at the edges of the work and leave a convex surface.

7. What is meant by the "taper" of files?
A. Files taper slightly, being slightly smaller in width toward the point (see Fig. 316). This does not apply to blunt files.

Fig. 316

8. How are files classified?
A. They may be divided into two classes, single cut (Fig. 317) and double cut (see Fig. 318).

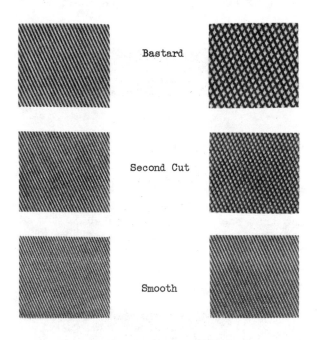

Bastard

Second Cut

Smooth

Fig. 317.
Types of Single Cut
Files

Fig. 318.
Types of Double Cut
Files

9. Describe the use of the following files: single cut (Fig. 317), double cut for general work (Fig. 318), and double cut for finishing work (Fig. 319).
A. The single cut mill files of Fig. 317 derive their name from the fact that they were first used for filing mill saws. They are also used for lathe filing, draw filing, and for finishing various compositions of brass and bronze. This type of file leaves a fine finish. It differs from other files in having single cut teeth.

The flat double cut files shown in Fig. 318 are used by machinists, machinery builders, ship and engine builders, repair men, and tool makers when a fast cutting file is needed. This type of file leaves a rough finish.

The hand finishing double cut file shown in Fig. 319 is used by machinists and tool makers when a smooth finish is desired.

Fig. 319. Double Cut Finish File

Double cut files are fast cutting but cannot be used for draw filing. The finish produced by a double cut file is not as smooth as that produced by a single cut file.

10. What is the meaning of such terms as rough, coarse, bastard, second cut, smooth, and dead smooth?
A. These terms refer to the distance apart of the parallel cuts. A numbering system is used for the same purpose on small files, ranging from 00 to 8, 00 being the coarsest. Bastard, second cut, and smooth files are illustrated in Figs. 317 and 318.

11. Is the pitch of a 16" second cut file the same as the pitch of a 6" second cut file?
A. No, the shorter the file the finer the pitch (as the illustrations in Fig. 320 show clearly).

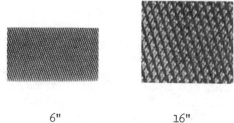

6"   16"
Fig. 320

12. How is the length of a file measured?
    A. From the point to the heel (see Fig. 313).

13. What are some of the different shapes of files?
    A. Cross sections of some of the most commonly used files are shown in Fig. 321.

Fig. 324

17. Describe the "three-square" file.
    A. The three-square file shown in Fig.

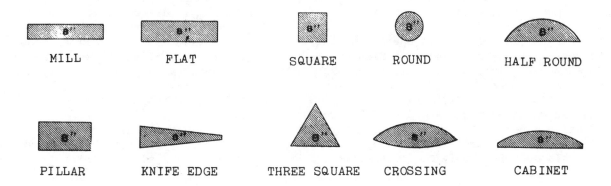

MILL   FLAT   SQUARE   ROUND   HALF ROUND

PILLAR   KNIFE EDGE   THREE SQUARE CROSSING   CABINET

Fig. 321

14. Describe the "pillar" file.

Fig. 322

A. The pillar file shown in Fig. 322 is similar to the hand file shown in Fig. 318 except that it is narrower and one or both edges are safe edges (have no teeth cut on them). The pillar file is used for filing slots and keyways and for filing against shoulders.

15. Give a description of the "square" file.
    A. The square file is similar to the mill file with the exception that its cross section is square and it has double cut teeth on all four sides (see Fig. 323). It is used for filing small square or rectangular holes, for finishing the bottoms of narrow slots, etc.

Fig. 323

16. Give a description of the "round" file.
    A. The round file shown in Fig. 324 is similar to the square file with the exception that the cross section is round. It is generally tapered and the small sizes are often called "rat-tail" files. It is used for enlarging round holes and for finishing fillets.

325 (commonly called the three-cornered file) is triangular in section, with angles of 60°. It tapers to the point, the corners are left sharp, and it is double cut on three sides and single cut on the edges. It is generally used for filing internal angles more acute than those of the rectangle, for clearing out square corners, and for filing taps, cutters, etc.

Fig. 325

18. Give a description of the "half-round" file.
    A. Fig. 326 shows the cross section of the "half-round" file (observe that it is not a half circle). It derives its name from the fact that one side is rounded while the other is flat. It can be used for various jobs on which no other type of file would be satisfactory.

Fig. 326

19. Describe the "crossing" file.

A. The crossing file shown in Fig. 327 has a double circular section, one side having the same radius as the half-round file and the other side having a flatter curve, or larger radius. It tapers to the point in both width and thickness and is double cut on both sides. This file is often used in place of the half-round file.

22. When should the coarser files be used?

A. A coarse file should be used only when a comparatively large amount of material is to be removed.

23. When should the finer files be used?

A. A fine file should be used only for finishing work.

Fig. 327   Fig. 328   Fig. 329

20. Describe the "knife" file.

A. The knife file shown in Fig. 328 is made from steel that is knife shaped, the included angle of the sharp edge being approximately 10°. This file tapers to the point in width and thickness, and is double cut on both flat sides and single cut on both edges. It is used for finishing the sharp corners of many kinds of slots and grooves.

21. What kind of a file should be used for filing lead or babbitt?

A. Use either the "lead float" file shown in Fig. 329 or the curved tooth "Vixen" file shown in Figs. 330 and 331.

24. What kind of a 12"-file should be used to remove stock rapidly?

A. A double cut bastard or double-cut coarse-tooth file should be used to remove stock rapidly.

25. What file should be used for finishing?

A. A single cut fine tooth file should be used for finishing.

26. What is the proper way to hold a file?

A. As illustrated in Fig. 332, grasp the handle in the right hand so that it rests against the palm of the hand, with the thumb placed on top. Place the left hand at the end of the file and let the fingers curl under it.

Fig. 330

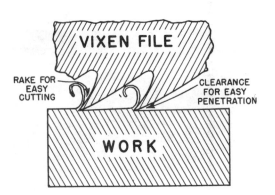

Fig. 331

Fig. 332

27. What should be the position of the body
    when filing?

A. It is important to have the body in
the correct position, as the muscles must
move freely. The left foot should point forward and the right foot be brought up close
enough to the left to give the necessary balance. When filing, the body should lean forward on the forward stroke for part of the
stroke and then return to the original position at the finish. The file must be held
straight or the surface will not be flat. The
strokes should not be too fast as this will
ruin the file and the work. Enough pressure
should be applied to make the file cut evenly.

28. What is meant by "draw filing"?

A. With the file in a position similar to
that shown in Fig. 333, grasp it with both
hands, the thumbs being one-half or three-
quarters of an inch from each side of the
work, and pull or push the file. By using a
single cut file in this manner, the teeth
will have a shearing action and give a smooth
finish to the work.

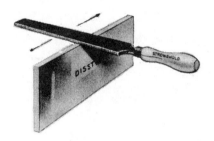

Fig. 333

29. What is meant by "crossing the stroke"?

A. Crossing the stroke means changing the
angle of the stroke to about 45 degrees from
the previous strokes. This will show the
high spots and also tends to keep the work
flat.

30. Should the file be lifted from the work
    on the return stroke? Explain thorough-
    ly.

A. Be sure to raise the file slightly
during the return stroke in order to clear
the work and avoid dulling the file by wearing away the back of the teeth, thus destroying the cutting edges. This procedure does
not hold true in the filing of soft metals,
such as lead or aluminum, as the file should
be drawn back along those metals on the return stroke, as an aid in cleaning the teeth.

31. What precaution should be taken before
    filing a cast iron casting?

A. Before attempting to file a cast iron
casting, the scale must be removed from its
surface. This can be done by chipping, using
the edge of the file, tumbling, sand blasting,
or pickling (a good pickling solution is 4 to
10 parts of water to 1 part of sulfuric acid).

32. How should a piece of filed work be tested
    to see if it is square?

A. Two surfaces may be tested for squareness by using a solid square.

33. What three factors affect the cutting ef-
    ficiency of a file?

A. The shape, sharpness, and hardness of
the teeth are the important factors.

34. What is meant by "pinning"?

A. When filing soft metals, narrow surfaces, or corners, small particles of the material being filed tend to become clogged in
the gullets between the teeth of the file,
causing scratches on the work. This is called
pinning.

35. What is the cause of pinning?

A. If the pressure applied to a file is
too great, the chisel cuts on the file will
clog, especially on fine cut files. In using
a new file see that the rough edges and burrs
are worn down slightly before taking a heavy
cut. Rubbing chalk on a file will help to
prevent pinning.

36. How would you clean a file that has become
    pinned?

A. Fig. 334 shows a file brush and the
method of using it to clean a file. The file

(a)

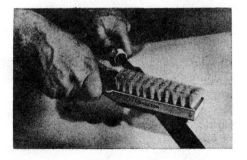

(b)

Fig. 334.   (a) File Brush, (b) Carding a File

brush is made of a rectangular piece of wood with bristles on one side and fine wires on the other. The card wire is fine enough to enter a fine tooth file and loosen chips and dirt while the brush is used to clean these out. The "scorer" in the handle of the file brush is used to remove burrs that cannot be loosened by the card. If a scorer cannot be obtained, a piece of soft metal can be used to remove material fastened in the gullets.

37. What are "increment cut" files?

A. Files were originally made by hand and the teeth were more or less irregular, according to the skill of the mechanic who cut them. Some of the teeth were higher than others, giving fewer contact points. This proved desirable as it increased the cutting life of the file, but the cost of the handmade files was necessarily great. Machines are now used to increase or decrease the cuts of the teeth, giving the same type of file as the earlier hand cut files. These are called increment cut files.

38. When does a file cut best?

A. After it has cut about 2500 strokes, or after it has removed about one cubic inch of material because at that time most of the cutting edges will be in contact with the work. It must be remembered however after continued use the worn down edges will continue to cut less and less until the life of the file is destroyed.

39. What two files are generally used in filing on the lathe?

A. The double cut flat file is used for rough filing and the single cut file is used for finish filing.

40. Why is the mill file best for filing on the lathe?

A. A mill file gives a better finish, as it is a single cut file with chisel cuts in only one direction, giving a shearing action.

41. What procedure should be used in filing on the lathe?

A. Use long, slow cuts. Be sure to bear down hard enough on the file to make it cut, but not hard enough to make it pin.

42. What is the result when short, quick strokes are taken to remove a small amount of material from a job by filing it on the lathe?

A. Either a series of small flats will be found on the periphery or the work will be out of round.

43. How much material should be left for finishing when filing on the lathe?

A. In general practice, not more than .003" is usually left for finish filing.

44. What is the effect of too much filing on the lathe?

A. The work will have a tendency to be out of round if too much stock is removed with a file.

45. What two files are most commonly used in the machine shop?

A. The flat file and the half-round file are most commonly used.

46. In placing an order for files, how should a required file be designated?

A. A file should be designated by the size, by the shape or cross section of the steel on which the teeth are cut, and by the spacing of the teeth (the size always refers to the length, which is measured from the point to the heel, and does not include the tang).

EXAMPLE: 12 - 6" Half-Round Mill Files

47. Is it ever permissible to use a file without a handle?

A. Never use a file without a handle. This is a safety rule. To do so is especially bad practice on a lathe.

48. Give two precautions that should never be overlooked when filing.

A. Never rub your hand over the work you are filing. Always make sure that filed surfaces are protected by soft material between the work and the hardened jaws of the vise.

49. Describe the use of the filing machine.

A. The filing machine shown in Fig. 335 saves a great deal of time and labor. Care must be used in setting the table to the correct clearance angle for dies, etc. Use the amount of force in setting the protractor against the file, to get the angle between the table and the file, that you would use in forcing the work against the file. The angle will vary with the amount of force used.

Fig. 336 shows the type of files used in the filing machine. In placing a file in the filing machine, the roller guide must be adjusted to give the proper amount of friction against the file. Failure to install the file correctly will cause it to break in the socket.

Fig. 335

Fig. 336

Chapter 10

SOLDERING

1. What is soldering?

   A. Soldering is the process of joining two metals by a third soft metal that is applied in the molten state.

2. What is the composition of solder?

   A. Solder consists of tin and lead. Bismuth and cadmium are frequently included to lower the melting point.

3. What is one of the most important operations in soldering?

   A. One of the most important operations, and one that is often overlooked, is that of cleaning the surface to be soldered.

4. Tell what flux is and explain why it is used.

   A. Flux is a cleanser and is used to remove and prevent the oxidation of the metals, allowing the solder to flow freely and unite more firmly with the surfaces to be joined.

5. What is a soldering copper? Make a sketch showing three different types.

   A. The soldering copper is a piece of copper attached to a steel rod with a handle. Soldering coppers are made in different lengths, forms, and weights. (See Fig. 337)

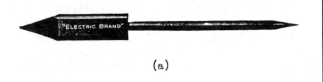

(a)

(b)

solder. The way to tin the point is to rub the clean, heated copper with sal ammoniac and then apply the solder. To have a good clean point, rub over it with a rag immediately after it has been tinned.

7. How should a surface be prepared for soldering?

   A. Use a file, scraper, or acid cleaner to clean the surface.

8. Why can we not solder two pieces of metal together without the aid of a flux after their surfaces have been cleaned?

   A. A cleaned metal surface oxidizes immediately upon exposure to the air. A thin coating of oxide is formed when the oxygen and metal combine. Solder will not unite with a metal that has a coating of oxide, grease, dirt, etc. A flux is used to remove the oxide the instant the solder comes in contact with the metal.

9. What is the most commonly used flux in the machine shop?

   A. Prepared soldering paste is the most commonly used flux in the shop, while tinsmiths favor "killed" muriatic acid (zinc chloride).

10. Describe "killed" muriatic acid.

    A. Muriatic (hydrochloric) acid is killed by adding small scraps of zinc, a few at a time, until the acid fails to eat the zinc. It is then called a saturated solution. The killed or cut acid is known as chloride of zinc.

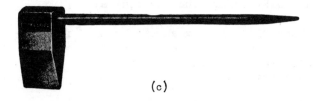

(c)

Fig. 337. (a) Square Point Copper, (b) Bottoming Copper, (c) Hatchet Copper.

6. Describe how to "tin" a soldering copper.

   A. Use a file to clean the copper back to the end of the bevel tip and heat the copper a little higher than the melting point of the

CAUTION. Do not kill the acid near any machines or tools, as the escaping fumes may cause the machines or tools to rust.

11. On what kind of metals is chloride of
    zinc used as a flux?
    A. Chloride of zinc is used as a flux on
steel, cast iron, brass, zinc, nickel, monel
metal, stainless steel, lead, tin, and galvanized iron.

12. For what metal is chloride of zinc flux
    diluted?
    A. When chloride of zinc is used as a
flux for soldering tin, it is diluted with
about 50% alcohol.

13. Is muriatic acid ever used in the raw
    state as a flux?
    A. Muriatic acid is frequently used in
the raw state as a flux for soldering galvanized iron. It is used in the raw state as
a cleaner on cast iron, galvanized iron, and
sheet steel.
        CAUTION. Where muriatic acid or zinc
chloride is used as a flux, the part should
be cleaned after soldering to prevent the
acid from eating the metal.

14. What kind of flux is used for soldering
    copper and brass?
    A. Zinc chloride or a prepared soldering
flux.

15. What kind of flux is used for soldering
    lead?
    A. Rosin, tallow, or zinc chloride may be
used as a flux for soldering lead.

16. What kind of flux is used for soldering
    sheet tin?
    A. Beeswax, rosin, or any of the prepared
fats, pastes, or liquids are good. Zinc
chloride can be used by diluting it with 50%
alcohol.

17. Tell how to solder cast iron.
    A. Cast iron may be soldered by first
scraping it, using raw muriatic acid to clean
the part, and then tinning the surface to be
soldered by using zinc chloride as a flux.

18. What kind of flux should be used for sol-
    dering wrought iron or steel?
    A. Zinc chloride is the best flux to use
for soldering wrought iron or steel.

19. Tell how to solder aluminum.
    A. Flux is molded in the special solder
used for aluminum. This solder requires a
greater heat to melt it than the lead-tin
solder. Another way to solder aluminum is to
use stearin as a flux with a solder made of
70% tin, 25% zinc, 3% aluminum, and 2% tin.

Paraffin or vaseline may also be used as a
flux.

20. What is meant by "sweating" parts together,
    and why do we use this operation?
    A. If the parts to be joined are perfect-
ly tinned, flux applied between them, and
then held together and heated, the parts will
become perfectly soldered. This is "sweating." Split bushings are sweated together so
they can be machined.

21. How can tin foil be used in soldering?
    A. Tin foil is used in the operation of
sweating parts together. The parts are
cleaned, flux is applied, a sheet of tin foil
is placed between the surfaces, and they are
clamped together. Enough heat is then applied
to fuse the tin foil.

22. Should a soft or hard solder be used to
    make a strong joint?
    A. Use a hard solder, for the higher the
melting point the stronger the solder.

23. What materials are used in making hard
    solder?
    A. Hard solder, usually called spelter,
is made of 4 parts of copper to 1 part of
zinc. Borax is used as a flux.

24. What materials are used in making soft
    solder?
    A. Soft solder is made of $1\frac{1}{2}$ parts of tin
and 1 part of lead. This melts at a temperature of 334°F.

25. What flux is used by canneries and packing
    houses because of its non-poisonous quali-
    ties?
    A. Rosin is generally used as a flux in
the sealing of foods because it is non-
poisonous. Palm oil or cocoa oil is also
used.

26. What kind of flux is best to use for com-
    mutator wires and electrical connections?
    A. Use an alcoholic solution of rosin for
commutator wires and electrical connections.
Never use an acid flux, as this will after-
wards cause a corrosive action.

27. How would you solder a horizontal piece
    of work if the solder had a tendency to
    run?
    A. In soldering a horizontal job, such as
a piece of pipe, tin it in the regular way and
make a small cup of clay to hold the solder.
Clay is also used for holding small parts to
be soldered.

28. What is the difference between soldering
    and brazing?
    A. Brazing requires a greater heat and
copper is used in the hard solder or spelter.
A torch is generally used to heat the spelter
and the part, instead of a copper as in soldering. A brazed joint is much stronger than
a soldered joint.

29. How does the addition of bismuth or cad-
    mium affect solder?
    A. The addition of a small amount of
either bismuth or cadmium will lower the
melting point of solder.

30. What peculiar quality does bismuth have?
    A. Bismuth expands when it cools.

31. What is the effect of mixing zinc with
    solder?
    A. The addition of zinc causes solder to
flow sluggishly. Aluminum has the same effect.

32. Why does solder contain lead?
    A. Lead is used because it cheapens the
mixture and because it has a low melting
point of 620.6°F.

33. What is the effect of phosphorus on a
    mixture of solder?
    A. The addition of a small percentage of
phosphorus causes solder to flow more freely.
It should be added in the form of phosphor-
tin.

34. What is the correct design of the top of
    the square point soldering copper and how
    is it used properly?
    A. Whatever the type of copper selected,
it must have adequate capacity for the work
it is to do. A perfectly soldered connection
can be obtained only when the surfaces to be
joined have absorbed enough heat to melt the
solder. For example, it is almost impossible
to solder a large vessel with a small copper,
because the large vessel will absorb all the
heat from the small copper and the part to be
soldered will not be heated sufficiently to
cause a good fusion. As a large copper will
carry more heat to the part being soldered, a
large copper should be used on large jobs. A
small copper should of course be used for
small or intricate jobs.

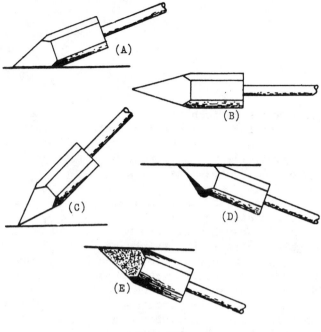

Fig. 338

The size of the copper is important,
but it must also have the proper shape and be
used efficiently. At A in Fig. 338, the heat
can be transmitted as rapidly as possible.
The copper at B is too pointed and the heat
cannot be delivered properly. The thick and
heavy-pointed copper at C is shaped properly
but it is being applied at the wrong angle to
pass the heat to the work. As solder will
not flow upward, an attempt to solder the
underside of a job by the method shown at D
simply causes the solder to flow away from
the joint. However, the following method can
be used to solder this job. Clean only one
side of the soldering copper, heat it, tin it,
and then apply it to the work. Solder can be
applied in this way because it will cling
only to the clean side of the copper. Be
sure that the dirty side shown at E and its
opposite side are left dirty so that the solder will not run off.

Therefore the proper design and application of the soldering copper is shown at A,
the point of the copper at B being too thin
and the two coppers at C and D being held in
wrong positions.

Chapter 11

SHOP REVIEW

1. What is an inch?

A. An inch is a standard unit of length or measurement. 12 inches = 1 foot, 3 feet = 1 yard, $16\frac{1}{2}$ feet = 1 rod, 5280 feet = 1 mile, 160 square rods = 1 acre.

2. What parts of an inch are generally used for measurements in the shop?

A. The usual units are halves, quarters, eighths, sixteenths, thirty-seconds, sixty-fourths, and thousandths. Ford Motor Company has adopted a decimal system for specifying divisions of an inch on new designs using tenths, hundredths, and thousandths of an inch as units of measurements. The rule shown in Fig. 339 is graduated into tenths and fiftieths of an inch.

used in the shop and drawing room and are written in the decimal form. For example, 1/2 = .500; 3/4 = .750; 13/16 = .8125.

7. Write the value of .6569 in words.

A. The value of .6569 is six hundred fifty-six and nine tenths thousandths.

8. How can a fraction be reduced to its lowest terms?

A. To reduce a fraction to its lowest terms divide both the numerator and the denominator by their greatest common divisor. The greatest common divisor (G.C.D.) of two or more numbers is the largest number that will exactly divide each of them. One method used to find the G.C.D. for a fraction is as

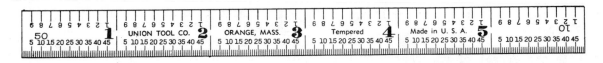

Fig. 339

3. How many eighths are there in an inch? In $\frac{1}{4}"$? In $\frac{1}{2}"$? In $\frac{3}{4}"$?

A. $1" \times \frac{8}{8} = \frac{8}{8}$; $\frac{1}{4}" \times \frac{2}{2} = \frac{2}{8}$;

$\frac{1}{2}" \times \frac{4}{4} = \frac{4}{8}$; $\frac{3}{4}" \times \frac{2}{2} = \frac{6}{8}$.

4. How many eighths are there in $\frac{1}{2}" + \frac{1}{8}"$?

How many sixteenths in $\frac{1}{2}" + \frac{1}{4}"$?

A. $\frac{1}{2}" \times \frac{4}{4} = \frac{4}{8}$; $\frac{4}{8} + \frac{1}{8} = \frac{5}{8}$.

A. $\frac{1}{2}" \times \frac{8}{8} = \frac{8}{16}$; $\frac{1}{4}" \times \frac{4}{4} = \frac{4}{16}$;

$\frac{8}{16} + \frac{4}{16} = \frac{12}{16}$.

5. How can a common fraction be changed to a decimal fraction?

A. A common fraction can be changed to a decimal fraction by dividing the numerator by the denominator. Example: $3/7 = 3 \div 7$ = .42857.

6. How are thousandths written in the shop and on mechanical drawings?

A. Thousandths are the unit of measurement

follows: Divide the larger term by the smaller term and find the remainder; divide the original divisor by the remainder; continue this process of dividing the remainder into the previous divisor until the remainder becomes zero. The last divisor is the G.C.D. If this number is one (1) the fraction cannot be reduced. This method can be applied to both proper and improper fractions.

Example: Reduce $\frac{6856}{20000}$ to its lowest terms.

6856	20000	2
6288	13712	1
568	6288	11
560	6248	14
8	40	5
	40	
	0	

Since 8 is the last divisor, it is the greatest common divisor.

$$\frac{6856 \div 8}{20000 \div 8} = \frac{857}{2500}$$

9. What is meant by the least common denominator of a group of fractions?

A. The least common denominator (L.C.D.) of a group of fractions is the least or smallest number which contains each of the denominators a whole number of times. Many times the least common denominator can be determined by inspection. When this is not possible it

can be found as follows:

Example: Find the L.C.D. of $\frac{1}{4}$, $\frac{2}{3}$, $\frac{5}{9}$, $\frac{3}{16}$.

First place the denominators in a row, separating them by commas, as shown at the right. Select the smallest prime number (a number that cannot be divided except by itself and one) that will exactly divide two or more of the denominators. In this case 2 will exactly divide 4 and 16. Write the quotients below, and bring down any numbers which are not divisible by this divisor.

2	4, 3, 9, 16
2	2, 3, 9, 8
3	1, 3, 9, 4
	1, 1, 3, 4

Proceed as before, again using the smallest prime number that will divide two or more of the numbers just obtained. Continue this process until no two of the remaining numbers can be divided by any number except 1. The product of all the divisors and all the numbers left in the last line of quotients is L.C.D. = 2 X 2 X 3 X 3 X 4 = 144, the L.C.D.

10. Find the sum of the following:
$1\frac{1}{2} + 2\frac{3}{4} + 1\frac{7}{8} + 3\frac{9}{16} + 1\frac{7}{32} + 2\frac{15}{64}$.

A. The least common denominator of all the fractions is 64; therefore, all the fractions must be changed to sixty-fourths before they can be added.

$$1\frac{32}{64} + 2\frac{48}{64} + 1\frac{56}{64} + 3\frac{36}{64} + 1\frac{14}{64} + 2\frac{15}{64} = 10\frac{201}{64}$$

$$= 13\frac{9}{64}.$$

11. Extract the square root of 1.0736 and of 125.38. Check the results.

```
      1. 0 3 6 1  (2)
     √1.07'36'00'00  (8)
      1
  203│  7 3 6
        6 0 9
   2066│ 1 2 7 0 0
        1 2 3 9 6
  20721│   3 0 4 0 0
          2 0 7 2 1
            9 6 7 9  (4)

      2² = 4      4 + 4 = 8
```

To check: Find the excess of nines in the root. This number is squared and added to the excess of nines in the remainder. Then find the excess of nines in this sum. The result should equal the excess of nines in the original number.

12. How can the square root of a fraction be found?

A. The square root of a fraction may be found either by extracting the square root of the numerator and denominator or by changing the fraction to a decimal and extracting the square root of the decimal.

Example: $\sqrt{\frac{5}{8}} = \frac{\sqrt{5}}{\sqrt{8}} = \frac{2.2361}{2.8284} = .7905$

$$\sqrt{\frac{5}{8}} = \sqrt{.625} = .7905.$$

13. What is a circle?

A. A circle is a closed plane bounded by a curved line, all points of which are equidistant from a fixed point called the center.

14. Draw a sketch of a circle showing the different parts.

A. The sketch (Fig. 340) shows a circle

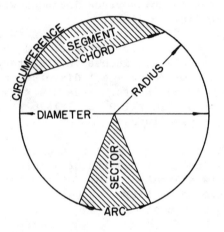

Fig. 340

```
       1 1. 1 9 7 3  (4)
      √1'25.38'00'00'00  (1)
       1
    21│ 25
        21
     221│ 4 3 8
          2 2 1
      2229│ 2 1 7 0 0        4² = 16     16 + 3 = 19
            2 0 0 6 1
       22387│  1 6 3 9 0 0    19 - 18 = 1
              1 5 6 7 0 9
        223943│   7 1 9 1 0 0
                6 7 1 8 2 9
                4 7 2 7 1  (3)
```

with its different parts. π(pi) = 3.1416, Circumference = 3.1416 X Diameter, Radius = Diameter ÷ 2, and Area = 3.1416 X Radius squared.

15. What is meant by the circumference of a circle?

A. The circumference is the distance around a circle. It is equal to 3.1416 (π) times the diameter.

16. What is meant by the diameter of a circle?

A. The diameter is the distance across the circle, measured through the center.

17. How is the diameter of a circle found if the circumference is given?

A. Divide the circumference by π (3.1416) to get the diameter.

18. What is the radius of a circle?

A. The radius is a straight line from the center to any point on the circumference, and is equal to one-half the diameter.

19. What is meant by area?

A. Area is the surface included within any given line or lines.

20. Solve for the circumference and the area of a circle 2.4 inches in diameter.

A. Circumference = π x diameter or 3.1416 x 2.4 = 7.53984 inches. Area = π x radius squared or 3.1416 x 1.2² or 3.1416 x 1.44 = 4.5239 square inches.

21. What is an arc? What is a chord?

A. An arc is a portion of a curved line. A chord is a straight line joining two points of a curve, especially two points on the circumference of a circle.

22. What is meant by a segment of a circle?

A. A segment is a part of a circle bounded by an arc and its chord.

23. What is a sector?

A. A sector is a figure bounded by the arc of a plane curve and the radii drawn from the extremities of the arc.

24. How many degrees are there in a circle? minutes in a degree? seconds in a minute?

A. There are 360 degrees in a circle; 60 minutes in one degree; 60 seconds in one minute; and 3600 seconds in one degree.

25. Give the symbols for degree, minute, and second.

A. The degree is indicated by °, the minute by ', and the second by ". For example, an angle of 23 degrees, 17 minutes, and 13 seconds is written 23°17'13".

26. Reduce 2°52'30" to degrees.

A. Change the minutes and seconds in the given angle to seconds, and divide by 3600". 52' = 52 x 60" = 3120". 3120"+ 30" = 3150". 3150" ÷ 3600 = .875°. Adding the original number of degrees, 2° + .875 = 2.875°. If the angle is given in degrees and minutes, divide the number of minutes by 60' to get the number of degrees.

27. Reduce 11.5625° to degrees, minutes, and seconds.

A. Multiply .5625 by 60' to obtain the number of minutes: .5625 x 60 = 33.75'. Multiply the decimal part of the minute by 60" to obtain the number of seconds: .75 x 60 = 45". Therefore, 11.5625° = 11°33'45".

28. What is meant by perimeter?

A. The perimeter is the distance around a body or figure. It is equal to the sum of the sides.

29. Use "perimeter" and "circumference" in sentences.

A. The perimeter of the rectangle is 24 inches. The circumference of the circle is 6.2832 inches.

30. What is a quadrilateral?

A. A quadrilateral is a figure bounded by four straight lines, as in the sketches in Fig. 341.

Fig. 341

31. What is a parallelogram?

A. A parallelogram is a quadrilateral with both pairs of opposite sides equal, as in the sketches in Fig. 342.

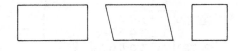

Fig. 342

32. What is a right angle? How many degrees does it contain?

A. Right angles are the angles formed when two perpendicular lines meet or intersect. Right angles contain 90 degrees.

33. What is a rectangle?

A. A rectangle is a parallelogram having four right angles, as in Fig. 343. The area of a rectangle is equal to the length multiplied by the width.

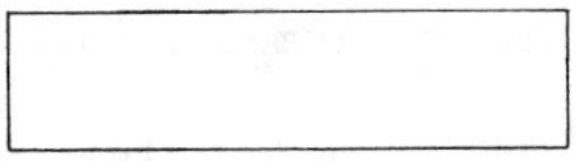

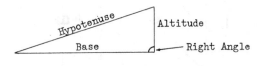

Fig. 343

34. What is a square? How is its area found?

A. A square is a rectangle having four equal sides and four right angles, as in Fig. 344. Its area is equal to the square of one side.

Fig. 344

35. What is a triangle? How is its area found?

A. A triangle is a plane figure bounded by three straight lines and having three angles. Its area is equal to one-half the base multiplied by the altitude.

36. What is a right triangle? What is the hypotenuse of a right triangle?

A. A right triangle is a triangle in which one of the angles is 90 degrees. The hypotenuse is the longest side of a right triangle.

37. Make a sketch of a right triangle, showing the different parts.
A. See Fig. 345.

Fig. 345

38. How can the length of the hypotenuse be found when the lengths of the other two sides are known?

A. The square of the hypotenuse is equal to the sum of the squares of the other two sides. For example, find the length of the hypotenuse of a right triangle if the base is four inches and the altitude is three inches. $3^2 + 4^2 = 9 + 16 = 25$. $\sqrt{25} = 5"$.

39. How many threads per inch are there on a one inch micrometer screw?

A. All micrometer screws, including the one inch, have 40 threads per inch.

40. What is the difference between a prick punch and center punch?

A. The included angle of the point of a prick punch is from $30°$ to $55°$ while the included angle of a center punch is approximately $90°$.

41. Give the uses of the prick punch and center punch.

A. The prick punch is used to mark scribed or layout lines. The center punch is used to mark the center of holes for drilling.

42. What is the difference between a heavy duty and a light duty screw driver?

A. The shank of the heavy duty screw driver is made square so that a wrench may be applied to it to assist in turning a screw. The shank of the light duty screw driver is usually round, because enough pressure to turn a screw can be applied at the handle.

43. Draw a sketch showing the correct way to grind a screw driver.

Right   Wrong

Fig. 346

A small land should be left on the tip of the blade (see Fig. 346). Never grind the tip of the blade as you would grind a chisel, because this will round the slot in the screw head. In grinding the point, be careful not to overheat and anneal it.

44. What angle should be ground on the point of a cold chisel?

A. The point on a chisel for cutting soft metal should be ground to an included angle of 55 to 70 degrees. If the chisel is to be used for cutting hard steel and chilled cast iron, the point should be ground to an included angle of 80 to 90 degrees.

45. What is meant by the word "lubricate"?

A. Lubricate means to apply a thin film of a friction-resisting substance (oil or grease) between moving parts.

46. What is the correct cutting angle to grind on a drill for ordinary work?

A. A cutting angle of 59 degrees (Fig. 347) should be ground on a drill used for ordinary purposes.

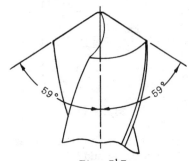

Fig. 347

47. What is the correct clearance angle to grind on a drill?

A. Grind a clearance angle of 9 to 12 degrees for hard steel and 12 to 15 degrees (Fig. 348) for soft steel. An angle of 12 degrees is generally used.

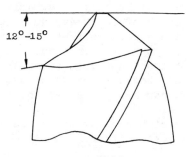

Fig. 348

48. What is the name of the taper shank used on drills?

A. Morse taper shanks are used on drills.

49. What two shapes of files are most commonly used in the machine shop?

A. The flat and the half-round files are most commonly used.

50. What kind of file should be used for filing soft metals?

A. A Vixen file should be used for filing soft metal.

51. When should a file be used without a handle?

A. Never use a file without a handle.

52. How is the single depth (Fig. 349) of an American National thread found?

A. Divide .6495 by the number of threads per inch.

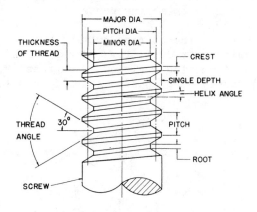

Fig. 349

53. How is the pitch diameter of an American National thread found?

A. Subtract the single depth from the major diameter (O.D.)

$$P.D. = \text{major diameter} - \frac{.6495}{\text{No. of threads per in.}}$$

54. What is the purpose of a thread micrometer?

A. The thread micrometer is used to measure the pitch diameter of threads.

55. What included angle is ground on tools for cutting American National threads?

A. A 60 degree angle should be ground on tools for cutting American National threads.

56. How is the tap drill size for a given thread determined?

A. Subtract twice the single depth, which is the double depth, from the major diameter and add one sixth of the double depth for clearance; or from the major diameter subtract $\frac{1.0825}{\text{No. of threads per inch}}$

57. What is the smallest diameter from which a 3/4" square can be cut? 3/4" hexagon?

A. See Figs. 350 and 351.

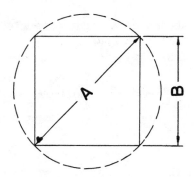

A = B X 1.4142

$\frac{3}{4}$ X 1.4142 = 1.0606

Fig. 350

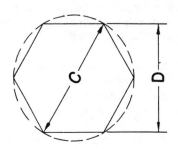

C = D X 1.1547

$\frac{3}{4}$ X 1.1547 = .86602

Fig. 351

58. Illustrate how dimensions are stated in ordering stock.

A. In writing requisitions for stock it is very important that dimensions be stated correctly and in the proper order.

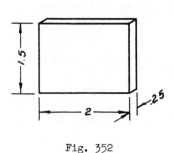

Fig. 352

In ordering rectangular stock the thickness is written first, then the width and the length. A piece of cast iron of the size shown in Fig. 352 would be stated as ¼" x 1½" x 2".

The size of the piece of cold rolled steel shown in Fig. 353 would be stated as ¼" x 1¼" x 4".

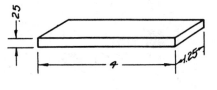

Fig. 353

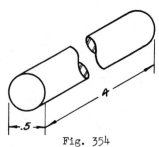 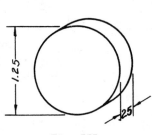 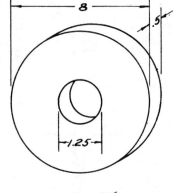

Fig. 354   Fig. 355   Fig. 356

In round stock the diameter is stated first, followed by the length or thickness. The size of a piece of cold rolled steel ½" in diameter and 4" long, as shown in Fig. 354, would be written as ½" x 4" C.R.S.

The dimensions of a piece of machine steel of the size shown in Fig. 355 would be written as 1¼" x ¼" M.S.

If round stock has a hole through it, as in milling machine cutters, grinding wheels, and cored stock, the diameter is stated first, the thickness, and then the size of hole. For example, a 46 H grinding wheel, 8" diameter and ½" thick, with a 1¼" hole (Fig. 356), should be stated as 8" x ½" x 1¼" 46 H.

59. What is bronze? brass?

A. Bronze is an alloy of copper and tin.

Brass is an alloy of copper and zinc.

60. What is steel?

A. Steel is an alloy of iron and carbon. Among the other elements found in steel are silicon, phosphorus, sulphur, manganese, and chromium.

61. How are the different types of steel designated?

A. In Ford Motor Co. some types of steel are designated by letter. For example, "A" steel is a low carbon alloy steel; "AAA" is a medium carbon alloy steel; "RR" is a high carbon tool steel; etc. Other types are designated by name such as: chrome non-shrink, Ford hot work, insert die steel, etc. Bar stock is identified by the colors painted on the end of the bar. Many other firms use the S.A.E. numbering system for identifying steel.

62. Give the percentage of carbon in each of the following steels: cold rolled, machine, "A", "AAA", "EE", "S", and tool steel.

A. The following list gives the percentage of carbon in each of these steels.

Cold rolled	.05% to .15%	carbon
Machine	.08% to .20%	"
"A"	.20% to .24%	"
"AAA"	.30% to .35%	"
"EE"	.35% to .40%	"
"S"	.60% to .70%	"
"RR" (Tool)	.95% to 1.05%	"

63. What kind of steel is usually used to make tool bits, milling cutters, reamers, drills, broaches, etc.?

A. High speed steel is usually used for making these tools.

64. Give a table of common types of screws and tell how their lengths are measured.

A. The table below gives information on standard screws of the National Coarse and National Fine Series. The length of flat head screws is measured over the entire length. The length of socket head, button head, and hexagon head screws is measured under the head.

SIZE OF SCREW	MAJOR DIA OF THD NC	THDS PER INCH NC	MINOR DIA OF THD NC	THDS PER INCH NF	MINOR DIA OF THD NF	FLAT HEAD		SOCKET HEAD		BUTTON HEAD		HEXAGON HEAD		
						A	B	A	B	A	B	A	B	C
2	.086	56	.064	64	.067	.164	.045	.132	.056	.154	.060	.205	.066	.187
3	.099	48	.073	56	.077	.190	.052	.153	.063	.178	.069	.205	.066	.187
4	.112	40	.081	48	.086	.216	.060	.175	.072	.202	.078	.275	.098	.250
5	.125	40	.094	44	.097	.242	.067	.196	.081	.226	.087	.344	.114	.312
6	.138	32	.099	40	.107	.268	.075	.217	.083	.250	.096	.344	.114	.312
8	.164	32	.125	36	.130	.320	.090	.260	.106	.298	.114	.378	.130	.344
10	.190	24	.139	32	.152	.372	.105	.303	.133	.346	.133	.413	.130	.375
12	.216	24	.165	28	.172	.424	.120	.345	.141	.394	.151	.482	.193	.438
1/4	.250	20	.185	28	.206	.468	.156	.375	.250	.438	.218	.482	.242	.438
5/16	.312	18	.240	24	.261	.625	.218	.438	.312	.562	.281	.557	.250	.500
3/8	.375	16	.294	24	.324	.750	.265	.562	.375	.625	.312	.628	.289	.562
7/16	.438	14	.345	20	.375	.812	.265	.625	.438	.750	.375	.698	.337	.625
1/2	.500	13	.400	20	.438	.875	.265	.750	.500	.812	.404	.840	.385	.750
9/16	.562	12	.460	18	.494	1.00	.293	.812	.562	.938	.468	.910	.433	.812
5/8	.625	11	.507	18	.557	1.12	.360	.875	.625	1.00	.500	.980	.481	.875
3/4	.750	10	.620	16	.673	1.37	.438	1.00	.750	1.25	.625	1.12	.576	1.00
7/8	.875	9	.730	14	.787			1.12	.875			1.26	.672	1.12
1	1.000	8	.837	14	.912			1.25	1.00			1.47	.768	1.31
1-1/8	1.125	7	.939	12	1.02							1.68	.863	1.50
1-1/4	1.250	7	1.06	12	1.15							1.89	.960	1.68

Chapter 12

DRILLS AND DRILLING OPERATIONS

The twist drill shown in Fig. 357 is a very efficient tool. It is generally formed by forging and twisting grooves in a flat strip of steel or by milling a cylindrical piece of steel. High speed steel is commonly used. This steel costs more but tools made of it withstand heat much better than those made of ordinary tool steel.

The twist drill may be divided into three principal parts: body, shank, and point. The "flutes" are the spiral grooves that are formed on the side of a drill. Drills are made with two, three, or four flutes. Those having three or four flutes are used for following smaller drills or for enlarging cored holes, and are not adapted for drilling into solid stock. In fact, a drill with three or four flutes might properly be classified as an end-cutting reamer.

1. Give four advantages of spiral flutes.
   A. (1) They give the correct rake to the lip of a drill.
   (2) They cause a chip to curl tightly within itself so that it occupys the minimum amount of space.
   (3) They form channels through which chips escape from the hole.
   (4) They allow the lubricant to flow easily down to the cutting edge.

2. What is the "dead center"?
   A. The dead center is the sharp edge at the extreme tip end of the drill (see Fig. 357). It is formed by the intersection of the cone-shaped surfaces of the point and should always be in the exact center of the axis of the drill.

3. Describe the "point."
   A. The point of the drill should not be confused with the dead center. The point is the entire cone-shaped surface at the cutting end of the drill (Fig. 357).

4. Describe the "heel."
   A. The heel is the portion of the point back of the lips or cutting edges.

5. What is the "lip clearance"?
   A. The lip clearance is the surface of the point that is ground away or relieved just back of the lips (Fig. 357).

6. Describe the "margin."
   A. The narrow strip between A and B in

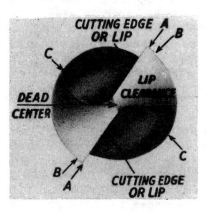

Fig. 357. Twist Drill and Point

Fig. 357 is called the margin. It is practically the full diameter of the drill and extends the entire length of the flute. Its
surface is part of a cylinder which is interrupted by the flutes and by what is known as
body clearance.

7. What is "body clearance"?

A. The portion from B to C in Fig. 357 is
of less diameter than the margin. This les-
sened diameter, called body clearance, reduces the friction between the drill and the
walls of the hole, while the margin between A
and B (which is the full diameter) insures
the hole being of accurate size. This diameter at the shank end of the drill is .0005"
to .002" smaller than the diameter at the
point. This clearance is called longitudinal
clearance and allows the drill to revolve
without binding in drilling deep holes.

8. Describe the "web" of the drill.

A. The web is the metal column which sep-
arates the flutes. It runs the entire length
of the drill between the flutes (see Fig.
358), and is the supporting section of the
drill, the drill's "backbone" in fact. It
gradually increases in thickness toward the
shank (see Fig. 359). This thickening of the
web gives additional rigidity to the drill.

Fig. 358

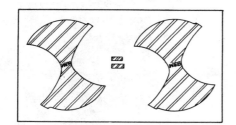

Fig. 359

The section on the left was cut from
a drill near the point while the section on the right was cut near the
shank. The difference in the thickness of the web at these two points is
shown by the length of the white lines
between the two sections in the illustration.

9. What is the "tang"?

A. The tang is usually found only on tapered shank tools (see Fig. 357). It is the
portion of the tool which fits into a slot in
the socket or spindle. It may bear a portion
of the driving strain but its principal use
is to make it easy to remove the drill from
the socket with the aid of a drill drift
(Fig. 360). In removing the drill, place the

round edge of the drift against the sleeve
and the flat edge against the top of the

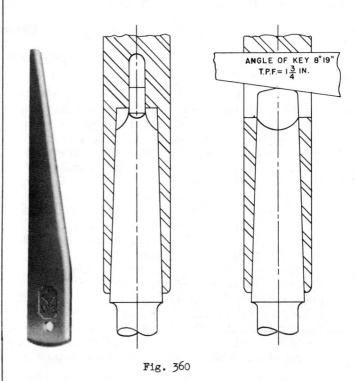

ANGLE OF KEY 8°19"
T.P.F.= 1 3/4 IN.

Fig. 360

drill. Never use a file or wedge as a drift.
Do not use a machinist's hammer to seat the
drill shank tightly in the sleeve or socket.
Instead, use a piece of wood or a rawhide or
lead hammer. In removing a drill from a
spindle or socket, take great care that the
point or tip does not hit the table of the
press or the work.

10. Give some causes of broken tangs.

A. Broken tangs result from a variety of
causes, such as the use of badly burred
sockets, drills whose shanks are so damaged
that they do not fit snugly into the sockets,
or drills with badly damaged tangs.

Although the tang of a drill is designed to carry a portion of the driving
strain, a tapered fit between the shank of
the drill and the wall of the socket should
carry a large share of this strain. A perfect fit is impossible if either the inner
wall of the socket or the surface of the
shank is damaged. Under such conditions the
whole driving strain is thrown on the tang,
and it may break through no defect of its
own, but rather through faulty treatment of
either the drill or the socket.

Before inserting a drill in a socket, rub off the shank to make certain it is
smooth and free from grit. Also inspect the
inside of the socket to be sure it is in the

same good condition.

11. What is the shank of the drill?

A. The shank is the end of the drill which fits into the socket, spindle, or chuck of the drill press. The Morse taper shank, which is the one commonly used, is shown in 357.

12. Why is the taper shank used on drills?

A. The taper shank is used so that drills may be easily forced out of a sleeve with a drill drift.

13. What makes a drill with a taper shank stick in a sleeve or spindle?

A. The taper shank is designed so that pressure at the point of the drill causes the shank to grip tighter in the sleeve, thereby keeping the drill from slipping.

14. Name the four most common shanks used on drills.

A. The four most common shanks (see Fig. 361) are the bit shank (a), straight shank (b), tapered shank (c), and ratchet shank (d).

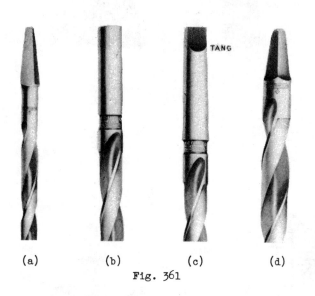

(a)   (b)   (c)   (d)

Fig. 361

15. Lips or cutting edges should be ground to what angle to the axis of a drill?

A. It has been found by experience that 59° is the best angle to grind a drill for work on steel or cast iron (see Fig. 362 in which the two lips are the same length and the same angle to the axis of the drill).

16. What happens in drilling steel or cast iron with a drill that has an angle greater or less than 59°?

A. If the angle is more than 59° the

drill will not center properly, because the cone-shaped point which should hold it in

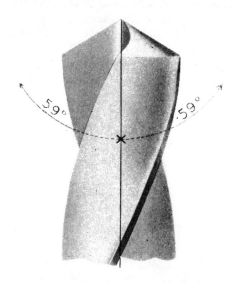

Fig. 362

position will be too flat to perform this work. If the angle is less than 59° the hole is drilled less rapidly and more power is required to drive the drill, because of the additional length of the cutting edges.

17. What happens if the point is on center but the cutting edges are ground at different angles?

A. The drill will bind on one side (see Fig. 363), only one lip or cutting edge will

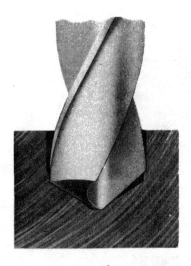

Fig. 363

do the work (resulting in rapid wear on that edge), and the hole will be larger than the drill.

18. What happens if the angles of the cutting edges are equal but the lips are of different lengths?

A. The result will be that both the point and the lip will necessarily be off center (see Fig. 364). This will cause the

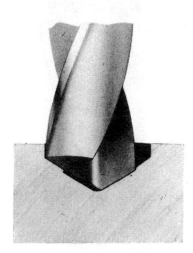

Fig. 364

hole to be larger than the drill. The effects of this condition are the same as the effects that would be obtained from a wheel with its axle placed at any point other than the exact center of the wheel. Such a condition in a drill obviously causes disastrous results. A tremendous strain is placed on the press, the spindle will weave

and wobble, the drill will wear away rapidly, and the machine will eventually break down because of strains on the spindle bearings and other parts.

19. What is the approximate angle to grind the lip clearance of a drill?

A. The heel (the surface of the point back of the cutting lip) should be ground away from the cutting lip at an angle of about 12° to 15°, as shown in Fig. 365 (in all cases, this angle of 12° to 15° is the angle at the circumference of the drill).

CAUTION: Do not overheat the point

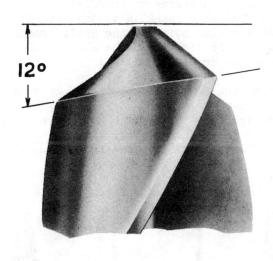

Fig. 365

Lip Clearance Angle Correctly Ground

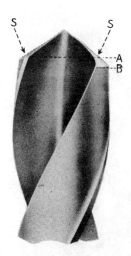

Fig. 366
Proper Lip Clearance. Note how much lower the heel line "B" is than the cutting lip line "A". This difference is the measure of the clearance.

Fig. 367
The cutting lip has already removed considerable metal ahead of the heel as indicated by the black portions of the hole on each side of the drill.

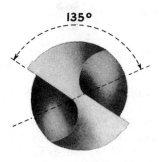

Fig. 368
Showing one way to gauge the correctness of your lip clearance angle.

of a drill in grinding it, as this will tend to anneal the cutting edge.

rolled helix.

Fig. 373 shows how chips will be re-

Lip Clearance Angle Incorrectly Ground and Results

Fig. 369
Showing a drill point without any clearance. Note that both the cutting lip and the heel "S" are in the same plane.

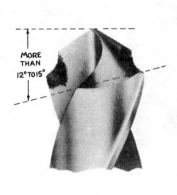

Fig. 370
Showing results of giving a drill too <u>great</u> lip clearance-- the edges of the cutting lips have broken down because of in- sufficient support.

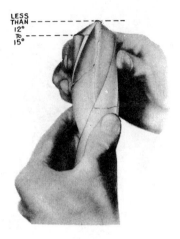

Fig. 371
Here the drill was given in- sufficient lip clearance. As a result there ceased to be any cutting edges whatsoever and, as the feed pressure was applied, the drill could not enter the work--as a result it "splits up the center."

20. What is the rake angle?

A. This is the angle of the flute in relation to the work (see Fig. 372). The rake angle is usually between 22° and 30°. If this angle were 90° or more, it would not give a good cutting edge. If the angle is ground too small, how- ever, it makes the cutting edge so thin that it breaks down un- der the strain of the work.

The rake angle also par- tially governs the tightness with which the chips curl (within themselves) and hence the amount of space which the chips occupy. Other conditions being the same, a very large rake angle makes a tightly rolled chip while a rather small rake angle gives a chip a tendency to curl into a more loosely

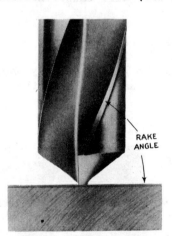

Fig. 372

moved from the job by a correctly ground drill.

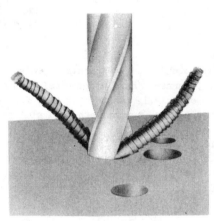

Fig. 373

21. What will happen to a drill if the speed is too great?

A. The outer corners of the drill will wear away quickly because the excessive speed will draw the temper (see Fig. 374).

22. Describe the grinding of a drill for working on brass, bronze, and cast iron.

A. Grind the lip or cutting edge so there

will be no rake, as shown in Fig. 375. When drilling brass or bronze, grind the cutting angle to 65° to 67° instead of 59°. Use a heavy feed.

Fig. 374   Fig. 375

23. Tell what "thinning the point" of a drill is, and explain why it is done.
    A. To strengthen a drill, the thickness of the web is increased as the flute approaches the shank (Fig. 376). As the point

Fig. 376

Shows "notched point" thinning. This is done with a sharp cornered hard emery wheel. This type of thinning is especially adapted for hand feed such as crankshaft drilling, turret lathe and similar work.

is ground back, the thicker web comes in contact with the work, causing greater wear and making penetration into the work more difficult. This condition may be remedied by

Fig. 377   Fig. 378

thinning the point, which may be done in various ways. The use of a convex grinding wheel (Fig. 377) is the most common method. Fig. 378 shows the result of this operation.

24. Describe the grinding of a drill for working on hard material.
    A. In drilling extremely hard material, using a light feed, the point should be ground to an angle of 68°, as shown in Fig. 379. The angle of lip clearance should be decreased to 9° at the periphery, as in Fig. 380. A point should be examined frequently

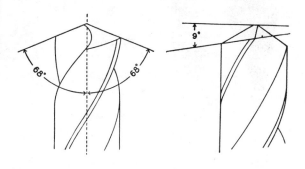

Fig. 379   Fig. 380

as it is being ground, as a guard against drill trouble. Use turpentine, kerosene, or soda water for cooling.

    NOTE. A high speed drill should be ground on a dry wheel of medium grain and soft grade. It must not be immersed in water afterward as this will cause the point of the drill to crack.

25. Name two things necessary to make a drill cut "up to size."
    A. The lips must be ground exactly the same length and at the same angle to the axis.

26. What is the advantage of rotating the work in drilling a deep hole?
    A. It is easier to drill a straight hole while the work is rotating, as the drill will operate like a boring bar.

27. How is a dull drill indicated?
    A. The hole will be rough, as shown in Fig. 381,

Fig. 381

and the drill will heat quickly.

28. Describe an oil hole drill.

A. The oil hole drill (Fig. 382) has oil holes through the body from the shank to the

Fig. 382

point, by which lubricant flows down to cool the point. This drill is generally used for deep hole drilling.

29. Why are flat drills used in laboratories instead of twist drills?

A. The flat drill (Fig. 383) has no rake, so that it breaks the chips into fine pieces

Fig. 383

suitable for laboratory work, but the twist drill causes the chips to come out of the hole in the shape of a curl or roll, as shown in Fig. 373.

30. Why is the old style flat drill, similar to the one shown in Fig. 383, better for some jobs than the modern twist drill?

A. The old style flat drill does not tend to "hog in" or "grab" soft material such as brass. Also, hard spots in steel will not throw a flat drill off center as easily as a twist drill.

31. How should the drill be started in the work?

A. After the job has been marked with a center punch (Fig. 384), the combined drill and countersink (Fig. 385) is placed in the

Fig. 384

drill chuck of a drill press and the center hole drilled.

In starting the drill, bring it down to the work by hand feed un-

Fig. 385

til it centers itself in the work.   Then throw in the power feed.

32. How should a job be laid out for drilling?

A. The laying out of the holes is done from a sketch or blue print and the work is then sent to the drilling machines, ready for the necessary operations. The surface of the work to be drilled is first coated with chalk or blue vitriol. Then the height gage shown in Fig. 386 or the surface gage shown in Fig. 387 is used to locate the center of the hole to be drilled (see Fig. 390). The center punch (Fig. 384) is used to mark the location of the center and assist in starting the center drill on it. A pair of dividers (Fig. 388) is used to scribe a circle having a diameter equal to that of the hole, as shown in Fig. 391.   The prick

Fig. 386

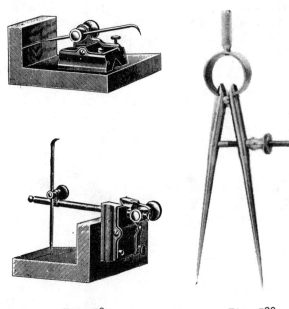

Fig. 387   Fig. 388

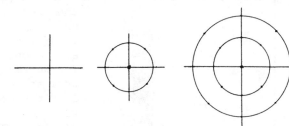

punch of Fig. 389 (notice that the angle of the point of the prick punch is much more acute than that of the center punch) is used to make small indentations at intervals about the circumference, as shown in Figs. 391 and 392.

Fig. 389   Fig. 390   Fig. 391   Fig. 392

33. Explain how a drill may be drawn back on center after it has moved.

A. As a drill begins to cut into a job, it forms a conical hole which is the shape of the drill point. If the hole is concentric with the layout, the drill has been started properly. A drill may start "off center," however, due to improper center drilling, careless starting of the drill, improper grinding of the drill point, hard spots in the metal, or to some other cause (Fig. 393a). To correct this condition, use a cape chisel which has been ground with a round nose (Fig. 393d) and cut a groove on the side of the hole toward which the center is to be drawn (Fig. 393b). The amount the center has been moved may be judged by comparing the edge of the hole with the "layout line." It may be necessary to move the center several times before the edge of the hole and the "layout line" are concentric (Fig. 393c). When the drill begins to cut its full diameter, the prick punch marks on the layout should be evenly cut at the centers.

34. Why is soda water used for cooling instead of plain water?

A. The addition of soda to plain water tends to reduce the amount of heat generated, improve the finish, and overcoming rusting.

35. Why does a drill sometimes squeak?

A. The drill may be ground improperly, too hot, or running without lubrication.

36. Describe the method used to drill glass.

A. Use a piece of copper or brass tubing as a drill. The tubing should have an outside diameter of the same size as the hole desired. An abrasive is used and the tubing run at a speed of about 100 feet per minute. A coarse grinding compound or emery dust mixed with gasoline or light oil makes a very good abrasive to apply between the end of the tubing and the glass. Rest the glass on a rubber or felt pad a little larger than the size of the hole being drilled. Drill about halfway through and then invert the glass and drill from the opposite side.

37. What is the feed of a drill?

A. The distance that the drill enters the work on each revolution, measured in fractions of an inch.

38. What is meant by the speed of a drill?

A. The speed of a drill is the speed of the circumference, called peripheral speed. Speed is the distance that a drill would roll if placed on its side and rolled for one minute at its given r.p.m. (revolutions per minute). The speed of a drill is usually not expressed in revolutions per minute but in feet per minute.

39. What approximate speed in feet per minute should be used with a H.S.S. drill for alloy steel, machine steel, cast iron,

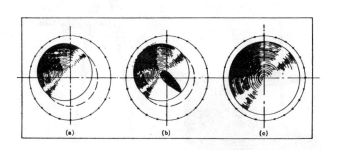

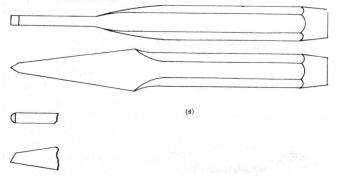

Fig. 393

and brass?

A. The following speeds are recommended by the Cleveland Twist Drill Co.

Alloy Steel	50 - 70 f.p.m.
Machine Steel	70 - 100 f.p.m.
Cast Iron	70 - 150 f.p.m.
Brass	200 - 300 f.p.m.

40. What methods are used to determine the approximate r.p.m. at which to run a $\frac{1}{4}$" drill in cast iron?

A. The correct r.p.m. may be determined mathematically as follows:

Speed for cast iron = 70 f.p.m.
70 f.p.m. = 840 inches per minute
π X diameter = circumference of drill
3.1416 x $\frac{1}{4}$ = .7854 distance per revolution
840 ÷ .7854 = 1069.51 r.p.m.

The correct r.p.m. may also be found from a table of cutting speeds. For example, if we look under 70' and opposite $\frac{1}{4}$" in the table given here we find 1070 r.p.m. Similar tables are found in various handbooks.

41. What different terms are used for sizes of wire drills and machinist's drills?

A. Wire drills are given in figure and letter sizes, as No. 46 or K. Machinist's drills are expressed in fractional sizes of an inch in diameter, as $\frac{1}{4}$" or $\frac{1}{2}$".

42. Besides the drilling of holes, what seven operations are commonly performed on the drilling machine?

A. Tapping, reaming, lapping, countersinking, counterboring, center drilling, and spot facing.

43. Describe what is meant by tapping, reaming, lapping, countersinking, counterboring, center drilling, and spot facing.

A. Taps operated by drilling machines are classified as straight, spiral, taper, pipe, and special. They are also known by several different names which indicate their use, as taper, plug, bottoming, etc.

Taps are internal threading tools. When a set of three taps (Fig. 394) is used, the taper tap makes the first cut; the plug tap (tapered only at the point) completes the thread nearly to the bottom of the hole; and the bottoming tap (which has no taper) finishes it. If the hole passes entirely through the piece and its whole length may be threaded, either the taper tap or the

Table of Cutting Speeds

(Fraction Size Drills)

Feet per Min.	30'	40'	50'	60'	70'	80'	90'	100'	110'	120'	130'	140'	150'
Diameter Inches	Revolutions per Minute												
$\frac{1}{16}$	1833	2445	3056	3667	4278	4889	5500	6111	6722	7334	7945	8556	9167
$\frac{1}{8}$	917	1222	1528	1833	2139	2445	2750	3056	3361	3667	3973	4278	4584
$\frac{3}{16}$	611	815	1019	1222	1426	1630	1833	2037	2241	2445	2648	2852	3056
$\frac{1}{4}$	458	611	764	917	1070	1222	1375	1528	1681	1833	1986	2139	2292
$\frac{5}{16}$	367	489	611	733	856	978	1100	1222	1345	1467	1589	1711	1833
$\frac{3}{8}$	306	407	509	611	713	815	917	1019	1120	1222	1324	1426	1528
$\frac{7}{16}$	262	349	437	524	611	698	786	873	960	1048	1135	1222	1310
$\frac{1}{2}$	229	306	382	458	535	611	688	764	840	917	993	1070	1146
$\frac{5}{8}$	183	244	306	367	428	489	550	611	672	733	794	856	917
$\frac{3}{4}$	153	203	255	306	357	407	458	509	560	611	662	713	764
$\frac{7}{8}$	131	175	218	262	306	349	393	436	480	524	568	611	655
1'	115	153	191	229	267	306	344	382	420	458	497	535	573

plug tap alone may be used.

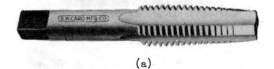

(a)

(b)

(c)
Fig. 394

Taps for internal threading: (a) taper; (b) plug; (c) bottoming.

Fig. 395 illustrates the operation of tapping on a drilling machine. A hole of the proper size must be drilled for a certain size tap, so that the thread will not be cut too deep or too shallow. A table of tap sizes and corresponding drills is given on page 83.

A tap that has been broken in a hole may be removed by a tap extractor, as shown in Fig. 396. Remove all loose particles

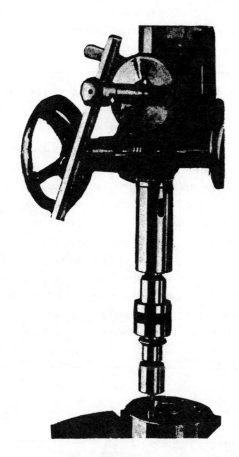

Fig. 395

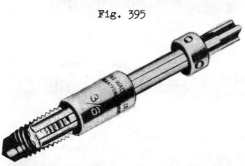

Fig. 396

before attempting to use an extractor, and
always use care and good judgment with it,
as it is a fragile tool.

If a screw or bolt has broken off in
a hole, it may be removed with an easy-out,
as in Fig. 397. Drill a hole in the broken
screw or bolt, a little smaller than its
minor diameter, and then insert the easy-out
and twist it as though you were tapping with
a left-hand tap. The twist causes the cork-
screw-like spirals of the easy-out to grip
the sides of the drilled hole, so that the
broken piece can be loosened and extracted.

It must be remembered that it is
practically impossible to drill a hole with

Fig. 397

a high degree of ac-
curacy. The reamer is
therefore used to re-
move a small amount of
material, leaving the
surface smooth and
within very close
limits.

Lapping may be
defined as a process
of finishing the sur-
face of a piece of
work by using another
piece of work (called
a lap) whose surface
is charged with an
abrasive (see Fig.
403). Only a small
amount of material (.0001" to .002") is left
to be removed by a lap.

Fig. 398. Straight Reamer.

Fig. 399. Taper Reamer.

Fig. 400. Expansion Reamer

Reamer Sizes		Limits of Expansion
$\frac{1}{4}$ to $\frac{15}{32}$		.005
$\frac{1}{2}$ to $\frac{31}{32}$		.008
1 to $1\frac{23}{32}$		.010
$1\frac{3}{4}$ to $2\frac{1}{2}$		.012

Fig. 401. Cherry Reamer.

Fig. 402
End View of Holder for Cherry Reamer

Fig. 403. Copper Head Laps

Fig. 404
Countersink

The countersink in Fig. 404 is used for beveling the mouth of a hole similar to the one shown in Fig. 405.

The counterbore in Fig. 408 is used to machine a larger hole on the same center line and on top or concentric with a hole previously drilled, as shown in Fig. 406.

Center drilling is the operation of drilling and countersinking a hole with a 60° countersink, both at the same operation (see Figs. 407 and 385).

Spot facing is the machining of a circular surface around the top

used for this operation.

The drilling machine is the second oldest known machine tool, having been invented shortly after the lathe, and is probably the most used of any machine. The drilling machine may be classified into three general types; Vertical Spindle, Multiple Spindle, and Radial Spindle machines.

The Vertical Spindle Drilling Machine comes in three types, Heavy Duty, plain, and Sensitive. The Heavy Duty Drilling Machine (Fig. 409) is a heavy and powerful machine for heavy drilling. It has an adjustable knee firmly gibbed to the front of the column and supported by an adjusting screw.

The Plain Vertical Spindle Drilling Machine (Fig. 410) is designed for lighter work and is more adaptable than the heavy duty machine. The spindle can be moved up and down on the column and the table can be adjusted to any desired height. These spindles are sometimes arranged in groups of three or more and called gang drills (Fig. 412). On the plain machine the table is located on an arm attached to the column and can be swung in a 180° arc at right angles to

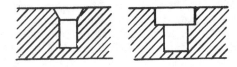

Fig. 405   Fig. 406

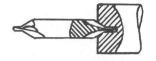

Fig. 407

of a hole to form a true bearing for a collar or washer. The counterbore shown in Fig. 408 is

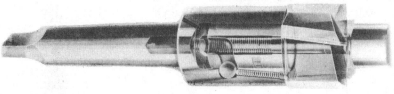

Fig. 408. Phantom View of Holder and Counterbore.

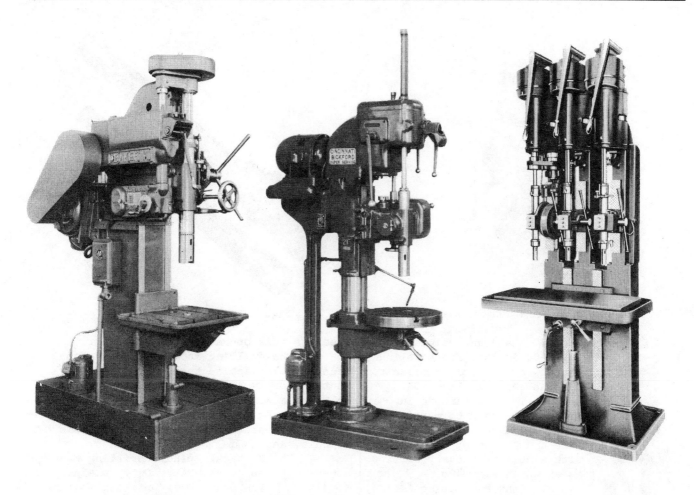

Fig. 409
Heavy Duty Drilling Machine

Fig. 410
Plain Drilling Machine

Fig. 411
Sensitive Drilling Machine

the column. The arm may be clamped in any position within the arc. The table proper is free to revolve 360° and may be clamped in any position, making it easy to locate a layout hole in line with the spindle.

The Sensitive Drilling Machine (Fig. 411) is a light drilling machine for drilling small holes, which have to be drilled at a high speed in light jobs. These machines have no feed mechanism and must be fed by hand. The table and spindle can be adjusted to the required height on the column.

The Multiple Spindle Drilling Machine (Fig. 413) may have any number of spindles arranged on the cross rail of the drilling machine, all spindles being driven from the same shaft by a worm and worm gear. The spindles are removable on the rail to accommodate different classes of work. The Multiple Spindle Drilling Machine should be distinguished from the Multiple Drill Head Machine (Fig. 414), which has any number of spindles from four to forty-eight, driven off the same spindle drive gear in the same head.

On the Radial Drilling Machine (Fig. 415), the spindle is movable and can be adjusted to the work instead of adjusting the work to fit the spindle. In cases where the work is too large and heavy to be set on the vertical spindle machine, or where a great many holes are required, the Radial Drilling Machine will be found a great convenience.

Fig. 412.   Gang Drills.

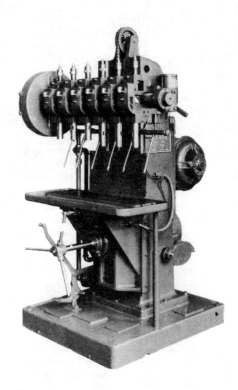

Fig. 413.   Multiple Spindle Drilling Machine.

Fig. 414.   Multiple Drill Head Machine.

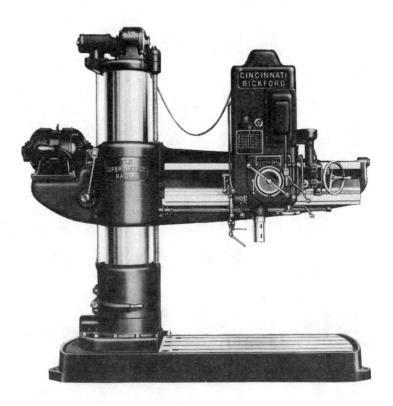

Fig. 415.   Radial Drilling Machine.

DECIMAL EQUIVALENTS OF DRILL SIZES

Inch	M.M.	Wire Gauge	Decimals of an Inch	Inch	M.M.	Letter Sizes	Decimals of an inch	Inch	M.M.	Letter Sizes	Decimals of an Inch
	4.2		.165354		5.9		.232283		8		.31496
		19	.166			A	.234			O	.316
	4.25		.167322	15/64			.234375		8.1		.318897
	4.3		.169291		6		.23622		8.2		.322834
		18	.1695			B	.238			P	.323
11/64			.171875		6.1		.240157		8.25		.324802
		17	.173			C	.242		8.3		.326771
	4.4		.173228		6.2		.244094	21/64			.328125
		16	.177			D	.246		8.4		.330708
	4.5		.177165		6.25		.246062			Q	.332
		15	.18		6.3		.248031		8.5		.334645
	4.6		.181102	1/4			.25		8.6		.338582
		14	.182		6.4		.251968			R	.339
		13	.185		6.5		.255905		8.7		.342519
	4.7		.185039			E	.257	11/32			.34375
3/16			.1875		6.6		.259842		8.75		.344487
	4.75		.187007			F	.261		8.8		.346456
	4.8		.188976		6.7		.263779			S	.348
		12	.189	17/64			.265625		8.9		.350393
		11	.191		6.75		.265747		9		.35433
	4.9		.192913			G	.266			T	.358
		10	.1935		6.8		.267716		9.1		.358267
		9	.196		6.9		.271653	23/64			.359375
	5		.19685			H	.272		9.2		.362204
		8	.199		7		.27559		9.25		.364172
	5.1		.200787			I	.277		9.3		.366141
		7	.201		7.1		.279527			U	.368
13/64			.203125	9/32			.28125		9.4		.370078
		6	.204			J	.281		9.5		.374015
	5.2		.204724		7.2		.283464	3/8			.375
		5	.2055		7.25		.285432			V	.377
	5.25		.206692		7.3		.287401		9.6		.377952
	5.3		.208661			K	.29		9.7		.381889
		4	.209		7.4		.291338		9.75		.383857
	5.4		.212598			L	.295		9.8		.385826
		3	.213		7.5		.295275			W	.386
	5.5		.216535	19/64			.296875		9.9		.389763
7/32			.21875		7.6		.299212	25/64			.390625
	5.6		.220472			M	.302		10.		.3937
		2	.221		7.7		.303149			X	.397
	5.7		.224409		7.75		.305117			Y	.404
	5.75		.226377		7.8		.307086	13/32			.40625
		1	.228		7.9		.311023			Z	.413
	5.8		.228346	5/16		N	.3125		10.5		.413385

DECIMAL EQUIVALENTS OF DRILL SIZES

Inch	M.M.	Wire Gauge	Decimals of an Inch	Inch	M.M.	Wire Gauge	Decimals of an Inch	Inch	M.M.	Wire Gauge	Decimals of an Inch
		80	.0135		1.25		.049212		2.5		.098425
		79	.0145		1.3		.051181			39	.0995
1/64			.015625			55	.052			38	.1015
	.4		.015748		1.35		.053149		2.6		.102362
		78	.016			54	.055			37	.104
		77	.018		1.4		.055118		2.7		.106299
	.5		.019685		1.45		.057086			36	.1065
		76	.02		1.5		.059055		2.75		.108267
		75	.021			53	.0595	7/64			.109375
	.55		.021653		1.55		.061023			35	.11
		74	.0225	1/16			.0625		2.8		.110236
	.6		.023622		1.6		.062992			34	.111
		73	.024			52	.0635			33	.113
		72	.025		1.65		.06496		2.9		.114173
	.65		.02559		1.7		.066929			32	.116
		71	.026			51	.067		3		.11811
	.7		.027559		1.75		.068897			31	.12
		70	.028			50	.07		3.1		.122047
		69	.02925		1.8		.070866	1/8			.125
	.75		.029527		1.85		.072834		3.2		.125984
		68	.031			49	.073		3.25		.127952
1/32			.03125		1.9		.074803			30	.1285
	.8		.031496			48	.076		3.3		.129921
		67	.032		1.95		.076771		3.4		.133858
		66	.033	5/64			.078125			29	.136
	.85		.033464			47	.0785		3.5		.137795
		65	.035		2		.07874			28	.1405
	.9		.035433		2.05		.080708	9/64			.140625
		64	.036			46	.081		3.6		.141732
		63	.037			45	.082			27	.144
	.95		.037401		2.1		.082677		3.7		.145669
		62	.038		2.15		.084645			26	.147
		61	.039			44	.086		3.75		.147637
	1		.03937		2.2		.086614			25	.1495
		60	.04		2.25		.088582		3.8		.149606
		59	.041			43	.089			24	.152
	1.05		.041338		2.3		.090551		3.9		.153543
		58	.042		2.35		.092519			23	.154
		57	.043			42	.0935	5/32			.15625
	1.1		.043307	3/32			.09375			22	.157
	1.15		.045275		2.4		.094488		4		.15748
		56	.0465			41	.096			21	.159
3/64			.046875		2.45		.096456			20	.161
	1.2		.047244			40	.098		4.1		.161417

Tap Drill Sizes

Size	Threads per Inch NC	NF	NS	Outside Diameter Inches	Pitch Diameter Inches	Root Diameter Inches	Tap Drill Approx. 75% Full Thread	Decimal Equivalent of Tap Drill
0		80		.0600	.0519	.0438	3/64	.0469
1	64			.0730	.0629	.0527	54	.0550
1		72		.0730	.0640	.0550	53	.0595
2	56			.0860	.0744	.0628	50	.0700
2		64		.0860	.0759	.0657	50	.0700
3	48			.0990	.0855	.0719	47	.0785
3		56		.0990	.0874	.0758	45	.0820
4	40			.1120	.0940	.0759	43	.0890
4		48		.1120	.0958	.0795	42	.0935
4			36	.1120	.0917	.0714	44	.0860
5	40			.1250	.1070	.0889	38	.1015
5		44		.1250	.1088	.0925	37	.1040
6	32			.1380	.1177	.0974	36	.1065
6		40		.1380	.1200	.1019	33	.1130
6			36	.1380	.1192	.1055	34	.1110
8	32			.1640	.1437	.1234	29	.1360
8		36		.1640	.1460	.1279	29	.1360
8			40	.1640	.1477	.1315	28	.1405
10	24			.1900	.1629	.1359	25	.1495
10		32		.1900	.1697	.1494	21	.1590
10			30	.1900	.1684	.1467	22	.1570
12	24			.2160	.1889	.1619	16	.1770
12		28		.2160	.1928	.1696	14	.1820
12			32	.2160	.1957	.1754	13	.1850
1/4	20			.2500	.2175	.1850	7	.2010
1/4		28		.2500	.2268	.2036	3	.2130
5/16	18			.3125	.2764	.2403	F	.2570
5/16		24		.3125	.2854	.2584	I	.2720
3/8	16			.3750	.3344	.2938	5/16	.3125
3/8		24		.3750	.3479	.3209	Q	.3320
7/16	14			.4375	.3911	.3447	U	.3680
7/16		20		.4375	.4050	.3726	25/64	.3906
1/2	13			.5000	.4500	.4001	27/64	.4219
1/2		20		.5000	.4675	.4351	29/64	.4531
9/16	12			.5625	.5084	.4542	31/64	.4844
9/16		18		.5625	.5264	.4903	33/64	.5156
5/8	11			.6250	.5660	.5069	17/32	.5312
5/8		18		.6250	.5889	.5528	37/64	.5781
3/4	10			.7500	.6850	.6201	21/32	.6562
3/4		16		.7500	.7094	.6688	11/16	.6875
7/8	9			.8750	.8028	.7307	49/64	.7656
7/8		14		.8750	.8286	.7822	13/16	.8125
1"	8			1.0000	.9188	.8376	7/8	.8750
1 1/8	7			1.1250	1.0322	.9072	63/64	.9844
1 1/8		12		1.1250	1.0709	1.0168	1 3/64	1.0469
1 1/4	7			1.2500	1.1572	1.0644	1 7/64	1.1094
1 1/4		12		1.2500	1.1959	1.1418	1 11/64	1.1719
1 3/8	6			1.3750	1.2667	1.1585	1 7/32	1.2187
1 3/8		12		1.3750	1.3209	1.2668	1 19/64	1.2969
1 1/2	6			1.5000	1.3917	1.2835	1 11/32	1.3437
1 1/2		12		1.5000	1.4459	1.3918	1 27/64	1.4219
1 3/4	5			1.7500	1.6201	1.4902	1 9/16	1.5625
2"	4 1/2			2.0000	1.8557	1.7113	1 25/32	1.7812
2 1/4	4 1/2			2.2500	2.1057	1.9613	2 1/32	2.0313
2 1/2	4			2.5000	2.3376	2.1752	2 1/4	2.2500
2 3/4	4			2.7500	2.5876	2.4252	2 1/2	2.5000
3	4			3.0000	2.8376	2.6752	2 3/4	2.7500
3 1/4	4			3.2500	3.0876	2.9252	3	3.0000
3 1/2	4			3.5000	3.3376	3.1752	3 1/4	3.2500
3 3/4	4		18	3.7500	3.5876	3.4252	3 1/2	3.5000

Table of Cutting Speeds

(Fraction Size Drills)

Revolutions per Minute

Feet per Min. Diameter Inches	30'	40'	50'	60'	70'	80'	90'	100'	110'	120'	130'	140'	150'
1/16	1833	2445	3056	3667	4278	4889	5500	6111	6722	7334	7945	8556	9167
1/8	917	1222	1528	1833	2139	2445	2750	3056	3361	3667	3973	4278	4584
3/16	611	815	1019	1222	1426	1630	1833	2037	2241	2445	2648	2852	3056
1/4	458	611	764	917	1070	1222	1375	1528	1681	1833	1986	2139	2292
5/16	367	489	611	733	856	978	1100	1222	1345	1467	1589	1711	1833
3/8	306	407	509	611	713	815	917	1019	1120	1222	1324	1426	1528
7/16	262	349	437	524	611	698	786	873	960	1048	1135	1222	1310
1/2	229	306	382	458	535	611	688	764	840	917	993	1070	1146
5/8	183	244	306	367	428	489	550	611	672	733	794	856	917
3/4	153	203	255	306	357	407	458	509	560	611	662	713	764
7/8	131	175	218	262	306	349	393	436	480	524	568	611	655
1"	115	153	191	229	267	306	344	382	420	458	497	535	573
1 1/8	102	136	170	204	238	272	306	340	373	407	441	475	509
1 1/4	92	122	153	183	214	244	275	306	336	367	397	428	458
1 3/8	83	111	139	167	194	222	250	278	306	333	361	389	417
1 1/2	76	102	127	153	178	204	229	255	280	306	331	357	382
1 5/8	70	94	117	141	165	188	212	235	259	282	306	329	353
1 3/4	65	87	109	131	153	175	196	218	240	262	284	306	327
1 7/8	61	81	102	122	143	163	183	204	224	244	265	285	306
2"	57	76	95	115	134	153	172	191	210	229	248	267	287
2 1/4	51	68	85	102	119	136	153	170	187	204	221	238	255
2 1/2	46	61	76	92	107	122	137	153	168	183	199	214	229
2 3/4	42	56	69	83	97	111	125	139	153	167	181	194	208
3	38	51	64	76	89	102	115	127	140	153	166	178	191

Chapter 13

TAPERS

Most machines in the machine shop are provided with revolving spindles having tapered holes into which the tapered shanks of drills, reamers, centers, etc., are fitted and securely held in place.

The term taper may be defined as the gradual lessening or increasing of the diameter or thickness of a piece of work toward one end. The amount of taper in any given job is found by subtracting the size at the small end from the size at the large end.

Since the taper attachment on tool room lathes and grinders is usually graduated to read in either degrees or taper per foot, it is frequently necessary to calculate one of these factors. This may be done by using similar figures and ratio and proportion. It must be remembered that when using this method for solving taper problems it is necessary to know either two lengths and a taper or two tapers and a length. On blue prints the taper is usually given as an angle or as the amount of taper per foot (T.P.F.) or taper per inch (T.P.I.).

1. (a) What taper is used on drills? (b) How are taper sizes designated?

   A. (a) Drills are usually made with the Morse standard taper shank. (b) Taper sizes are designated by a numbering system, the number of the taper increasing with the size of the drill, as 1, 2, 3, 4, etc.

2. What taper is used on milling machines for the shanks of arbors, collets, end mills, etc.?

   A. The Brown and Sharpe standard taper is used for milling machine spindle shanks.

3. What standard taper does the Ford Motor Company use for the shanks of lathe centers?

   A. The Ford Motor Company uses the Jarno standard taper.

Fig. 416. Grinding the Taper on a Milling Machine Arbor

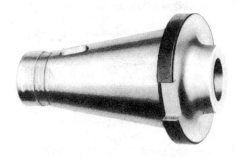

Fig. 417. Tapered Face Milling Cutter Arbor

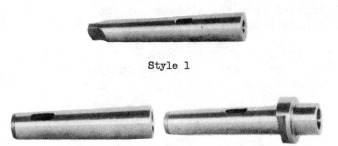

Style 1

Style 2   Style 3

Fig. 418. Tapered Sleeve

Fig. 419. Collets for Holding Tools

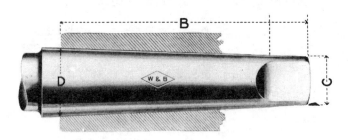

Fig. 420. Tapered Arbor for Screw Cutter

Fig. 421. Morse Standard Taper Reamer

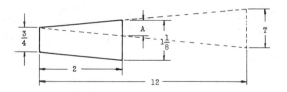

Taper $\frac{1}{4}$ Inch per Foot

Fig. 422. Standard Taper Pin Reamer

4. What taper system is the simplest? Why?

A. The Jarno system is the simplest, because of the easy method of finding the various taper dimensions.

5. Give the Jarno taper formulas.

A.

(1) Diameter of large end = $\dfrac{\text{No. of Taper}}{8}$

(2) Diameter of small end = $\dfrac{\text{No. of Taper}}{10}$

(3) Length of taper = $\dfrac{\text{No. of Taper}}{2}$

(4) T.P.F. = .600 and T.P.I. = .050.

6. How may the accuracy and efficiency of tapers be preserved?

A. The accuracy and efficiency of tapers are preserved by keeping them free from dirt, chips, nicks, and burrs.

7. How should a taper ring gage be checked with a taper plug gage?

A. The taper plug gage should be covered with Prussian blue and then inserted into the ring gage, giving the plug a gentle twisting motion. Any irregularities of the bearing surface will be shown by blue marks inside the ring gage.

8. How can ratio and proportion be used in solving taper problems?

A. By using similar figures (figures which have the same characteristics but are not the same size). A ratio is an indicated division, 1/4, for example, being the ratio between 1 and 4. Where specific terms are used the ratio must be expressed between two terms of like denomination. A proportion is a statement of equality between two given ratios.

9. Find the taper per foot of the job shown in Fig. 423.

Fig. 423

The diameter of the small end subtracted from the diameter of the large end equals the amount of taper in the length of the work. Then 1-1/8 - 3/4 = 3/8, the taper in the length of the work or the distance A. Using ratio and proportion the problem may be stated as follows:

$$\frac{\text{Length}}{\text{Length}} = \frac{\text{Taper}}{\text{Taper}} \quad \text{or} \quad \frac{2}{12} = \frac{\frac{3}{8}}{T}$$

Cross multiplying, $2T = 12 \times \frac{3}{8}$

Clearing fractions, $16T = 12 \times 3$ or 36

Then, $T = \frac{36}{16}$ or $2\frac{1}{4}$ inches taper per foot

It must be remembered that if the first couplet starts with a small number the second couplet must do the same, or vice versa.

10. Find the taper per foot in Fig. 424.

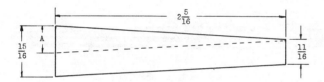

Fig. 424

$$A = \frac{15}{16} - \frac{11}{16} \quad \text{or} \quad \frac{1}{4}$$

$$\frac{2\frac{5}{16}}{12} = \frac{\frac{1}{4}}{T} \qquad 2\frac{5}{16}\,T = \frac{12}{4}$$

$$T = \frac{12}{4} \div 2\frac{5}{16}$$

$$T = 1.2972 \text{ inches taper per foot}$$

11. Find the diameter A in Fig. 425.

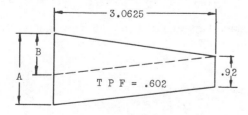

Fig. 425

$$\frac{B}{.602} = \frac{3.0625}{12}$$

$12B = 3.0625 \times .602$

$$B = \frac{3.0625 \times .602}{12}$$

$B = .15363$

$A = .15363 + .920$

$A = 1.0736.$

12. Find the distance A in Fig. 426.

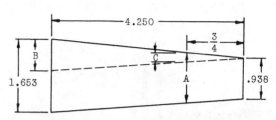

Fig. 426

$B = 1.653 - .938 = .715$

$$\frac{C}{.715} = \frac{\frac{3}{4}}{4.250}$$

$$4.250C = \frac{3}{4} \times .715$$

$$C = \frac{\frac{3}{4} \times .715}{4.250}$$

$C = .12617$

$A = .12617 + .938$

$A = 1.0641.$

13. Determine the taper per foot in Fig. 427.

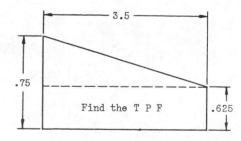

Find the T P F

Fig. 427

$\text{Taper} = .75 - .625 = .125$

$$\frac{12}{3.5} = \frac{T}{.125}$$

$3.5T = .125 \times 12$

$3.5T = 1.5$

$$T = \frac{1.5}{3.5} = .42857.$$

14. Determine the diameter A in Fig. 428.

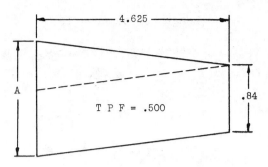

Fig. 428

$$\frac{4.625}{12} = \frac{X}{.500}$$

$$12X = 4.625 \times .500$$

$$12X = 2.3125$$

$$X = \frac{2.3125}{12} = .1927$$

$$A = .840 + .1927 = 1.0327.$$

15. Determine the diameter A in Fig. 429.

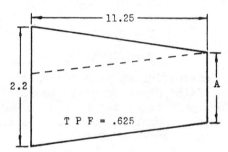

Fig. 429

Let X = taper of job

$$\frac{11.25}{12} = \frac{X}{.625}$$

$$12X = 11.25 \times .625$$

$$12X = 7.0312$$

$$X = \frac{7.0312}{12} \text{ or } .58593$$

$$A = 2.2 - .58593$$

$$A = 1.6141.$$

16. Determine the length A in Fig. 430.

Let X = taper of job

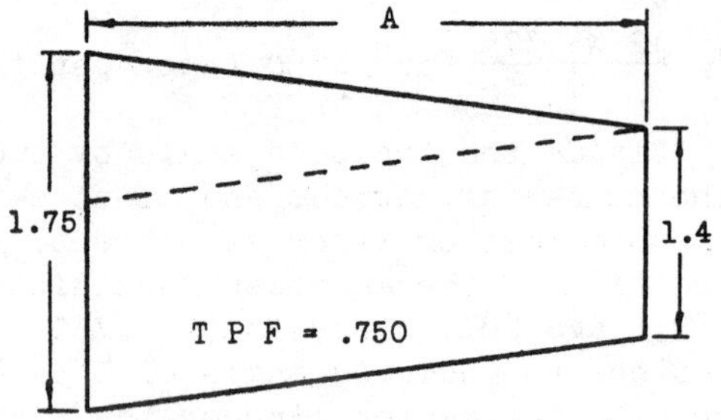

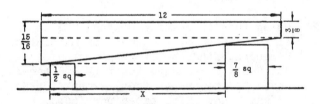

Fig. 430

$$X = 1.75 - 1.4 \text{ or } .350$$

$$\frac{.750}{.350} = \frac{12}{A}$$

$$.750A = .350 \times 12$$

$$.750A = 4.200$$

$$A = \frac{4.200}{.750}$$

$$A = 5.600.$$

17. Find the distance X in Fig. 431.

Fig. 431

$$\frac{15}{16} - \frac{3}{8} = \frac{9}{16}$$

$$\frac{7}{8} - \frac{1}{2} = \frac{3}{8}$$

$$\frac{\frac{3}{8}}{\frac{9}{16}} = \frac{X}{12}$$

$$\frac{\frac{3}{8} \times 12}{\frac{9}{16}} = 8 \qquad X = 8.$$

18. At what angle should the compound rest or taper attachment be set over if the small diameter of the work = 1.500, large diameter = 1.875, and length of taper = 6.000?

A. 1.875 - 1.500 = .375 taper in six inches

$$\frac{6}{12} = \frac{.375}{T}$$

$$T = \frac{12 \times .375}{6} \text{ or } .750 \text{ taper per foot}$$

Divide the included angle by two to find the number of degrees and minutes to set the compound rest or taper attachment. (The sixth edition of the American Machinists' Handbook, page 681, shows that a .750 taper per foot has an included angle of $3°34'44''$. The angle to the center line or the angle to set the taper attachment is therefore $1°47'22''$, which is one-half the included angle.)

19. Find the angle to set the taper attachment on a job which has a small diameter of .36875, a large diameter of .475, and a length of 2-1/8.

A. By ratio and proportion the taper per foot is found to equal .600. The taper to the center of the work will equal .600 ÷ 2 or .300. The length of 12" and the taper to the center line form a right angle triangle. Using trigonometry to find the cotangent of the angle, 12.000 ÷ .300 = 40. The cotangent of 40 equals $1°25'6''$, the angle to set the taper attachment.

20. Using the information on page 672 in the sixth edition of the American Machinists' Handbook, find the large diameter of a No. 5 B & S standard taper (B & S tapers have .500 taper per foot, except for No. 10 which has .5161 taper per foot).

A. Diameter at small end = .450

Length of taper = $2\frac{1}{8}$

Taper per foot = .500

Let X = taper for $2\frac{1}{8}$

$$\frac{2\frac{1}{8}}{12} = \frac{X}{\frac{1}{2}}$$

$$12X = 2\frac{1}{8} \times \frac{1}{2}$$

$$X = \frac{1.0625}{12} \text{ or } .0885$$

Diameter of plug at large end = .450 + .0885 or .5385.

21. Find the diameter at the large end of a plug with a No. 5 B & S milling machine standard taper.

A. Diameter at small end = .450

Length of taper = $1\frac{3}{4}$

Taper per foot = .500

$$\frac{1\frac{3}{4}}{12} = \frac{T}{\frac{1}{2}}$$

$$12T = 1\frac{3}{4} \times \frac{1}{2} = .875$$

$$T = \frac{.875}{12} = .0729$$

Diameter of plug at large end = .450 + .0729 = .5229.

22. If a taper plug has 3/16 of an inch taper in 4 inches, what is the taper per foot? At what angle should the taper attachment be set?

A. The equation for finding the taper per foot is

$$\frac{4}{12} = \frac{\frac{3}{16}}{T}$$

$$4T = 12 \times \frac{3}{16}$$

$$4T = 2.250$$

$$T = .5625.$$

Information on page 681 of the American Machinists' Handbook gives the included angle for .5625 taper per foot as $2°41'4''$. The taper attachment would be set at one-half of this angle, or at $1°20'32''$.

23. Give the dimensions for a No. 9 Jarno taper (see question 5).
A.

Large diameter = $\frac{9}{8}$ = 1.125

Small diameter = $\frac{9}{10}$ = .900

Length = $\frac{9}{2}$ = 4.500

Taper per foot = .600.

24. If .015 grinding stock is left on a #7 B & S taper plug gage, how far will it come from reaching the end of the ring gage (see Fig. 432)?
A. The taper per foot for a #7 B & S taper is .500.

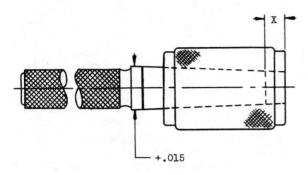

Fig. 432

$$\frac{.015}{.500} = \frac{X}{12}$$

$$.500X = .015 \times 12$$

$$.500X = .180$$

$$X = \frac{.180}{.500} = .360.$$

25. Give the amount of taper per foot for three tapers in common use.

A.

Brown Sharpe	- .500 T.P.F. (except #10 which has .5161 T.P.F.)
Jarno	- .600 T.P.F.
Morse	- .600 - .630 T.P.F. (depending on the number of taper)

26. What is a taper dowel pin? What is the T.P.F. for a taper dowel pin?

A. A taper dowel pin is a pin used to hold jigs, fixtures, brackets, machine parts, etc., in alignment after they have been located properly. These pins have a taper of 1/4" per foot.

27. How may tapers be produced on a lathe?

A. Tapers may be produced on a lathe by setting over the tailstock, by setting the compound at an angle and feeding the tool by hand, by using a taper attachment, and by using a square nose tool.

28. How is a taper hole produced with a taper reamer?

A. To produce a taper hole with a reamer, first drill a hole large enough to allow the small end of the taper to enter. Then ream out by hand with a taper reamer until the proper size is reached.

29. How far must a boring tool be fed into the work to allow a .750 T.P.F. plug gage to advance .062 into a ring gage?

A. Solve for the amount of taper in .062 and divide the result by 2.

$$\frac{.750}{X} = \frac{12}{.062}$$

$$12X = .750 \times .062$$

$$X = \frac{.750 \times .062}{12} = .003875$$

$$.003875 \div 2 = .00194.$$

Chapter 14

THREADS

TERMS RELATING TO SCREW THREADS

1. Screw Thread. A ridge of uniform section in the form of a helix on the surface of a cylinder or cone.

2. External and Internal Threads. An external thread is a thread on the outside of a member. Example: A threaded plug. An internal thread is a thread on the inside of a member. Example: A threaded hole.

3. Major Diameter (formerly known as "outside diameter"). The largest diameter of the thread of the screw or nut. The term "major diameter" replaces the term "outside diameter" as applied to the thread of a screw and also the term "full diameter" as applied to the thread of a nut.

4. Minor Diameter (formerly known as "core diameter" or "root diameter"). The smallest diameter of the thread of the screw or nut. The term "minor diameter" replaces the term "core diameter" as applied to the thread of a screw and also the term "inside diameter" as applied to the thread of a nut.

5. Pitch Diameter. On a straight screw thread, the diameter of an imaginary cylinder, the surface of which would pass through the threads at such points as to make equal the width of the threads and the width of the spaces cut by the surface of the cylinder. On a taper screw thread, the diameter, at a given distance from a reference plane perpendicular to the axis of an imaginary cone, the surface of which would pass through the threads at such points as to make equal the width of the threads and the width of the spaces cut by the surface of the cone.

6. Pitch. The distance from a point on a screw thread to a corresponding point on the next thread measured parallel to the axis.

$$\text{Pitch in inches} = \frac{1.0000}{\text{Number of threads per inch}}$$

7. Lead. The distance a screw thread advances axially in one turn. On a single-thread screw, the lead and pitch are identical; on a double-thread screw the lead is twice the pitch; on a triple-thread screw, the lead is three times the pitch, etc.

8. Angle of Thread. The angle included between the sides of the thread measured in an axial plane.

9. Helix Angle. The angle made by the helix of the thread at the pitch diameter with a plane perpendicular to the axis.

10. Crest. The top surface joining the two sides of a thread.

11. Root. The bottom surface joining the sides of two adjacent threads.

12. Side. The surface of the threads which connect the crest with the root.

13. Axis of a Screw. The longitudinal central line through the screw.

14. Base of Thread. The bottom section of the thread, the greatest section between the two adjacent roots.

15. Depth of Thread. The distance between the crest and the base of the thread measured normal to the axis.

16. Number of Threads. Number of threads in one inch of length.

17. Length of Engagement. The length of contact between two mating parts, measured axially.

18. Depth of Engagement. The depth of thread contact of two mating parts, measured radially.

19. Pitch Line. An element of the imaginary cylinder or cone specified in Definition 5.

20. Thickness of Thread. The distance between the adjacent sides of the thread measured along or parallel to the pitch line.

TERMS RELATING TO CLASSIFICATION
AND TOLERANCE

Five distinct classes of screw thread fits have been established by the National Screw Thread Commission for the purpose of insuring the interchangeable manufacture of screw thread parts throughout the country. The numbers and corresponding fits are as follows: #1 - loose fit, #2 - free fit, #3 - medium fit, #4 - close fit, #5 - wrench fit.

21. Allowance. An intentional difference in the dimensions of mating parts. It is the minimum clearance or the maximum interference which is intended between mating parts. It represents the condition of the tightest permissible fit, or the largest internal member mated with the smallest external member. This is illustrated by the two following examples.

One-half inch, class 1, loose fit, American National coarse thread series:
     Minimum pitch diameter of nut   - .4500
     Maximum pitch diameter of screw - .4478
                 Allowance (positive) - .0022

One-half inch, class 4, close fit, American National coarse thread series:
     Minimum pitch diameter of nut   - .4500
     Maximum pitch diameter of screw - .4504
                 Allowance (negative) - .0004

22. Tolerance. The amount of variation permitted in the size of a part. Example:

One-half inch screw, class 1, loose fit, American National coarse thread series:
          Maximum pitch diameter - - .4478
          Minimum pitch diameter - - .4404
                      Tolerance - - .0074

23. Basic Size. The theoretical or nominal standard size from which all variations are made.

24. Crest Allowance. Defined on a screw form as the space between the crest of a thread and the root of its mating thread.

25. Finish. The character of the surface of a screw thread or other product.

26. Fit. The relation between two mating parts with reference to the conditions of assembly, as wrench fit, close fit, medium fit, free fit, and loose fit. The quality of fit depends upon both the relative size and finish of the mating parts.

27. Neutral Zone. A positive allowance (see "Allowance").

28. Limits. The extreme permissible dimensions of a part. Example:

One-half inch screw, class 1, loose fit, American National coarse thread series:
     Maximum pitch diameter - .4478 These are
     Minimum pitch diameter - .4404 the limits.

The old U.S.S. thread has been changed to American National coarse thread series.

The old S.A.E. thread has been changed to American National fine thread series.

American National Form of Threads

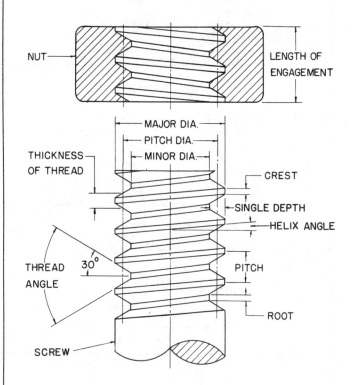

Fig. 433

N = Number of Threads per Inch

$$\text{Pitch (P)} = \frac{1.000}{N} \qquad \text{Root} = \frac{P}{8}$$

$$\text{Crest} = \frac{P}{8} \qquad \text{Single Depth} = \frac{.6495}{N}$$

$$\text{Double Depth (D.D.)} = \frac{1.299}{N}$$

$$\text{Clearance} = \frac{D.D.}{6}$$

Pitch Diameter = Major Diameter - Single Depth
Minor Diameter = Major Diameter - Double Depth

STANDARD THREADS

AMERICAN NATIONAL STANDARD THREAD

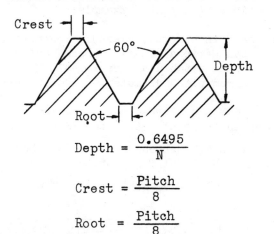

$$\text{Depth} = \frac{0.6495}{N}$$

$$\text{Crest} = \frac{\text{Pitch}}{8}$$

$$\text{Root} = \frac{\text{Pitch}}{8}$$

(a)

SHARP "V" THREAD

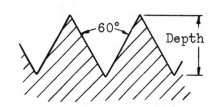

$$\text{Depth} = \frac{0.8660}{N}$$

(b)

ACME THREAD

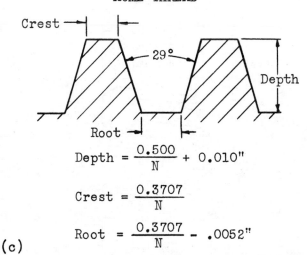

$$\text{Depth} = \frac{0.500}{N} + 0.010"$$

$$\text{Crest} = \frac{0.3707}{N}$$

$$\text{Root} = \frac{0.3707}{N} - .0052"$$

(c)

BROWN & SHARPE WORM THREAD

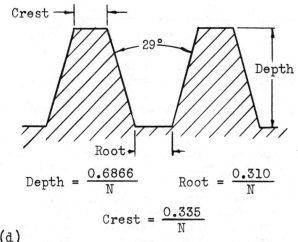

$$\text{Depth} = \frac{0.6866}{N} \qquad \text{Root} = \frac{0.310}{N}$$

$$\text{Crest} = \frac{0.335}{N}$$

(d)

SQUARE THREAD

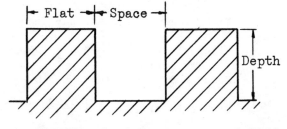

$$\text{Depth} = \frac{0.5000}{N} \qquad \text{Width of flat} = \frac{0.5000}{N}$$

$$\text{Width of space} = \frac{0.5000}{N}$$

(e)

BUTTRESS THREAD

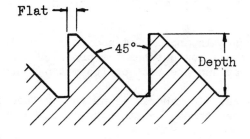

$$\text{Depth} = \frac{0.7500}{N}$$

$$\text{Width of flat} = \frac{\text{Pitch}}{8}$$

(f)

Fig. 435

American National Form of Threads

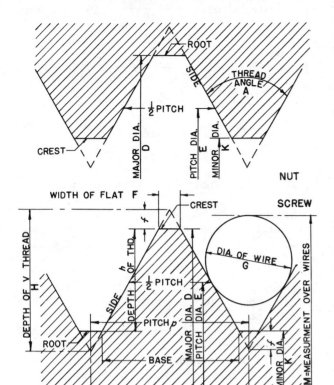

Fig. 434

Size to Bore or Drill a Hole for Internal
Threading = Major Dia. $- \dfrac{1.0825}{N}$

Tangent of Helix Angle $= \dfrac{\text{Lead}}{\text{Pitch Dia.} \times 3.1416}$

M (measurement over three wires) = Major Dia.
$+ 3G - \dfrac{1.5155}{N}$

Best Wire $= \dfrac{.57735}{N}$ Min. Wire $= \dfrac{.56}{N}$

Max. Wire $= \dfrac{.90}{N}$

29. Describe the difference between the lead
    and the pitch of a screw.
    A. The pitch is the distance between a
point on one thread and the corresponding
point on the next, measured parallel to the
axis. The lead is the distance a nut will
advance on a screw, parallel to its axis, in
one revolution (Fig. 436). On a single
thread screw the lead equals the pitch, on a
double thread screw it equals twice the
pitch, etc.

30. Describe the tap and its use.
    A. A tap is a cylindrical bar of steel
with threads formed around it and grooves or
"flutes" running lengthwise in it, intersecting with the threads to form cutting edges.
These cutting edges cut the threads in the
nut.

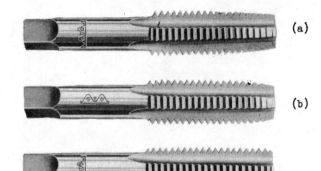

Fig. 437. Taps: (a) taper; (b) plug;
(c) bottoming.

31. Name three different kinds of taps.
    A. Taper, plug, and bottoming taps (shown
in Fig. 437).

32. State how to find the size of hole to
    drill for tapping a National Form thread.
    A. Divide 1.0825 by the number of threads
per inch and subtract this number from the
major diameter.

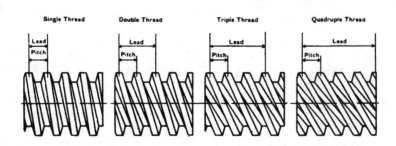

Fig. 436. Relation of Lead and Pitch of Multiple Threads

To check the squareness of the tap with the hole, place the blade of a square against the tap, permitting the beam of the square to rest on the surface of the work, as shown in Fig. 438.

Fig. 438. Testing Tap for Squareness with Work

33. Describe a die and its use.

A. A die is usually a flat piece of steel, internally threaded, with grooves or flutes intersecting the threads to form cutting edges (Fig. 439. These cut the external threads on the screw.

Fig. 439. Die

34. Find the size of the hole to drill to tap the following American National threads: 3/4 NC; 1/2 NF, and 3/4 - 12N. Use the following formula (N = number of threads per inch):

$$\text{Size of Hole} = \text{Major Diameter} - \frac{1.0825}{N}$$

A.

(a) .750 major dia. $- \dfrac{1.0825}{10} = .750 - .10825$

$\qquad\qquad\qquad = .64175$

(b) .500 major dia. $- \dfrac{1.0825}{20} = .500 - .0541$

$\qquad\qquad\qquad = .4459$

(c) .750 major dia. $- \dfrac{1.0825}{12} = .750 - .0902$

$\qquad\qquad\qquad = .6598.$

35. Why is the Sharp "V" thread almost obsolete?

A. The sharp top of the "V" thread is very easily nicked or burred, and the taps and dies will not stand up long because of the sharp thread.

36. After the first cut is taken on a lathe, how should the correct number of threads per inch be checked?

A. Place a rule on the thread parallel with the center line and count the number of threads in one inch (or the number of threads in one-half inch and multiply by two), or check with a pitch gage.

37. What is meant by "rolled threads"?

A. Most of the external commercial threads made today are rolled, that is, the part to be threaded is rolled between grooved dies in a special machine.

38. How is the major diameter (O.D.) of the blanks for rolled threads determined?

A. The major diameter of a blank is equal to the pitch diameter of the finished thread.

39. Figure the size of the blank stock that should be used for 3/4" American National coarse rolled threads.

A. A 3/4" dia. thread has ten threads per inch. We use the following formula to find the pitch diameter: Pitch Diameter = Major Diameter - Single Depth.

$$\text{Single Depth} = \frac{0.6495}{10} = .06495$$

Major Diameter - - - - - - - - 0.75000
Single Depth - - - - - - - - - .06495
Pitch Dia. of Finished Thread - .68505

40. Find the number of threads per inch on a screw with a pitch of .750.

A.

$$\frac{1.0000}{.750} = 1\tfrac{1}{3} \text{ threads per inch.}$$

41. Find the lead of a screw which has 16 threads per inch.
   A.

$$\text{Lead} = \frac{1.000}{16} = 0.0625$$

   The lead and the pitch are the same for single threads. If this were a double thread, the lead would be 0.125.

42. Figure the minor diameter of a Brown & Sharpe triple worm thread with 3/4" lead and 3.000" major diameter.
   A. The pitch equals the lead divided by the number of threads, which is 3 (Triple). 0.75 ÷ 3 = .250. The number of threads per inch (N) equals 1.0000 divided by the pitch. 1.0000 ÷ .250 = 4. The single depth equals .6866 divided by the number of threads per inch. .6866 ÷ 4 = .1716. The double depth equals .1716 multiplied by two, which is .3432.

   Major Diameter - 3.0000
   Double Depth   -   .3432
   Minor Diameter - 2.6568

43. What kind of lubricant should be used for general threading?
   A. Mineral lard oil is a very good lubricant for threading. It is made of white lead, graphite, and fatty oil. The Geometric Tool Company advises the use of the following compound for screw thread cutting.
   40 gallons of water
   10 gallons of mineral lard oil
   2½ pounds of soda (exact quantity)

   In some cases the addition of sulfur to a cutting compound proves very helpful.

44. How should the compound be set for cutting screw threads?
   A. Do not force the tool into the work at 90 degrees with the center line, but swing the compound around one-half the included angle of the thread. For example, set the compound at an angle of 30 degrees for cutting American National threads, and at an angle of 14½ degrees for cutting Brown & Sharpe worm threads.

45. In setting the cutting tool to cut pipe threads, should the tool be set at 90 degrees with the center line of the work or at 90 degrees with the taper?
   A. Set the tool at 90 degrees with the center line of the work.

46. Is the profile of the space the same as that of the thread? Give a reason for your answer.

   A. Yes. A nut with a one inch major diameter and eight square threads per inch would not fit a one inch American National standard coarse thread screw which also had eight threads per inch.

47. What is the purpose of the pitch diameter?
   A. The pitch diameter determines the fit of the thread.

48. Name the tools used in a Ford standard tap set.
   A. Let us take a one-half inch tap set as an example.

   ½ inch body drill
   13/32 tap drill
   ½ x 3/4 counterbore
   ½-13 std. taper tap
   ½-13 std. plug tap
   ½-13 std. bottoming tap

49. Name some methods of producing a thread.
   A. Threads may be produced by revolving the work in the lathe against a proper cutting tool, by rolling the part to be threaded between dies in a special thread milling machine, by hand dies, and by cutting with a revolving cutter in a thread milling machine.

50. How is the size of the tap drill for a specified size determined?
   A. Divide 1.0825 by the number of threads per inch and subtract this result from the major diameter.

51. Why was the Acme thread designed?
   A. Square threads were formerly used to a great extent on adjusting and power screws. The Acme thread was designed because of the difficulty of cutting square threads with taps and dies. It is much easier to produce accurately than the square thread. The Brown & Sharpe worm thread is deeper than the Acme thread.

52. Why are fine threads used on automobile parts instead of coarse threads?
   A. Threads on automobile parts are cut in hard and tough materials, and need not be as coarse as threads cut in cast iron. A screw or bolt of a given size has a greater minor diameter and consequently a greater strength if the pitch is fine rather than coarse. A fine pitch screw or nut may be set tighter and does not shake loose as readily as one of a coarse pitch, although it must be remembered that a fine pitch thread will strip or shear easier than one of coarse pitch.

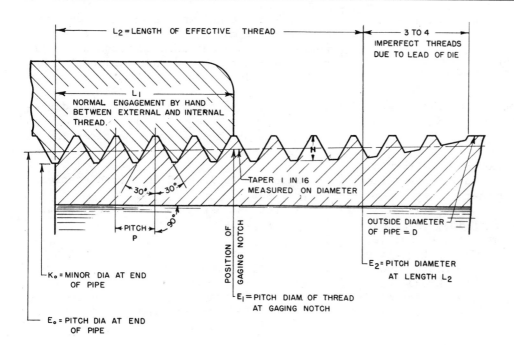

American National Taper
Pipe Thread Notation

Notation

$$E_0 = D - (0.005D + 1.1)p$$
$$E_1 = E_0 + 0.0625\,L_1$$
$$L_2 = p(0.8D + 6.8)$$
$$h = 0.8p.$$

NATIONAL SCREW THREAD COMMISSION

Dimensions of national (American Briggs') taper pipe threads

Nominal size of pipe in inches	Number of threads per inch, n	Pitch, p	Depth of thread	Outside diameter of pipe, D	Length of normal engagement by hand, L_1	Length of effective thread, L_2	Increase in diameter per thread, $\frac{0.0625}{n}$	At end of pipe, or at length L_1 from end of coupling, $E_0 = D - \frac{0.05D+1.1}{n}$ Basic	At length L_1 on pipe, or at end of coupling, $E_1 = E_0 + \frac{L_1}{16}$ Maximum	Basic	Minimum
1	2	3	4	5	6	7	8	9	10	11	12
		Inch	Inch	Inches	Inches	Inches	Inch	Inches	Inches	Inches	Inches
⅛	27	0.03704	0.02963	0.405	0.180	0.26385	0.00231	0.36351	0.37823	0.37476	0.37129
¼	18	.05556	.04444	.540	.200	.40178	.00347	.47739	.49510	.48989	.48468
⅜	18	.05556	.04444	.675	.240	.40778	.00347	.61201	.63222	.62701	.62180
½	14	.07143	.05714	.840	.320	.53371	.00446	.75843	.78513	.77843	.77173
¾	14	.07143	.05714	1.050	.339	.54571	.00446	.96768	.99557	.98887	.98217
1	11½	.08696	.06957	1.315	.400	.68278	.00543	1.21363	1.24678	1.23863	1.23048
1¼	11½	.08696	.06957	1.660	.420	.70678	.00543	1.55713	1.69153	1.68338	1.67523
1½	11½	.08696	.06957	1.900	.420	.72348	.00543	1.79609	1.83049	1.82234	1.81419
2	11½	.08696	.06957	2.375	.436	.75652	.00543	2.26902	2.30442	2.29627	2.28812
2½	8	.12500	.10000	2.875	.682	1.13750	.00781	2.71953	2.77388	2.76316	2.75044
3	8	.12500	.10000	3.500	.766	1.20000	.00781	3.34062	3.40022	3.38850	3.37678
3½	8	.12500	.10000	4.000	.821	1.25000	.00781	3.83750	3.90053	3.88881	3.87709
4	8	.12500	.10000	4.500	.844	1.30000	.00781	4.33438	4.39684	4.38712	4.37540
4½	8	.12500	.10000	5.000	.875	1.35000	.00781	4.83125	4.89766	4.88594	4.87422
5	8	.12500	.10000	5.563	.937	1.40630	.00781	5.39073	5.46101	5.44929	5.43757
6	8	.12500	.10000	6.625	.958	1.51250	.00781	6.44609	6.51769	6.50597	6.49425
7	8	.12500	.10000	7.625	1.000	1.61280	.00781	7.43984	7.51406	7.50234	7.49062
8	8	.12500	.10000	8.625	1.063	1.71280	.00781	8.43359	8.51175	8.50003	8.48831
9	8	.12500	.10000	9.625	1.130	1.81250	.00781	9.42734	9.50969	9.49797	9.48625
10	8	.12500	.10000	10.750	1.210	1.92500	.00781	10.54531	10.63266	10.62094	10.60922
11	8	.12500	.10000	11.750	1.285	2.02500	.00781	11.53906	11.63110	11.61938	11.60766
12	8	.12500	.10000	12.750	1.360	2.12500	.00781	12.53281	12.62953	12.61781	12.60609
14 O. D.	8	.12500	.10000	14.000	1.562	2.25000	.00781	13.77500	13.88434	13.87262	13.86090
15 O. D.	8	.12500	.10000	15.000	1.687	2.35000	.00781	14.76875	14.88591	14.87419	14.86247
16 O. D.	8	.12500	.10000	16.000	1.812	2.45000	.00781	15.76250	15.88747	15.87575	15.86403
17 O. D.	8	.12500	.10000	17.000	1.900	2.55000	.00781	16.75625	16.88672	16.87500	16.86328
18 O. D.	8	.12500	.10000	18.000	2.000	2.65000	.00781	17.75000	17.88672	17.87500	17.86328
20 O. D.	8	.12500	.10000	20.000	2.125	2.85000	.00781	19.73750	19.88203	19.87031	19.85859
22 O. D.	8	.12500	.10000	22.000	2.350	3.05000	.00781	21.72500	21.87734	21.86562	21.85390
24 O. D.	8	.12500	.10000	24.000	2.375	3.25000	.00781	23.71250	23.87266	23.86094	23.84922
26 O. D.	8	.12500	.10000	26.000	2.500	3.45000	.00781	25.70000	25.86797	25.85625	25.84453
28 O. D.	8	.12500	.10000	28.000	2.625	3.65000	.00781	27.68750	27.86328	27.85156	27.83984
30 O. D.	8	.12500	.10000	30.000	2.750	3.85000	.00781	29.67500	29.85860	29.84688	29.83516

53. Where is the Buttress thread used?

A. The Buttress thread is used where a thread requiring great strength in one direction is required, as on vises, jacks, and power press screws.

54. What material is used between the threads of pipe and pipe coupling to prevent leakage?

A. Either red lead or white lead is generally used to seal the pipe couplings.

55. How can you distinguish a right-hand thread from a left-hand thread?

A. If the top of a bolt must be revolved to the right to be screwed into the receiving member, it is termed a right-hand thread, or if it must be revolved to the left, it is known as a left-hand thread.

56. Why is the O.D. of a 3/4 inch pipe much greater than 3/4 of an inch?

A. The inside diameter of a pipe is always as large or a little larger than the nominal size. For example, the inside diameter of a 3/4 inch pipe is .824 and the inside diameter of a one inch pipe is 1.049.

57. Describe the gaging of external and internal pipe threads.

A. In gaging external or male threads, the ring gage (Fig. 440), should screw tight by hand on the pipe or male thread until the small end of the gage is flush with the end of the thread.

In gaging internal or female threads, the plug gage (Fig. 441), should screw tight by hand into the fitting or coupling until the notch is flush with the face. When the thread is chamfered, the notch should be flush with the bottom of the chamfer.

Fig. 440. Ring Gage

Fig. 441. Plug Gage

CAUTION. As the pipe tap is tapered, the hole is tapped for a short length only,

before gaging with pipe tap gage. Never attempt to pass the pipe tap through the hole when tapping.

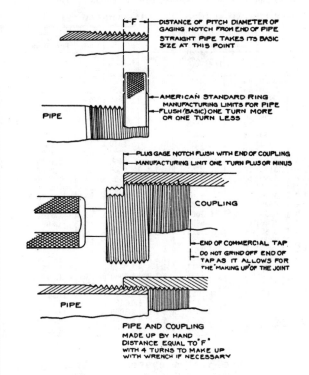

Fig. 442. Use of Taper Pipe Gages

58. What measurement is checked with a thread micrometer?

A. The thread micrometer (Fig. 443) gages the thread on the pitch diameter.

Capacity		Range		
1 in.	8	to	13	threads
1 "	14	"	20	"
1 "	22	"	30	"
1 "	32	"	40	"
2 "	$4\frac{1}{2}$	"	7	"
2 "	8	"	13	"
2 "	14	"	20	"
2 "	22	"	30	"

Fig. 443. Thread Micrometer

Capacity and Range of Thread Micrometers

Pitch Diameter = Major (Outside) Diameter—Depth of Thread

COARSE THREAD SERIES				FINE THREAD SERIES			
Size		Threads Per Inch	Cal. Reading or Pitch Diam.	Size		Threads Per Inch	Cal. Reading or Pitch Diam.
Number or Fraction	Decimal	N	$D - \dfrac{.649519}{N}$	Number or Fraction	Decimal	N	$D - \dfrac{.649519}{N}$
1	.0730	64	.0629	0	.0600	80	.0519
2	.0860	56	.0744	1	.0730	72	.0640
3	.0990	48	.0855	2	.0860	64	.0759
4	.1120	40	.0958	3	.0990	56	.0874
5	.1250	40	.1088	4	.1120	48	.0985
6	.1380	32	.1177	5	.1250	44	.1102
8	.1640	32	.1437	6	.1380	40	.1218
10	.1900	24	.1629	8	.1640	36	.1460
12	.2160	24	.1889	10	.1900	32	.1697
1-4	.2500	20	.2175	12	.2160	28	.1928
5-16	.3125	18	.2764	1-4	.2500	28	.2268
3-8	.3750	16	.3344	5-16	.3125	24	.2854
7-16	.4375	14	.3911	3-8	.3750	24	.3479
1-2	.5000	13	.4500	7-16	.4375	20	.4050
9-16	.5625	12	.5084	1-2	.5000	20	.4675
5-8	.6250	11	.5660	9-16	.5625	18	.5264
3-4	.7500	10	.6850	5-8	.6250	18	.5889
7-8	.8750	9	.8028	3-4	.7500	16	.7094
1	1.0000	8	.9188	7-8	.8750	14	.8286
1 1-8	1.1250	7	1.0322	1	1.0000	14	.9536
1 1-4	1.2500	7	1.1572	1 1-8	1.1250	12	1.0709
1 1-2	1.5000	6	1.3917	1 1-4	1.2500	12	1.1959
1 3-4	1.7500	5	1.6201	1 1-2	1.5000	12	1.4459
2	2.0000	4 1-2	1.8557	1 3-4	1.7500	12	†1.6959
2 1-4	2.2500	4 1-2	2.1057	2	2.0000	12	†1.9459
2 1-2	2.5000	4	2.3376	2 1-4	2.2500	12	†2.1959
2 3-4	2.7500	4	2.5876	2 1-2	2.5000	12	†2.4459
3	3.0000	*4	2.8376	2 3-4	2.7500	12	†2.6959
3	3.0000	3 1-2	2.8144	3	3.0000	10	†2.9350

59. When is the 3-wire system of checking the pitch diameter of screw threads used?

A. Three wires are laid in the thread groove and measured with an outside micrometer as shown in Figs. 444 and 445, when accurate pitch diameters on thread gages, taps, or pitch diameters beyond the range of available thread micrometers are checked.

Fig. 444

A set of wires consists of three wires having the same diameter within .00003", and a common diameter equal to .57735 divided by the number of threads per inch. Wires of this size will come in contact with the thread on the pitch diameter.

60. Find the micrometer reading over 3 wires (M) for a 3/4 N.C. thread.

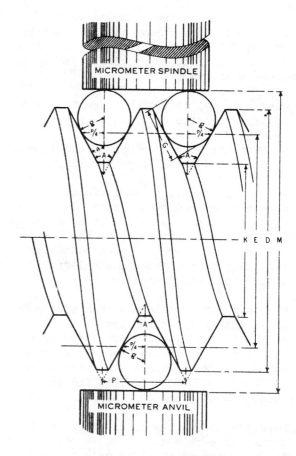

Fig. 445. Three-Wire System of Measuring Pitch Diameter.

M = measurement over wires
K = root diameter
A = angle of thread
D = major diameter of thread
G = diameter of wire
E = pitch diameter

A. Use the formula $M = D + 3G - \dfrac{1.5155}{N}$

To find the wire size to use, read down the column Threads per Inch to the required number of threads; then follow horizontally to the number in the column headed Diameter of Best Size Wires.

M = Measurement over the wires.
G = Diameter of the wire.
D = Major diameter of the thread.
N = Number of threads per inch.

$$M = D + 3G - \frac{1.5155}{N}$$

$$= .750 + 3 \times .05773 - \frac{1.5155}{10}$$

$$= .750 + .1732 - .15155$$

$$= .7717.$$

Threads per Inch	Pitch	Diameter of "Best-Size" Wires	Single Depth Nat. Form Thread	Width of Flat on Crest and Root NC and NF	Single Depth V-Thread	*
4	.250000	.1443375	.162379	.0312	.216506	.2344
4½	.222222	.1282998	.144337	.0278	.192449	.2187
5	.200000	.1154700	.129903	.0250	.173205	.1875
5½	.181818	.1049726	.118094	.0227	.157458	.1719
6	.166666	.0962246	.108253	.0208	.144336	.1562
7	.142857	.0824784	.092788	.0179	.123717	.1406
8	.125000	.0721687	.081189	.0156	.108253	.1250
9	.111111	.0641499	.072168	.0139	.096224	.1094
10	.100000	.0577350	.064951	.0125	.086602	.0938
11	.090909	.0524863	.059047	.0114	.078729	.0938
11½	.086956	.0502040	.056479	.0108	.075306	.....
12	.083333	.0481123	.054126	.0104	.072168	.0781
13	.076923	.0444114	.049963	.0096	.066617	.0781
14	.071428	.0412389	.046394	.0089	.061858	.0726
16	.062500	.0360843	.040594	.0078	.054126	.0625
18	.055555	.0320746	.036084	.0069	.048112	.0563
19	.052631	.0303865	.034185	.0065	.045579	.....
20	.050000	.0288675	.032475	.0062	.043301	.0469
22	.045454	.0262428	.029523	.0057	.039364	.0469
24	.041666	.0240558	.027063	.0052	.036083	.0422
27	.037037	.0213833	.024056	.0046	.032074	.0375
28	.035714	.0206194	.023197	.0045	.030929	.0360
30	.033333	.0192448	.021650	.0042	.028867	.0313
32	.031250	.0180421	.020297	.0039	.027063	.0313
34	.029411	.0169804	.019103	.0037	.025470	.....
36	.027777	.0160370	.018042	.0035	.024055	.....
40	.025000	.0144337	.016237	.0031	.021650	.....
44	.022727	.0131214	.014761	.0028	.019682	.....
48	.020833	.0120279	.013531	.0026	.018041	.....
50	.020000	.0115470	.012990	.0025	.017320	.....
56	.017857	.0103097	.011598	.0022	.015464	.....
64	.015625	.0090210	.010148	.0020	.013531	.....
72	.013888	.0080182	.009021	.0017	.012027	.....
80	.012500	.0072168	.008118	.0016	.010825	.....

*The figures in this column, when subtracted from basic major diameter, will give the nearest commercial standard drill size.

62. Describe the method of inspecting with a bench comparator.

A. Gaging with the bench comparator is a method of obtaining a very accurate comparison between a template and a job, Fig. 446 shows a Jones & Lamson standard bench comparator--inspecting a hob. A great variety of objects; such as hobs, taps, small tools, instrument gages, etc., can be inspected for accuracy by this method.

NUT THREADS

The difference in the size of the minor diameter of the nut and the minor diameter of the screw is called the clearance (see Fig. 447). The allowance for clearance is made in the nut.

$$\text{Clearance} = \frac{\text{Double Depth}}{6}$$

$$\text{Double Depth} = \text{Single Depth} \times 2 = \frac{1.299}{N}$$

$$\text{Single Depth} = \frac{.6495}{N}$$

61. When the size of wire required to gage the thread on the pitch diameter cannot be obtained, state the method of finding the commercial size of wire that can be successfully used to check threads by the three-wire system.

A. When wires are used in conjunction with a micrometer for measuring screw threads, the minimum wire diameter must be such that the wires extend beyond the top of the screw to prevent the micrometer from bearing on the threads instead of on the wires, and the maximum limit must be such that the wires bear on the sides of the thread and not on the crest. The following formulas for determining the wire diameters do not give the extreme theoretical limits, but the smallest and largest sizes that are practicable.

The smallest diameter of wire to use is equal to .56 divided by the number of threads per inch.

The largest diameter of wire to use is equal to .90 divided by the number of threads per inch.

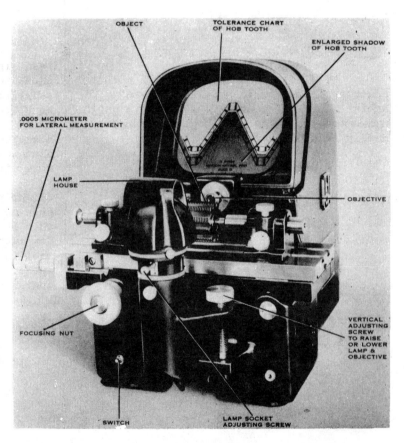

Fig. 446. Bench Comparator--Inspecting a Hob

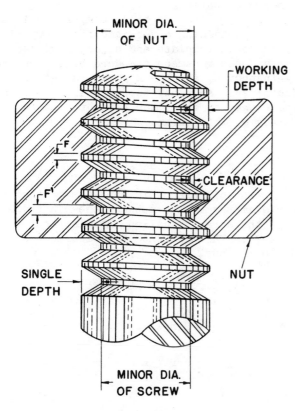

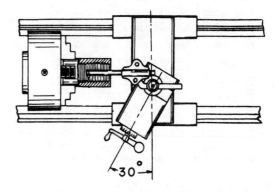

Fig. 449. External Threading

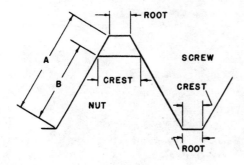

Fig. 447

$$\text{Working Depth} = \frac{.54125}{N}$$

The size to bore or drill a hole for an internal thread is equal to the major diameter minus the quantity, the double depth minus the clearance. This gives approximately an 80% thread.

$$\text{Major Dia.} - \left(\frac{1.299}{N} - \frac{1.299}{6N}\right)$$

$$= \text{Major Dia.} - \frac{1.0825}{N}$$

Fig. 448

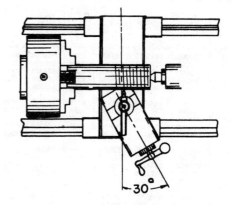

Fig. 450. Internal Threading

The crest of the thread in the nut (see Fig. 448) is larger than the root because the clearance is cut in the nut (see F and F' in Fig. 447). When the clearance is equal to DD ÷ 6, the crest F' is equal to P ÷ 4. When the compound is set at 30°, A = .750 ÷ N and B = .625 ÷ N. This is the distance to force the threading tool into the work with the compound feed.

In practice the compound is set at slightly less than 30°, so that there is clearance on the back edge of the threading tool when it is fed in by the compound. The last few cuts are taken by feeding the threading tool into the work with the cross feed of the lathe. The threading tool must be set on center (see page 140, Lathe Section).

Chapter 15

GEARING

Spur Gearing

A gear is a toothed wheel used to transmit positive and uniform rotary motion from one shaft to another.

Gears are divided into groups and named according to the position which the teeth occupy with respect to the axis of rotation of the gear body. Among the more common types of gears in use we find the SPUR GEAR, BEVEL GEAR, HELICAL GEAR, WORM and WORM GEAR, HERRINGBONE GEAR, and SPIRAL GEAR. Some special gears are used, such as square gears, elliptical gears, planetary gears, and intermittant gears. These special gears all have a definite purpose and are suitable for certain conditions, but are not used nearly as much as the first mentioned group.

SPUR GEARS, Fig. 451, are gears having straight teeth cut parallel with the

Fig. 451

axis of rotation of the gear body. They are used to connect shafts whose axis are parallel.

BEVEL GEARS, Fig. 452, are gears having teeth cut so that they radiate from the apex of a cone and lie on the conical

surface. They are used to transmit motion

Fig. 452

between shafts whose center lines intersect. Miter gears are mating bevel gears having the same number of teeth and whose pitch cone angles are 45°.

HELICAL GEARS, Fig. 453, are gears having teeth cut on a cylinder and at an angle with the axis of rotation of the gear body. These gears are sometimes called spiral gears by mistake. They are used in machines where the shafts do not intersect and where quiet, smooth operation is required. Since the contact action of these gears is sliding rather than a rolling action, they should be run in an oil bath.

Fig. 453

A WORM GEAR is a gear having the teeth cut angular with the axis of rotation of the gear body and radially in the gear

101

face. A worm gear is driven by a WORM which resembles a large screw thread. These gears give very quiet operation and have a long life. They are used extensively as speed reducers. Fig. 454 illustrates a worm and a worm gear.

Fig. 454

HERRINGBONE GEARS, Fig. 455, are gears having helical teeth which diverge from the center of the face toward the sides of the gear body. They have a particular advantage over the other types of gears, in that tooth load or shock is neutralized when it is transmitted from one tooth to another, thus eliminating the necessity of thrust bearings. They also give a more continuous action which is entirely free from tooth vibration such as is found in straight-tooth gears. Herringbone gears are used where high speeds and high gear ratios are necessary.

Fig. 455

SPIRAL GEARS are gears with teeth cut on a conical surface so that they curve continually toward or away from the apex of the cone upon which they are cut. See Fig. 456. These gears closely resemble bevel gears and are frequently called spiral bevel gears. This type of gear can be used satisfactorily for ratios as high as 6 to 1 and give quiet, efficient operation. They are used exten-

sively in the rear axles of automobiles.

Fig. 456

In the formation of gear teeth, two systems are in general use, the INVOLUTE and the INVOLUTE STUB. The involute tooth has a pressure angle of $14\frac{1}{2}°$ and is the form most commonly used. The involute stub, commonly known as the STUB TOOTH, has a standard pressure angle of $20°$; however, some deviations are made from this standard value. The stub tooth has two pitch sizes given in the form of a fraction. The numerator of the fraction indicates the diametral pitch to be used in calculating the chordal thickness and the pitch diameter. The denominator of the fraction is the diametral pitch to be used in calculating the addendum, dedendum, and whole depth.

The pressure angle is, as the name implies, the angle at which the tooth pressure is applied and distributed. Through experience it was found that for maximum efficiency the pressure angle should be $14\frac{1}{2}°$. As this angle of pressure becomes greater the outward strain upon the shaft bearings becomes greater, which increases the friction on the bearings with a consequent shortening of bearing life. It can be readily understood that, where a stub tooth gear is desirable for longer tooth life, special bearings must be installed to compensate for the added friction and to maintain a correct center distance. In Fig. 457 "Z" indicates the pressure angle.

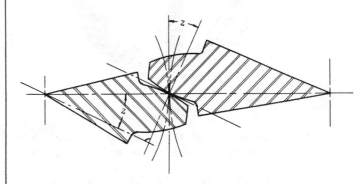

Fig. 457

In order to calculate gearing properly, a thorough knowledge of the terms used is necessary. For example, one should know exactly what the term PITCH means with respect to gears. In this case pitch means frequency. The DIAMETRAL PITCH means the frequency with which the teeth occur in 3.1416 inches measured on the pitch circumference. It should be remembered that the term DIAMETRAL PITCH means a quantity; whereas PITCH DIAMETER is a distance which depends on the value of the diametral pitch. If any of these terms are confused it may result in the use of the wrong formula; consequently the calculations will be in error. When the CENTER DISTANCE is spoken of, one should think of two circles either externally or internally tangent to each other. The distance between the centers of these circles is the CENTER DISTANCE. This dimension is very important, because the gears must operate freely and quietly at exactly this distance.

The following definitions cover the terms most frequently used in spur gearing:

The PITCH CIRCLE of a gear is an imaginary circle located about half-way down the teeth, where the teeth of both gears contact each other.

The PITCH DIAMETER is the diameter of the pitch circle, and is important because it regulates the pitch circle or the point of contact between two gears.

The DIAMETRAL PITCH is the number of teeth in 3.1416 inches measured on the pitch circle, or the number of teeth for each inch of pitch diameter. It regulates the size of the tooth.

The CIRCULAR PITCH is the distance from the center of one tooth to the center of the next consecutive tooth measured on the pitch circle.

The ADDENDUM is that portion of the tooth which projects above or outside of the pitch circle.

The DEDENDUM is that portion of the tooth between the pitch circle and the root circle and is equal to one addendum plus the clearance.

The ROOT CIRCLE is the circle formed by the bottoms of the teeth.

The WHOLE DEPTH is the distance from the top of the tooth to the bottom and consists of two addendums plus the clearance.

The BASE CIRCLE is, as the name implies, the basic circle upon which the teeth are mechanically constructed or drawn in the drafting room. It also forms an important part in calculating the pressure angle. To find the radius of the base circle multiply the pitch radius by the Cosine of the pressure angle.

It must be remembered that the diametral pitch regulates the size of the tooth much the same as the number of threads per inch regulates the size of a screw thread. The diametral pitch is always given in the form of a number as, 2, 3, 4, 8, etc.; and as this number gets larger the size of the tooth becomes smaller. For example a 2 diametral pitch gear would have a much larger tooth than a 10 diametral pitch gear, because in the case of the 10 diametral pitch more teeth are being crowded into 3.1416 inches. (TWO GEARS MUST BE OF THE SAME DIAMETRAL PITCH BEFORE THEY CAN MESH) For example two 7 diametral pitch spur gears will mesh, but a 7 diametral pitch gear will not mesh with a 10 diametral pitch gear.

When motion is transmitted from one shaft to another through the medium of gear teeth, it is of the utmost importance that they be actually as well as theoretically correct. To check the size of a gear tooth to any degree of accuracy, it must be measured on a straight line from where the pitch circle touches the tooth (a) on one side (see Fig. 458) to where it leaves the tooth

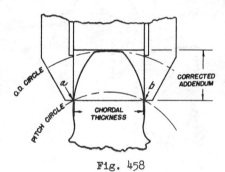

Fig. 458

on the other side (b). This thickness is known as the CHORDAL THICKNESS, because geometrically it is the length of a chord. In order to measure the true length of this chord the corrected addendum must be used.

The CORRECTED ADDENDUM is the perpendicular distance from the chord at the pitch circle to the top of the gear tooth. Why this corrected addendum must be used can be readily understood from the sketch in Fig. 458. A calculated distance equal to the height of the arc above the chord when added to "S" gives the distance "H". When the addendum beam on the gear tooth vernier caliper is set at this distance (H), it permits the jaws to contact the gear tooth exactly at the pitch line. The height of the

arc above the chord for the different diametral pitches and for varying numbers of
teeth have been calculated and charted. It
is convenient to use the values from these
charts when checking gear tooth parts.

        To test the thickness of a gear
tooth with a gear tooth vernier caliper, the
addendum beam is set for the corrected addendum. Then, after the burrs have been
stoned from the face of the tooth, the addendum beam is set on top of the tooth as
shown in Fig. 459. The caliper jaws are ad-

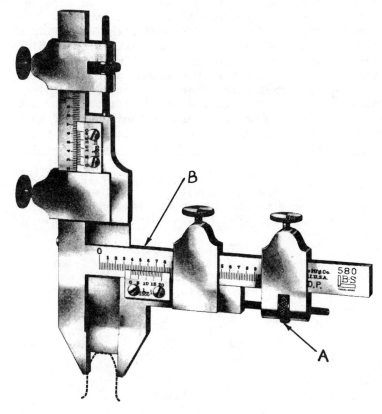

Fig. 459

justed to the tooth thickness by means of
the knurled nut "A". The chordal thickness
can then be read from the "B" scale.
        CAUTION: Make certain that the
vernier graduations are read correctly. The
smaller sizes of calipers have graduations
equal to .020 of an inch, while the larger
sizes have graduations equal to .025 of an
inch.
        It must be remembered that this tool
checks the corrected addendum and chordal
thickness but does not check the pitch diameter of a gear. To check the pitch diameter of a gear accurately the diameters of
plug gages are calculated so that they

will be tangent to the teeth at the pitch
circle. Then by means of trigonometry the
distance over the outside of these plugs is
calculated and measured with a micrometer.
        The following notations and abbrevia-
tions will be used in the formulas for spur
gear calculations:

$$Ng = \text{Number of teeth in the gear}$$
$$Np = \text{Number of teeth in pinion}$$
$$P = \text{Diametral Pitch}$$
$$P' = \text{Linear or Circular Pitch}$$
$$D = \text{Pitch Diameter}$$
$$O = \text{Outside Diameter}$$
$$H = \text{Corrected Addendum}$$
$$I = \text{Inside Diameter}$$
$$S = \text{Addendum}$$
$$S + F = \text{Dedendum}$$
$$F = \text{Clearance}$$
$$W = \text{Whole Depth}$$
$$C = \text{Center Distance}$$
$$T = \text{Chordal Thickness}$$
$$L = \text{Length of Rack}$$
$$Z = \text{Pressure Angle}$$

        The following rules and form-
ulas will be used in calculating the
dimensions for spur gears:

To Find	Formula
Diametral Pitch	$P = \dfrac{3.1416}{P'}$
Diametral Pitch	$P = \dfrac{Ng + Np}{2C}$
Diametral Pitch	$P = \dfrac{N + 2}{O}$
Diametral Pitch	$P = \dfrac{N}{D}$
Circular Pitch	$P' = \dfrac{3.1416}{P}$
Pitch Diameter	$D = \dfrac{N}{P}$
Pitch Diameter	$D = O - 2S$
Outside Diameter	$O = \dfrac{N + 2}{P}$
Outside Diameter	$O = D + 2S$
Number of Teeth	$N = P \times D$
Addendum	$S = \dfrac{1}{P}$
Addendum	$S = \dfrac{P'}{3.1416}$
Dedendum	$S + F = \dfrac{1.157}{P}$
Clearance	$F = \dfrac{.157}{P}$
Clearance	$F = \dfrac{3.1416}{20 \times P}$

To Find	Formula
Whole Depth	$W = \dfrac{2.157}{P}$
Whole Depth	$W = .6866 \times P'$
Center Distance	$C = \dfrac{Ng + Np}{2P}$
Length of Rack	$L = NP'$

For internal spur gears use the following formulas:

To Find	Formula
Inside Diameter of Internal Gears	$I = \dfrac{N - 2}{P}$
Center Distance of Internal Gears	$C = \dfrac{Ng - Np}{2P}$
Diametral Pitch of Internal Gears	$P = \dfrac{Ng - Np}{2C}$

In the following gear tooth parts chart, "N" is the number of teeth in the gear to be cut, "T" is the chordal thickness, and "H" is the corrected addendum of a ONE diametral pitch gear.

N	T	H
8	1.5607	1.0769
9	1.5628	1.0648
10	1.5643	1.0616
11	1.5654	1.0559
12	1.5663	1.0514
14	1.5675	1.0444
17	1.5686	1.0362
21	1.5694	1.0294
26	1.5698	1.0237
35	1.5702	1.0176
55	1.5706	1.0112
135	1.5707	1.0047

RACK TOOTH

The Diametral Pitch (Pitch or P.) indicates the size of the tooth.

$$\text{Addendum} = \frac{1}{P} \qquad \text{Dedendum} = \frac{1.157}{P}$$
$$\text{Linear Pitch} = \frac{\pi}{P}$$

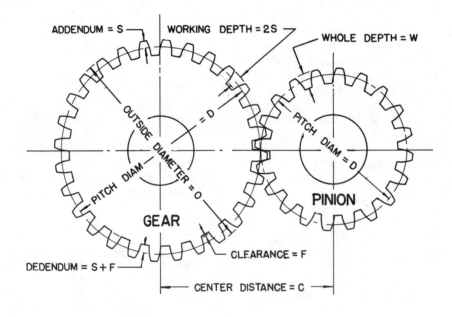

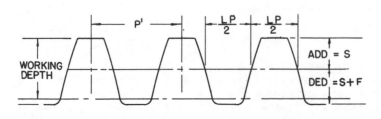

Fig. 460

Working Depth $= \dfrac{2}{P}$ Whole Depth $= \dfrac{2.157}{P}$

Linear or Circular Pitch = P'

To find the chordal thickness or corrected addendum of a gear, divide the number under "T" or "H", corresponding to the nearest number of teeth in the gear, by the diametral pitch. Example: Find the chordal thickness and corrected addendum for a gear having 74 teeth and a diametral pitch of 8. Referring to the above chart, find the number of teeth nearest to 74, which is 55. Opposite 55 under "T" find the value 1.5708, which is the chordal thickness for a ONE diametral pitch gear. Divide 1.5708 by 8 to get .1963, which is the chordal thickness required. Opposite 55 under "H" find 1.0112 and divide it by 8 to get .1264, which is the corrected addendum required. Therefore an 8 diametral pitch gear having 74 teeth will have a chordal thickness of .1963 and a corrected addendum of .1264.

In using a milling machine for cutting the teeth in a gear, it is necessary to use a rotary gear cutter. This rotary cutter must be selected so that it will give the gear teeth the correct curvature. The cutters are selected according to numbers, depending on the number of teeth in the gear to be cut and the diametral pitch. The cutter number merely represents the curvature on the sides of the teeth. Therefore when milling the teeth in a gear with a rotary milling cutter it is necessary to give the cutter number on the tool crib requisition.

The following chart gives the numbers of rotary cutters which will cut from 12 teeth up to and including a rack:

```
No. 1 for 135 teeth to a rack
 "  2  "   55   "     " 134 teeth
 "  3  "   35   "     "  54   "
 "  4  "   26   "     "  34   "
 "  5  "   21   "     "  25   "
 "  6  "   17   "     "  20   "
 "  7  "   14   "     "  16   "
No. 8 for  12 teeth to  13 teeth
```

The following problems illustrate the use of the notations and formulas given on pages 104 and 105.

PROBLEM:

Two gears, A and B, have 48 and 27 teeth respectively and are of 12 diametral pitch. In gear A find 1) the outside diameter, 2) the chordal thickness, 3) the corrected addendum. In gear B find 4) the circular pitch, 5) the pitch diameter, 6) the

number of the rotary gear cutter used for milling the teeth.

Procedure:

According to the formulas:

1. $0 = \dfrac{N + 2}{P}$; $0 = \dfrac{48 + 2}{12}$; $0 = \dfrac{50}{12}$;

    $0 = 4.1666.$

2. In solving for the chordal thickness, "T", refer to the table of gear tooth parts on page 105. The given number of teeth for which the chordal thickness must be found is 48; the nearest number of teeth to 48 is 55. Taking the value for "T" opposite 55 and dividing by the diametral pitch, 12, gives .1309 as the chordal thickness of a 12 pitch gear having 48 teeth.

3. To solve for the corrected addendum refer to the table of gear tooth parts on page 105. The number of teeth nearest to 48 is 55; opposite 55, under "H" find the value 1.0112. Divide this value by the diametral pitch, 12, which gives .0842 as the corrected addendum of a 12 diametral pitch gear having 48 teeth.

4. $P' = \dfrac{3.1416}{P}$; $P' = \dfrac{3.1416}{12}$; $P' = .2618$

5. $D = \dfrac{N}{P}$; $D = \dfrac{27}{12}$; $D = 2.25$

6. Referring to chart, find what number corresponds to a 26 tooth gear. In this case it is found that a #4 cutter will cut from 26 to 34 teeth; therefore a #4 rotary cutter of 12 pitch should be taken from the crib to cut the number of teeth in the pinion.

PROBLEM:

A gear has 53 teeth and a pitch diameter of 3.7857". Calculate the following: 1) Find P; 2) find 0; 3) find W; 4) find T; 5) find P'; 6) find S; 7) find H; 8) find F; 9) find S + F; 10) What number of cutter should be used for cutting this gear?

Procedure:

1) $D = \dfrac{N}{P}$; $3.7857 = \dfrac{53}{P}$; $3.7857\,P = 53$;

   $P = \dfrac{53}{3.7857}$; $P = 14$

2) $0 = \dfrac{N + 2}{P}$; $0 = \dfrac{53 + 2}{14}$; $0 = \dfrac{55}{14}$;

   $0 = 3.9285$

3) $W = \frac{2.157}{P}$; $W = \frac{2.157}{14}$; $W = .15407$

4) $T = \frac{1.5708}{P}$; $T = \frac{1.5708}{14}$; $T = .11213$

5) $P' = \frac{3.1416}{P}$; $P' = \frac{3.1416}{14}$; $P' = .2244$

6) $S = \frac{1}{P}$; $S = \frac{1}{14}$; $S = .0714$

7) $H = \frac{1.011}{P}$; $H = \frac{1.011}{14}$; $H = .0722$

8) $F = \frac{.157}{P}$; $F = \frac{.157}{14}$; $F = .0112$

9) $S + F = \frac{1.157}{P}$; $S + F = \frac{1.157}{14}$;

   $S + F = .08264$

10) For 53 teeth select a #3 cutter.

PROBLEM:

Two gears, A and B, are of 9 diametral pitch. Gear A has an outside diameter of 8.1111". Gear B has 38 teeth. Calculate (1) the number of teeth for gear A, (2) the center distance, (3) the circular pitch, (4) the whole depth, (5) the pitch diameter of gear A, (6) the outside diameter of gear B, (7) the dedendum, (8) the pitch diameter of gear B, (9) the cutter number for gear A, (10) the cutter number for gear B.

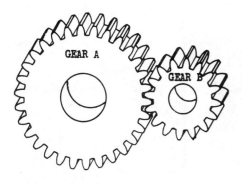

GEAR A

GEAR B

Fig. 461

Procedure:

1) $O = \frac{N + 2}{P}$; $8.1111 = \frac{N + 2}{9}$; $N + 2$

   $= 72.9999$; $N = 72.9999 - 2$ or $N = 71$ teeth

2) $C = \frac{Ng + Np}{2P}$; $C = \frac{71 + 38}{2 \times 9}$; $C = \frac{109}{18}$;

   $C = 6.0555$

3) $P' = \frac{3.1416}{P}$; $P' = \frac{3.1416}{9}$; $P' = .34906$

4) $W = \frac{2.157}{P}$; $W = \frac{2.157}{9}$; $W = .23966$

5) $D = \frac{Ng}{P}$; $D = \frac{71}{9}$; $D = 7.8888$

6) $O = \frac{Np + 2}{P}$; $O = \frac{38 + 2}{9}$; $O = 4.4444$

7) $S + F = \frac{1.157}{P}$; $S + F = \frac{1.157}{9}$;

   $S + F = .12855$

8) $D = \frac{Np}{P}$; $D = \frac{38}{9}$; $D = 4.2222$

9) For 71 teeth select a #2 cutter.

10) For 38 teeth select a #3 cutter.

The following problems are typical of those encountered in the shop and drafting rooms and should be worked out very carefully:

1. Make a working drawing of each gear of a pair of spur gears, having a 2 to 1 ratio and a 3" center distance. Choose bore and pitch to suit.

2. Make a working drawing of each gear of a pair of spur gears, having a 2 to 1 ratio and a 2 17/32" center distance. Choose bore and pitch to suit.

3. A rack measures approximately 29 1/32" over 37 teeth. Calculate all dimensions necessary for drawing a gear that will move the rack 25 1/8" when the gear makes one revolution.

4. A spur gear having a 1" face, a 2.083" outside diameter, and 23 teeth is mounted on a pump shaft which is run at 300 R.P.M. This gear is driving a second gear running at 100 R.P.M. Calculate the dimensions necessary for both gears.

5. Calculate the dimensions necessary to make drawings of each gear of a pair of stub tooth gears, 4-6 pitch, 2 to 1 ratio, if the center distance is 4.5".

6. A gear revolving through an angle of $112\frac{1}{2}°$ causes a rack to move 4 11/64". Calculate the dimensions necessary for drawing the gear.

7. A gear having 75 teeth and an outside diameter of 9.625" runs at 100 R.P.M. The pinion runs at 312.5 R.P.M. Calculate all the dimensions for both gears.

8. What diametral pitch is equivalent to .52432" circular pitch?

9. What number of teeth should a gear contain if the outside diameter is 18" and the diametral pitch is 8?

10. What is the outside diameter of a gear having 70 teeth and an 8 diametral pitch?

11. Find the chordal thickness and corrected addendum of a gear having 36 teeth and 5 diametral pitch.

12. The center line of a shaft running at 700 R.P.M. is located 6½" from the center of another shaft that is to run at 300 R.P.M. Make a detail drawing of each gear of a pair of spur gears to connect shafts. Find necessary dimensions.

13. What is the diametral pitch of a gear if the circular pitch is .7854?

14. How many teeth should there be in a spur gear if the outside diameter is 18" and the diametral pitch is 6?

15. What is the outside diameter of an 8 diametral pitch spur gear having 70 teeth?

16. Two 8 diametral pitch spur gears have 24 and 72 teeth respectively, what is their center distance?

17. The ratio of two spur gears is 4 to 5 and the center distance is 27". What number of teeth does each gear have if the diametral pitch is 5?

18. The speed ratio of two spur gears is 3 to 4, the diametral pitch is 10 and the center distance is 21". Find the outside diameter, the pitch diameter, and the number of teeth in each gear.

19. Find the pitch of a spur gear which has 80 teeth and an outside diameter of 16.400".

20. If a spur gear has 140 teeth and an outside diameter of 10 1/7", what is its diametral pitch?

Figures 462 to 465 show some of the methods of producing gears.

Fig. 462. Gear Hobbing Machine for Spur and Helical Gears.

Fig. 463. Hobbing Spur Gears.   Fig. 464. Hobbing Helical Gears.   Fig. 465. Hobbing Sprockets

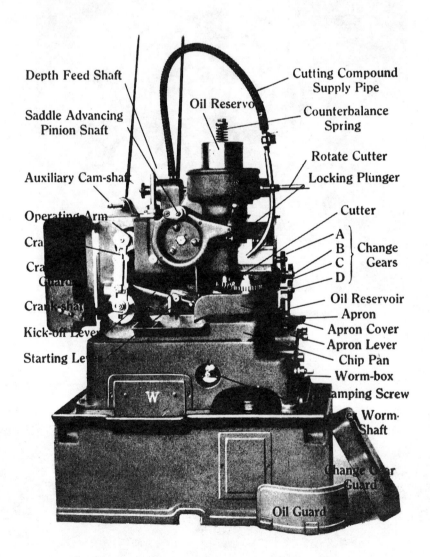

Fig. 466. Front View of Fellows Gear Shaper with All Guards Removed.

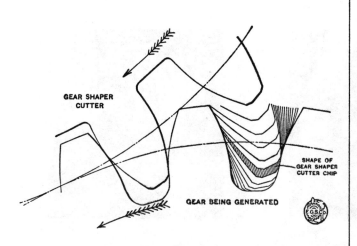

Fig. 467
Shows the action of the cutter and work.

Fig. 466 shows the principal operating members of a Fellows Gear Shaper. This machine is designed to cut one gear at a time using high reciprocating cutter speeds combined with a short cutting stroke.

The Gear Shaper Cutter, Fig. 467, can take a coarser feed than the milling machine type of cutter. The cutter and work are shown rotating together in the direction of the arrows. The cutter is given a reciprocating motion at right angles to the illustration, similar to that of a planning tool.

The outlines show the various positions which the cutting edge will occupy for each successive stroke. The distance between any two adjacent outlines at any point is the thickness of the chip at that point.

Fig. 468 shows the Gleason Bevel Gear Generator used for machining the teeth on bevel gears, a two-tool machine for rough and

Fig. 468.. 12" Straight Bevel Gear Generator.

Fig. 469   Fig. 470   Fig. 471

finish cutting.

Fig. 469 shows an Involute Form cutter used on a milling machine for cutting gear teeth. Fig. 470 shows the kind of cutter used for producing gear teeth by the gear shaper method, while in Fig. 471 is a Hob, used for producing gear teeth by the hobbing method.

Gearing Questions

1. What is a gear?
2. From what factors do gears derive their names?
3. Name six kinds of gears that are commonly used.
4. Name three kinds of special gears.
5. What is a spur gear?
6. What are bevel gears? For what are they used?
7. Are all bevel gears miter gears? Explain.
8. What is the difference between a spiral gear and a helical gear?
9. What advantages do helical and spiral gears have over spur and bevel gears?
10. What is a worm gear?
11. What is the purpose of a worm and worm gear?
12. What is a herringbone gear?
13. What particular advantages do herringbone gears possess?
14. When is it advisable to use herringbone gears?
15. What are spiral gears?
16. How do spiral gears differ from helical gears?
17. Where are spiral gears generally used?
18. What is meant by the involute form of gear tooth?
19. Why are involute gear teeth used most?
20. What is a stub tooth gear?
21. How is the pitch of a stub tooth gear generally specified?
22. Under what conditions is it advisable to use a stub tooth gear?
23. What are the disadvantages of the stub tooth gear?
24. What is meant by the pressure angle of a gear?
25. Why is the pressure angle important?
26. What is meant by the pitch of a gear?
27. Why is it necessary to have a good understanding of the terms used in gearing?
28. What is the difference between diametral pitch and circular pitch?
29. Could gears of different diametral pitches be made to run with each other? Why?
30. Why is the pitch diameter of a gear important?
31. How does the length of the addendum affect the chordal thickness?
32. Under what conditions would it be necessary to know the diameter of the base circle?
33. Why is the center distance important?
34. How are diametral and circular pitch sizes generally specified?
35. Could a 4 diametral pitch gear by made to run with an 8 diametral pitch gear? Why?
36. What is the difference between the addendum and the corrected addendum?
37. Explain how to check the chordal thickness of a gear tooth with a gear tooth vernier caliper.
38. Under what conditions would it be possible to have the chordal thickness correct and yet not have the gears run together properly?
39. For precision gear work how should the gear be checked?
40. How is the chordal thickness determined for a gear having a definite number of teeth and diametral pitch?
41. How is the corrected addendum determined for a gear having a definite number of teeth and diametral pitch?
42. In using the milling machine for cutting the teeth in a gear what kind of a cutter is used?
43. What two things govern the selection of the cutter?
44. Why is it necessary to have a great many cutters for milling the teeth in gears of the same diametral pitch?
45. Give three methods of cutting gears.

Chapter 16

CUTTING TOOLS

The successful operation of a lathe, shaper, planer, etc., turning out work of good quality, depends to a great extent on the skill of the operator in grinding his tools. Dull and improperly ground tools throw a heavy strain on the feed mechanism, causing the work to spring and the machine to chatter.

Cutting tools are made of carbon steel, high speed steel, and alloys such as stellite and tungsten carbide. The stellite and tungsten carbide tools are used only on high production work because their cost is too great for universal use. Some carbon steel tools are still used but the general practice is to use high speed steel tool bits in holders. A quick and simple method to tell whether a tool is carbon or high speed steel is to grind the end and watch the sparks. The wheel will throw light-colored sparks on carbon steel and dark red sparks on high speed steel.

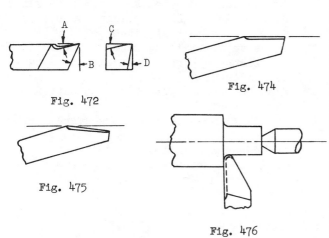

Fig. 473

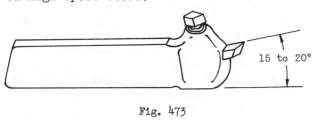

Fig. 472

Fig. 474

Fig. 475

Fig. 476

All cutting tools should be ground with a wet grinding wheel, or if ground dry, they should be frequently dipped in water to prevent annealing of the cutting edge.

Four angles are important in grinding tools. These angles, shown in Fig. 472, are the top rake (A), side rake (C), front clearance (B), and side clearance (D). The angles vary with the material being machined. (NOTE. The rake angles deal with the top of the tool bit and the clearance angles deal with the side and front.)

The top rake is usually provided for in the tool holder by setting the tool on an angle. This angle is usually 15° but it may be as high as 20°, as shown in Fig. 473. These rake angles are correct for the machining of steel and cast iron, but on forged tools it is necessary to grind the top rake in the tool. This top rake can be varied some to suit the material being turned by adjusting the tool in the post through the rocker. The softer the material the less the top rake should be, as there is a tendency for the tool to dig in if the rake is too great.

There should be no top rake for turning bronze, as shown in Fig. 474, and the cutting edge of the tool should be about horizontal. A negative rake (Fig. 475) is often used for turning soft copper, babbitt, and some die casting alloys.

The side rake (Fig. 472, C) also varies with the material being machined. With this angle large enough, the tool will drag the carriage along by feeding into the work of its own accord, especially if the material is soft. The tool would not cut without the side rake and the feed mechanism would be under excessive strain. The proper angle is from 6° for soft material to 15° for steel. A tool is ground with side clearance (Fig. 472, D) to take care of the feed advance and prevent the tool from dragging on the shoulder formed by the cut (see Fig. 476). This angle is about 6° from the vertical and is constant.

The front clearance (Fig. 472, B) depends somewhat on the diameter of the work to be turned. If a tool were ground square, without any front clearance, it would not cut, as the material being turned would rub on the cutting tool just below the cutting edge. The front clearance is necessary to eliminate this interference. This clearance, which should be less for small diameters than for large diameters, may range from 8° to 15°. Do not grind more front clearance than is necessary, as this takes away the support from the cutting edge of the tool.

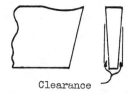

Clearance

Fig. 477

Tool bits should not be ground in their holders.

After a tool has been ground on a grinding wheel, it will produce better work and last longer if the cutting edge is stoned with an oil stone. This takes away the wheel marks and gives a smooth cutting edge. In grinding cutting off tools, be careful to see that both sides of a tool have the necessary clearance. A tool of this kind cuts better if the lip is ground back of the cutting edge (see Fig. 477), to curl the chip as it comes off the work.

The cutting edge of a tool must be set on the center line of the work, as shown in Fig. 478. This illustration also shows a tool properly ground and set for cutting the harder metals, such as tool steel and high carbon steel. Note the top and side rake.

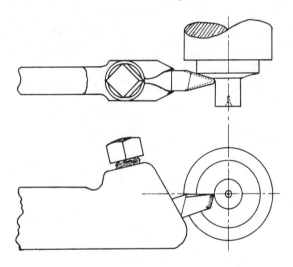

Fig. 478

The helix angle is the angle made by the helix of the thread at the pitch diameter with a plane perpendicular to the axis. Because of the helix angle, the square threading tool cannot be ground like the cutting off tool. The part of the tool bit which supports the cutting edge must be ground to the helix angle so that the bottom of the tool will not rub against the sides of the thread. The helix angle is found by the method shown in Fig. 480. The line FG equals the lead of the thread. EF equals the circumference of the pitch diameter and is drawn at right angles to FG. Angle B is the helix angle to which the tool must be ground. The tool must also be given clearance on the sides, as shown in Fig. 480.

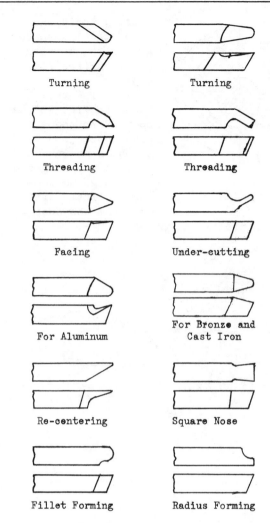

Turning   Turning

Threading   Threading

Facing   Under-cutting

For Aluminum   For Bronze and Cast Iron

Re-centering   Square Nose

Fillet Forming   Radius Forming

Fig. 479. Shapes of Turning Tools

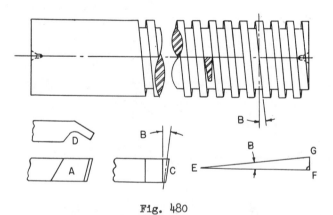

Fig. 480

In grinding boring tools see that the front clearance is sufficient to prevent the tool from rubbing in the hole and dragging at the point shown in Fig. 481. The boring tool should be ground as shown in Fig. 482. The front clearance angle of shaper and planer

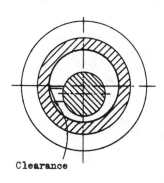

Fig. 481

Clearance

cutting tools should be 3° to 4° when they are in the cutting position (see Fig. 483).

The tool bits shown in Fig. 484 can be used to shape and plane most jobs. If a job is an unusual one, it may be necessary to grind the form a little different from the ones shown in this illustration.

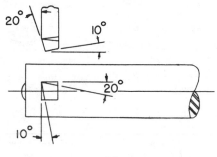

Fig. 482

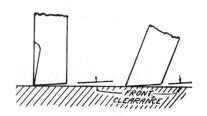

Fig. 483

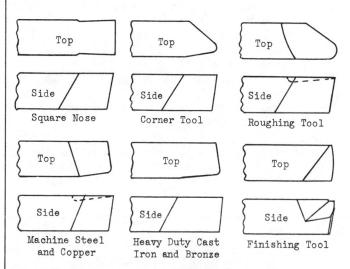

Fig. 484. Common Types of Shaper and Planer Bits

Chapter 17

SHAPER

1. What is a shaper?

A. A shaper is a machine that has a ram with a reciprocating motion. The ram holds the apron and tool. A chip is peeled off the work on the forward stroke. An adjust-able table with T slots holds the work, vise, and other fixtures for holding the work. Fig. 485 shows one of the types of shapers used in the Trade School shop.

Shapers are divided into two distinct

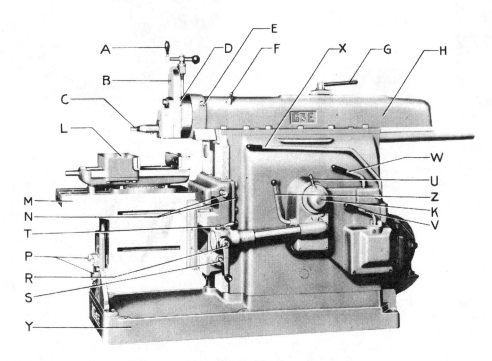

Fig. 485
Nomenclature of Gould and Eberhardt Shaper

A - TOOL HEAD SLIDE CONTROL - For positioning slide vertically.

B - TOOL HEAD SLIDE

C - TOOL POST

D - TOOL SLIDE LOCK - For clamping slide in vertically adjusted position.

E - TOOL HEAD LOCK - For clamping head in adjusted position.

F - RAM POSITIONING CONTROL

G - RAM CLAMP - Clamp ram to link and lever after positioning.

H - RAM

K - STROKE LENGTH CONTROL - Automatically locked. Positions crank pin in crank plate to obtain stroke lengths.

L - VISE

M - REGULAR BOX WORK TABLE - Called a Universal Table when constructed to swivel 360° and to have one surface that tilts.

N - RAIL CLAMP - For locking cross rail in adjusted position.

P - TABLE SUPPORT CLAMP AND LOCK

R - TABLE HORIZONTAL POSITION CONTROL - For positioning and feeding table manually.

S - TABLE VERTICAL POSITION CONTROL - For raising and lowering work table.

T - FEED DIRECTION CONTROL - Neutral position disengages feed.

U - FEED SELECTOR

V - TRANSMISSION SPEED SELECTOR - For obtaining eight ram speeds.

W - BACK GEAR SELECTOR - Used in conjunction with V.

X - CLUTCH CONTROL - For starting and stopping. Operates brake when stopping machine.

Y - BASE

Z - POWER RAPID TRAVERSE CONTROL - Moves table horizontally in direction opposite to feed set.

classes, namely, crank and geared. The most
common type found in the toolroom is the

have a tendency to hinge in a direction away
from the cut.

Fig. 487

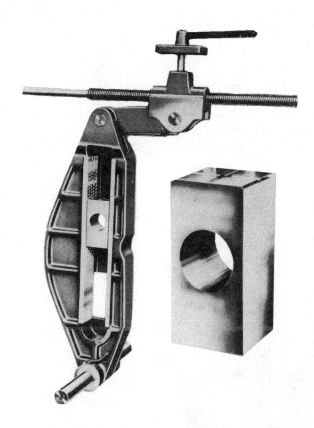

Fig. 486

crank shaper, shown on page 115. Figs. 486
and 487 show the rocker arm which drives the
ram, and the mechanism for regulating the
length of the stroke. The ram supports the
tool head. The head carries the downfeed
mechanism and will swivel from side to side
to permit the cutting of angles. This is
generally a hand feed, but some shapers are
equipped with a power downfeed in addition to
the regular hand feed.

     The table of the shaper is of box
form with "T" slots on the top and sides, for
clamping work. Some shapers have a universal
table, shown in Fig. 489, which can be
swiveled through an arc of about 180°. This
type of table also has a top which can be
tilted approximately 15°.

     The apron is fastened to the front of
the shaper head and consists of the clapper
box, tool block, hinge pin, and tool post.
It can be made to swivel, and when taking
vertical or angular cuts the upper part of
the apron is turned in a direction away from
the surface of the cut. In this position the
tool will not dig into the work, but will

The cross rail is bolted directly to
the frame or column of the shaper with bolts.
When these bolts are released, the table may
be adjusted up or down to suit the job being
machined. When the table adjustment is made
it is also necessary to readjust the supporting leg or table support from the base
of the machine to the table. This supports
the weight of the table.

     The table is operated horizontally,
either by hand or power, by the feed screw
which runs through the cross rail. When the
hand feed is desired, the feed screw is operated by a crank which is placed on the
squared end of the screw. The automatic
feed or power feed is obtained by a pawl
which engages in a notched wheel or ratchet.
This is either fastened to the feed screw or
transmits its motion to the screw through
gears. This mechanism causes the feed screw
to make part of a revolution each time the
ram makes a stroke. On the crank shaper the
feed screw should always be made to operate
on the return stroke of the ram.

2. How is the size of the shaper determined?
   A. By the size cube that it will shape.
A 24-inch shaper, for example, has a stroke
and transversal table feed of 24 inches.
The downfeed, however, is only a few inches
and the cube must therefore be turned each
time a new surface is shaped.

3. What is the difference between a Plain
   and a Universal shaper?

A. The main difference is the type of table each has. Fig. 488 shows the table of the plain shaper. This table cannot be swiveled or tilted. Fig. 489 shows the table

machining. The job must be machined to the dimensions given on the blue print. If the metal is a casting on which the line has been scribed from the outline of a template,

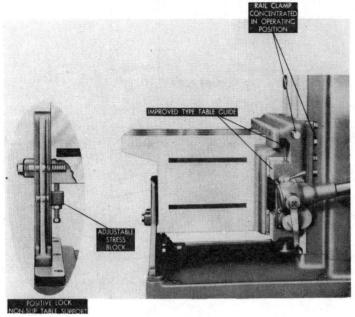

Fig. 488

Fig. 489

of a universal shaper. This table can be tilted to 15° and swiveled through an arc of 180°. The universal type is commonly used in toolrooms for machining angular cuts.

4. What are some of the first things to do before starting a cut on a shaper?
   A. Be sure that you thoroughly understand the blue print and work; see that all tools are clear of working or moving parts, that the job is held securely, that the machine is well oiled and in condition to operate, that parallels are both of the same height, and that you have enough stock.

5. What are three essential things to remember while operating a shaper?
   A. Wear your goggles, always be careful, and keep your mind on your work.

6. What is the first step in machining a job having several cuts and angles?
   A. Check the sketch or blue print and see if the job is understood clearly.

7. What is the purpose of layout lines on a job?
   A. Layout lines are generally used as a guide to show the amount of material left for

then in machining the tool point must split the layout line. On a job of this type, consult your instructor or foreman.

8. Sketch some shaper and planer tool bits.
   A. See Fig. 490.

9. In what respect does a shaper or planer tool differ from a lathe tool?
   A. Owing to the fact that the shaper tool feeds into the work on the return stroke only, the side clearance angle need not be nearly as great as that of the lathe tool, which continually feeds into the work.

10. How should the tool holder be set in the tool post?
    A. Set the tool holder so that the tool bit does not extend farther than two inches from the tool post, to avoid chattering.

11. How much clearance should there be between the work in the vise and the ram?
    A. About two inches, or enough to clear the hand.

12. What tools are used to hold work squarely in the vise?
    A. Use parallels or hold-downs. Place true side of work against the solid jaw, and insert a parallel or hold-down between the

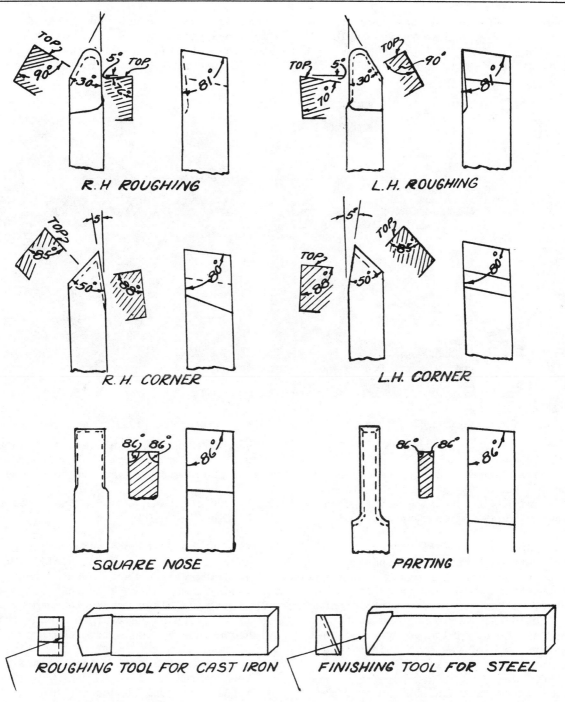

R.H ROUGHING   L.H. ROUGHING

R.H. CORNER   L.H. CORNER

SQUARE NOSE   PARTING

ROUGHING TOOL FOR CAST IRON   FINISHING TOOL FOR STEEL

Tool with broad cutting edge for   Tool with slightly rounded cutting
cutting cast iron, using coarse   edge for cutting steel, using coarse
feed. Wide cutting edge gives a   feed. Rounded cutting edge gives a
smooth finish.   shearing cut.

Fig. 490

other side of work and movable jaw.

13. State the operations in shaping a rec-
    tangular job.
    A. Shape the thickness first, then the
width, and then the length. After shaping

the thickness, it is necessary to shape the
width square with the finished surface. Place
job in horizontal plane, one side of the
thickness against the solid jaw. Then place
parallel or hold-down against the other side
and tighten vise. After the width has been

squared with the thickness, place the finished width on parallels and hold with hold-
down, forcing job against solid jaw.

     Next, the job must be squared and
shaped to length. Swivel vise to 90° with
stroke of ram and place job in end of vise.
The end to be shaped should be placed in
the horizontal plane. The other end of the
vise must be blocked with a piece of material of like thickness, to maintain parallelism of vise jaws and to hold the work
more securely. Place width of job parallel
with stroke of ram; the solid jaw will hold
job square in this direction. With the aid
of the solid square placed on the bottom of
the vise and against the work, true work and
clamp tightly. This will bring end of job
square with thickness.

14. How should the tool be started when tak-
     ing the first cut?
     A. The shaper is started and the length
of the stroke adjusted for length and position, to give the tool a movement somewhat
greater than total length of work. The tool
is fed downward into the work and the latter
is moved crosswise by hand until the desired
depth is started. The automatic feed is
then engaged by dropping the feed-pawl into
mesh with the ratchet gear on lead screw.

15. On what stroke should the table feed?

     A. The table should feed on the return
stroke.

16. What are the most common causes of chat-
     tering?
     A. Tool clearance too great, tool sus-
pended too far from tool holder, work not
held rigidly in vise, and ram gibs not in
proper adjustment.

17. How can chattering be eliminated?
     A. By either of the following adjust-
ments: Sharpen the tool with from 2° to 3°
clearance and clamp the work rigidly in the
vise, or see that the gibs are properly adjusted.

18. Explain how to set the stroke of ram for
     length and for position.
     A. When setting length of stroke, bring
ram to the extreme back position and set for
length of work plus about 5/8 of an inch as
shown on the graduated scale on machine.
This will allow for about 1/8 of an inch
clearance in the front of the work and 1/2 of
an inch in the back. For position of stroke,
bring ram to extreme forward position and adjust it so that front of tool bit will clear
front of job by about 1/8 of an inch.

19. How may the solid jaw of the vise be
     checked for being square and parallel

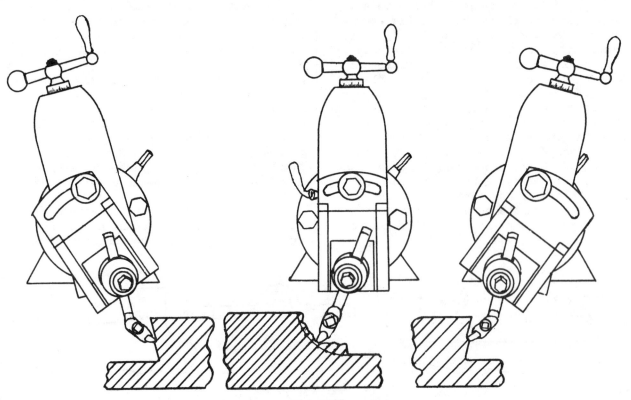

Fig. 491

with stroke of ram?

A. To check solid jaw for being square with stroke of ram, place an indicator in tool holder with point of indicator on solid jaw, and move the vise at right angles to the ram with cross feed. To check solid jaw for being parallel with stroke of the ram, place an indicator in the tool holder with point of the indicator against solid jaw and move ram forward and backward.

20. What should be done if work is not parallel within reasonable limits after a cut has been taken from each side?

A. Inspect the vise thoroughly for dirt and chips and make sure that it is clean. If this does not correct the condition check the vise with an indicator.

21. Draw sketches showing how to set head and clapper box when taking vertical and angular cuts.

A. When making angular or vertical cuts on a shaper the clapper box must be set at an angle so that the tool will swing away from the work on the return stroke of the ram. See Fig. 491.

22. Explain how a dovetail bearing may be cut on a shaper.

A. The toolhead of the shaper should be set over at an angle the same as the angle of the dovetail to be cut. When dovetail bearings such as those shown in Fig. 492 are to be cut the work should not be disturbed in shaping the angular and flat surfaces of the dovetail Aa, Bb, and Cc. Surfaces Bb and Cc should be machined before completing the angular surfaces Aa. A right-hand tool and a left-hand tool are used to machine the angular sides, as shown in the illustration. A roughing and a finishing tool should be used if considerable stock is to be removed. In using two tools and moving the tool head from one side of the center line to the other if there is any variation in the angular setting of the head, a variation in the angular sides of the dovetail will result.

Another way of cutting a dovetail, when the sides of the work are parallel and the solid jaw of the vise is parallel with the stroke of the ram, requires only one tool. First, rough out the sides of the dovetail to within 1/64 or 1/32 of the finished size. Next, take a light cut on one side, reverse the work in the vise but do not disturb the setting of the table, and take a light cut off the other side. Then check for size and repeat the process until the finished size is reached. Using this method the dovetail will be held central with the work and the angles will be the same.

In shaping dovetail bearings it is very important to incline the clapper box in the proper direction so that the tool will swing away from the work on the return stroke of the ram. The beginner should pay strict attention to this point because sometimes the setting may not be correct although it may appear to be. Remember that the top of the clapper box must be set in a direction away from the surface being machined (see question 21).

23. How should a dovetail be checked for size?

A. A dovetail slide bearing, as the name implies, derives its name from the shape of a dove's tail. It is used a great deal in the modern toolroom on parts of machines such as the ram and cross slide on the shaper, compound slide, and taper attachment on the lathe, etc. Dovetail bearings are usually machined on a shaper or planer but may also be made on a milling machine. When a gib (a piece located alongside a sliding member to take up the wear) is used, the accuracy required in planning is not as great as when the two pieces are fitted together. In either case, however, a smooth cut is necessary and one or two thousandths of an inch should be left for scraping. It is good practice to lay out the dovetail before starting the work.

In order to check the size of a dovetail to determine the amount necessary to machine off or to check the finished product

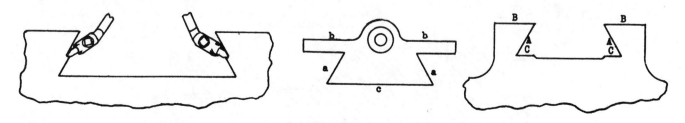

Fig. 492

accurately, two plug gages are placed in the angles of the dovetail and a measurement is taken either over or between the plugs. This measurement may be determined by applying the rule from trigonometry which states that the length of the side adjacent in a right triangle is equal to the length of the side opposite multiplied by the cotangent of the given angle. The side adjacent is that side which with the hypotenuse makes the acute angle in question. The side opposite is the side which is opposite the angle in question. The cotangent is the ratio existing between the side opposite and the side adjacent. The example below shows the method used to find the required measurement.

24. When cutting an angle on a large job, using a shaper which has a universal table, should the table be tilted, or the tool head swung to the required angle?

A. The universal table should be tilted because this makes it possible to use the automatic table feed. If the tool head were set on the required angle and did not have an automatic down feed, it would be necessary to use the hand down feed.

25. Explain the method used to shape a "V" or keyway central in a block.

A. One method used is to lay out the job and shape to scribed layout lines. A more accurate method of shaping a "V" block is to

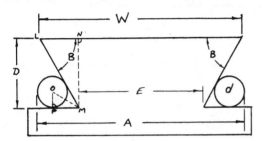

Fig. 493. Internal Dovetail

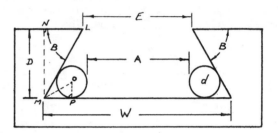

Fig. 494. External Dovetail

To solve for A (Fig. 493).

    E = W - 2(LN)
    A = E + 2MP + d

Example

    W = 4.5
    d = .250
    D = .750
    B = 60°

Solution

    LN = .75 x cot 60° = .43301
     E = 4.5 - (2 x .43301) =
         4.5 - .86602 = 3.6339
    OP = .250 ÷ 2 = .125
    MP = .125 x cot 30° =
         .125 x 1.732 = .2165
     A = 3.6339 + (2 x .2165) + .250 =
         3.6339 + .433 + .250 = 4.3169

To solve for A when E is given (Fig. 494).

    W = E + 2(LN)
    A = W - (2MP. + d)

Example

    E = 3.634
    d = .250
    D = .750
    B = 60°

Solution

    LN = .75 x cot 60° = .43301
     W = 3.634 + (2 x .43301) =
         3.634 + .86602 = 4.5
    OP = .250 ÷ 2 = .125
    MP = .125 x cot 30° =
         .125 x 1.732 = .2165
     A = 4.5 - (2 x .2165) - .250 =
         4.5 - .683 = 3.817

Distance measured over plugs = A + 2d

The following formulas and table of values are given for the most commonly used forms of dovetails.

$$MP = OP \times Cot \frac{B}{2}$$

$$E = W - 2(D \times Cot\ B)$$

$$A = E + 2MP + d$$

Angle B	15°	17½°	20°	22½°	25°	27½°	30°	35°	40°	45°	50°	55°	60°
Cotangent	3.732	3.1716	2.7475	2.4142	2.1445	1.9210	1.732	1.4281	1.1917	1.0000	.83910	.70021	.57735

set the vise jaws parallel to the stroke, set the tool head at the required angle, and rough out to the layout lines. Next, take a cut from one side of the "V", then reverse the job in the vise. With the table set in the same position, take a cut off the opposite side. Continue this until both sides of the "V" are cleaned up and the proper depth has been reached. The "V" will then be in the center of the block.

26. Tell how to machine a job similar to the one shown in the sketch at the right.

   A. First machine the six sides following the instructions given in question 13. Next, lay out the angle and the step according to the required dimensions. See Fig. 495. Place the work in the vise on parallels, swivel the vise to 15°, and rough out.

Complete the machining by setting the tool head at 7°, feeding the tool down to the layout on this angle and finishing the thickness of the step in one operation.

27. How may a keyway be cut on a shaper when the keyway does not extend the entire length of the shaft?

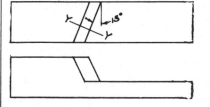
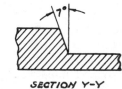

Fig. 495

Fig. 496
Vertical Shaper

A. Drill a hole slightly wider and deeper than the width and depth of the keyway at the place where the keyway ends. Set the position of the shaper stroke so that the tool will stop in center of the drilled hole at the end of the forward stroke. Then cut the keyway to size in the usual manner.

28. How should a thin job be machined to prevent warping?

A. First a light cut should be taken off each side to relieve the internal strain. Then light cuts should be taken off each side alternately until the correct thickness is reached.

29. Why should the speed of the shaper be increased when taking a finishing cut?

A. The speed should be increased because this gives a better finish and shortens the length of time for machining.

30. What tool is the most important for checking the squareness of shaper work?

A. The solid square is the most important.

31. Why is it unnecessary to hammer the shaper vise handle when clamping work?

A. The vise handle is long enough so that the leverage gained when pressure is exerted on the end of it is sufficient to clamp the jaws tightly on the work.

32. Describe the slotter.

A. Vertical slotters were originally designed for simple operations of slotting or key seating. They are still simple rugged machines primarily adapted for such simple operations.

33. Describe the vertical shaper.

A. The vertical shaper (shown in Fig. 496) is more commonly used than the slotter. It is a machine having all the design features of the vertical slotter and in addition has a swiveling tool holder and an angularly adjustable ram. By using this adjustable ram cuts can be taken at an angle with the vertical. This feature is used mostly for shaping clearance on dies, etc.

Fig. 497. Tools and Holder

Fig. 498
Vise

Fig. 499
Angle Iron

Fig. 500
Showing Angular Adjustment of Ram

Fig. 501.   Internal Splining

Fig. 502
Machining a punch for producing laminations.

Fig. 503
Facing a casting on both sides without a
change in set up.

Fig. 504.   Machining A Box Jig.

34. What is the function of the planer?

A. The planer (shown in Fig. 505) is for producing flat surfaces on work that is too large or otherwise impracticable or impossible to machine on the milling machine or shaper.

35. What is the main difference between a shaper and a planer?

A. With the planer the table or platen has a reciprocating motion past the tool head, while on the shaper the table is stationary and the tool head, which is fastened to the ram has the reciprocating motion.

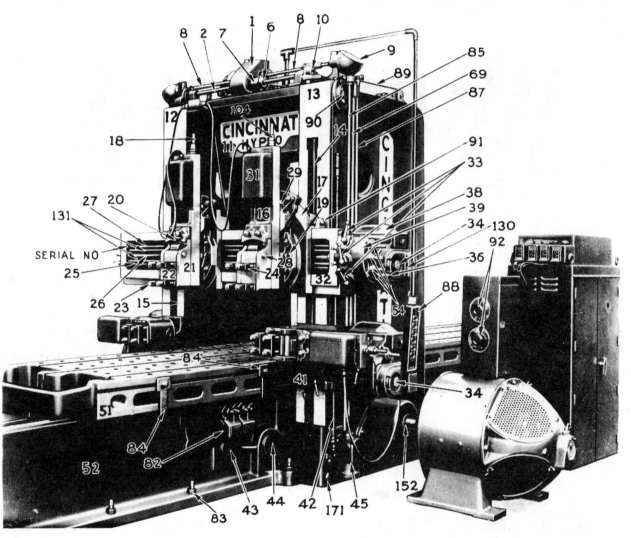

Fig. 505. Planer

1. Elevating Device
2. Horitonzal Elevating Shaft
6. Elevating Shifter Fork
7. Elevating Clutch
8. Horizontal Rapid Traverse Shaft
9. Rapid Traverse Bracket
10. Elevating Bracket
11. Arch
12. Left-Hand Housing
13. Right-Hand Housing
14. Right-Hand Elevating Screw
15. Left-Hand Elevating Screw
16. Slide
17. Saddle
18. Down Feed Screw

19. Harp
20. Clapper Box Clamp
21. Clapper Box
22. Tool Block
23. Tool Block Stud
24. Tool Block Clamp
25. Rail Screw, Right-Hand Head
26. Feed Rod
27. Rail Screw, Left-Hand Head
28. Clapper Box Taper Pin
29. Harp Gib
31. Right-Hand Head complete
32. Rail
33. Rapid Traverse and Feed Engaging Lever
34. Feed Changing Knob

36. Worm Shaft or Rail Clamping Shaft
38. Rail Clamp Feed Cover
39. Rail Clamp Cover
41. Side Head complete
42. Side Head Screw
43. Bull Wheel Shaft
44. Bull Pinion Shaft
45. Feed Drive Box
51. Table
52. Bed
54. Feed Reverse Knob
59. Feed Shaft
82. Table Reversing Master Switch
83. Leveling Screw

84. Table Dogs
85. Rapid Traverse Shaft
87. Rod, Interlocking Device
88. Pendant Station
89. Counterweight Cable
90. Counterweight Sheave Wheel
91. Top Saddle Gib
92. Cut and Return Speed Rheostats
104. Down Feed Screw Micrometer Collar
130. Cover Measuring Unit
131. Shifter Rod
152. Coupling Main Motor
171. Housing Bolt

36. How is the size of the planer designated?

A. The size of a planer is designated by the distance between the uprights, the distance between the platen and the cross rail at its highest position, and the maximum stroke. For example, 24" x 24" x 6'.

37. Describe a method of ascertaining if the cross rail is parallel to the platen.

A. Clamp an indicator on the tool box, with the point touching the platen, and note the indicator reading as the head is moved the length of the cross rail.

pin is provided for locating the chuck either parallel or at right angles with the base. The base has a slot at each side for securing it to the platen of the planer.

39. Describe the planer centers.

A. Planer centers are provided with an index consisting of circular plates or disks which have notches accurately spaced on their periphery and give a wide range of divisions. See Fig. 508.

The work, together with the index, is

Fig. 506. Planer Tools

1. Right-Hand Rougher
2. Left-Hand Rougher
3. Round Nose
4. Square Nose
5. Square Nose

6. 3/8 in. Square Nose
7. Gooseneck Finisher
8. Right Side Tool
9. Left Side Tool
10. Right Side Rougher

11. Left Side Rougher
12. Right Angle or Side Tool
14. Left Angle or Side Tool
16. Right Angle Tool
18. Left Angle Tool

38. The planer chuck shown in Fig. 507 is mounted upon a swivel base which is provided with a split clamping ring for securing chuck in any position. A tapered

revolved by means of a worm and gear, which can readily be disengaged when not required.

Tongues are inserted in the bottom of the headstock and tailstock, which can be taken out and fitted to the slot in any planer by planing or filing off one side, without destroying the alignment. The footstock is inserted in a block which is adjustable up and down for convenience in planing bevels or tapers.

Fig. 507. Planer Chuck

Fig. 508
(Left) Headstock; (Right) Footstock

40. Give some hints on clamping work to the
    planer table.

A. In clamping work on the planer table
the operator must take great care to see
that the work is fastened securely by clamps,
bolts, poppets, toe-dogs, etc., (see Figs.
509 to 516).

Clamps should not be placed on a finished part of the work unless it is protected with copper, brass, paper, etc.

Flat thin work is held down best by
toe-dogs and poppets.

See that the work does not spring

when tightening clamps or toe-dogs.

A washer must be used with every bolt
and nut. Select bolts of the proper length.

An open end wrench of the correct
size should be used on all square and hexagon
nuts.

When placing the work in position see
that it does not mar the finished surface of
the table. Allow enough clearance for the
work to pass under the cross rail and by the
uprights.

Place thin paper under the work to
seat it properly.

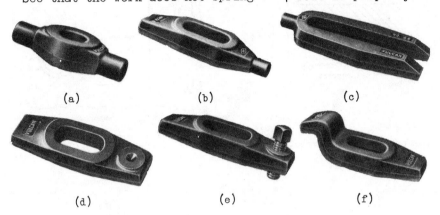

(a)   (b)   (c)

(d)   (e)   (f)

Fig. 509. Clamps: (a) Finger, Double
End; (b) Finger, Single End; (c) "U"-Pattern;
(d) Plain Slot; (e) Adjustable Step; (f) Gooseneck.

Fig. 510. Step Block

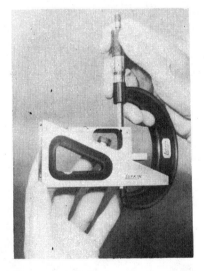

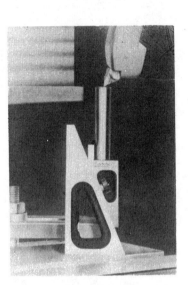

Fig. 511. Planer Jack   Fig. 512. Setting Planer Gage   Fig. 513. Use of Planer Gage

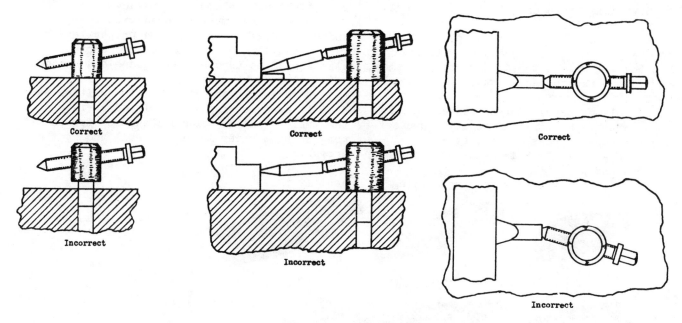

Fig. 514.   Correct and Incorrect Use of Toe-Dogs and Poppets

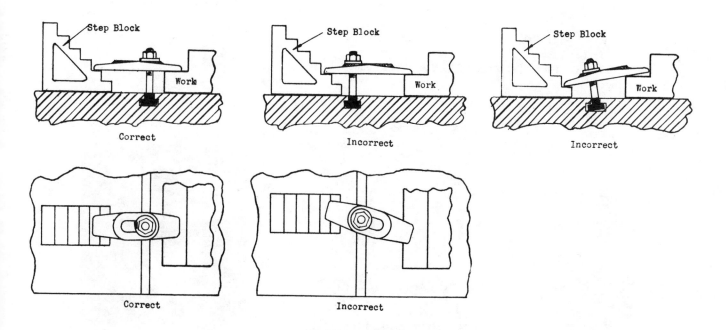

Fig. 515.   Correct and Incorrect Methods of Clamping

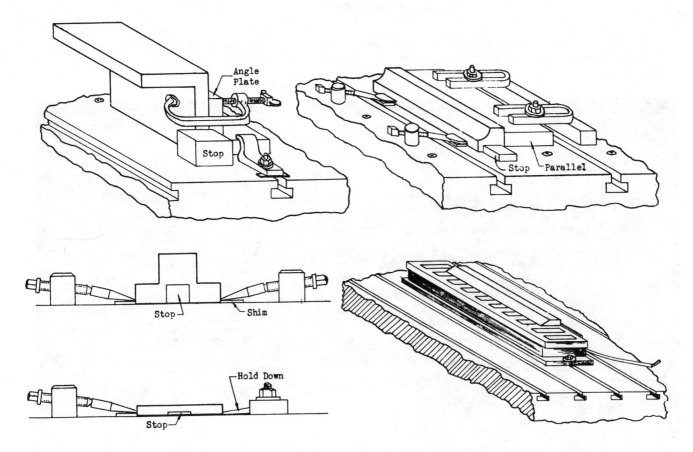

Fig. 516. Planer Set-Ups

Chapter 19

LATHE

The lathe is one of the most important machines in the machine shop. It removes material from revolving work by using suitably formed cutting tools of hardened and tempered steel or of alloy metals such as Carboloy, Stellite, etc.

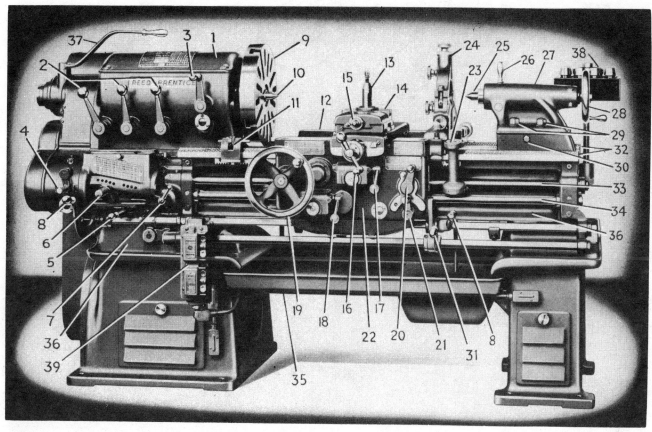

Fig. 517

1. Headstock
2. Speed Change Levers
3. Lever which controls back gears in headstock
4. Lever for controlling series of feed change
5. 2nd Lever for   "   "   "   "   "
6. Sliding Tumbler Gear Lever
7. Lever for controlling feed to lead screw or feed rod
8. Spindle Start, Stop and Brake Lever
9. Face Plate
10. Live Center
11. Micrometer Carriage Stop
12. Saddle or Carriage
13. Tool Post
14. Compound Rest
15. Compound Rest Feed Screw Handle
16. Handle for Operating Cross Feed by Hand
17. Lever for Controlling Power Longitudinal Feed Clutch
18. Lever for   "   "   Cross Feed Clutch
19. Hand Wheel for Longitudinal Carriage Travel

20. Lead Screw Split Nut Lever
21. Feed Rod and Lead Screw Safety Lever (Prevents engaging both feeds)
22. Apron
23. Thread Chasing Dial
24. Steady Rest
25. Dead Center
26. Tailstock Spindle Binding Lever
27. Tailstock
28. Tailstock Handwheel
29. Tailstock Clamping Screws
30. Screw for "Setting Over" Tailstock (For Taper Turning)
31. Lead Screw Reverse Lever
32. Bed Ways
33. Lead Screw
34. Feed Rod
35. Chip and Oil Pan
36. Bed
37. Collet Attachment Lever
38. Collet Cabinet
39. Electric Motor Control Switch

1. What are the different types of lathes?

A. The engine lathe shown in Fig. 517 is used mostly in tool rooms. Because of its adaptability to various kinds of work it requires a great degree of skill to operate.

The vertical turret lathe is the fastest lathe for short or heavy work.

The horizontal turret lathe, commonly called the screw machine, is used extensively in the production of duplicate parts. Most of these machines are equipped with a pump and a metal basin so that a lubricant or coolant can be used.

The production lathe and the automatic screw machine are used for the production of small duplicate parts.

2. How is the size of the engine lathe determined?

A. Engine lathes vary in size, ranging from the small bench lathe of only a few inches to one many feet in length. The size of the engine lathe is based upon two measurements--the approximate largest diameter that can be revolved over the ways and the total length of the bed. The actual size of the maximum swing is usually somewhat greater than the nominal size listed. For example, an 18 inch lathe may actually swing $18\frac{1}{2}$ inches over the ways of the bed.

3. Name six important parts of the lathe.

A. Bed, carriage, headstock, tailstock, thread cutting mechanism, and feeding mechanism.

4. Describe the bed.

A. The bed consists of two heavy metal sides located lengthwise, with ways or V's formed upon them. The bed is rigidly supported by cross-girths.

5. What is the carriage and what is its function?

A. The carriage, consisting of the saddle and apron, is the movable part which slides over the ways between the headstock and tailstock. The saddle has the form of the letter H, being bridged across the lathe bed to carry the cross-slide and tool rest, fitted to the outside ways, and gibbed to the bed. The apron (shown in the two illustrations of Fig. 518) contains the gears and clutches for transmitting motion from the feed rod to the carriage and the split nut which engages with the lead screw in cutting threads. The function of the carriage is to carry the compound rest.

6. Describe the headstock.

A. The complete headstock (Fig. 519) consists of the headstock casting, which is located on the ways at the left of the operator, the spindle in which the live center is

Fig. 519

(a)

(b)

Fig. 518. Apron: (a) front, (b) rear.

rigidly held by a taper, and the necessary gears and mechanism for obtaining the various spindle speeds.

16 Speed Index Plate

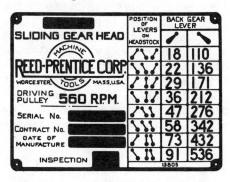

Fig. 520. Typical instruction plate for engaging gears in headstock.

7. Describe the tailstock.
   A. The tailstock (Fig. 521) is a movable

Fig. 521

Fig. 522

casting located opposite the headstock on the ways. It contains the dead center, the adjusting screw, and the hand wheel.

8. What are the thread cutting and feeding mechanisms?
   A. The thread cutting mechanism includes any gears or mechanism which transmits motion from the main spindle to the lead screw. The carriage movement is adjusted by the split nut on the lead screw, using the lever on the outside of the apron to move the split nut (see Fig. 518). The lead screw is used only for thread cutting, except on a lathe that does not have a feed rod. The apron on such a lathe is moved along by a spline in the lead screw instead of by using the thread.

   The feeding mechanism is a train of gears (series of gears in mesh) which transmits motion from the headstock or main spindle to the feed rod, and also to the lead screw when it is being used. The motion is then transmitted from the feed rod to various gears in the apron. The feed gears are generally controlled by friction, through small knobs located on the front of the apron. Fig. 522 shows the quick change gears used in securing the correct number of threads per inch in thread cutting or for varying the feed to suit different conditions. The plate (Fig. 523) indicates the position of the levers for producing the desired feed in thousandths or threads per inch.

9. What is the purpose of the compound rest?
   A. The compound rest supports the cutting tool in its various positions. It may be swiveled on the cross-slide to any angle in the horizontal plane. Its base is graduated to 180°, or 90° each way from the center, and the feed screw collar is graduated in thousandths of an inch. A compound rest is essential in turning and boring short taper work and in turning angles and forms on forming tools.

THREADS PER INCH										END GEARS
48	52	56	64	72	C	80	88	92	96	COMPOUNDED
24	26	28	32	36	B	40	44	46	48	
12	13	14	16	18	C	20	22	23	24	DIRECT
6	6½	7	8	9	B	10	11	11½	12	
3	3¼	3½	4	4½	A	5	5½	5¾	6	

FEEDS IN THOUSANDTHS										
7	6½	6	5¼	4¾	C	4¼	3	3.6	3½	COMPOUNDED
14	13	12	10½	9½	B	8½	7¾	7½	7	
28	26	24	21	19	C	17	15½	14¾	14	DIRECT
56	52	48	42	37½	B	33¾	31	29½	28	
112	104	96	84	75	A	67½	61½	59	56	

Fig. 523

In cutting American National and Sharp V threads the compound rest is swiveled to 30° from the center, which is one-half the included angle of these threads. The reason for this is that the cutting tool acts similarly to a facing tool when it is forced into the work at an angle of 30° instead of 90°. This causes the tool to cut only on one side, giving a better finish to the thread because the chips do not have a tendency to clog up the point.

CAUTION. Take great pains not to force the corner of the compound rest into the lathe chuck. This is done too frequently. Always keep the dovetail slide covered with the compound rest, to exclude all dirt and protect the slide from being nicked or marred by tools falling on it.

When facing the thickness of a piece of work, such as a spacer, to a size which is held to close limits, the compound rest may be swiveled at an angle of 30° from the face of the work and the tool fed in with the compound feed screw. The distance this screw advances is shown on a dial graduated in thousandths. Fig. 524 shows the compound rest set at a 30° angle. The tool is fed in along the hypotenuse of triangle ABC. From a theorem in geometry (which states that in a right triangle the side opposite the 30° angle is equal to one-half the hypotenuse) it can be seen that the thickness of the work is reduced one-half thousandth for every thousandth the tool is advanced with the compound feed screw.

In a 30° right triangle the side opposite the 30° angle is always one-half the hypotenuse. In Fig. 524 BC equals one-half of AB. Then if point B is moved in .002" along the line AB, BC will move in .001" against the face.

10. Name and describe attachments used on the lathe.

A. Taper attachment, relieving or backing off attachment, milling attachment, steady rest, and follower rest.

The taper attachment is used to turn and bore tapers. This is considered to

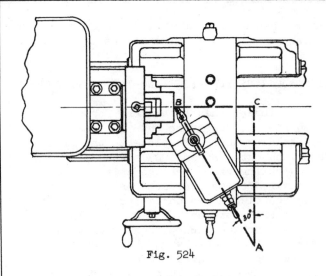

Fig. 524

be the best method of machining tapers. Fig. 525 shows a taper attachment with guide bar set in position to machine a taper.

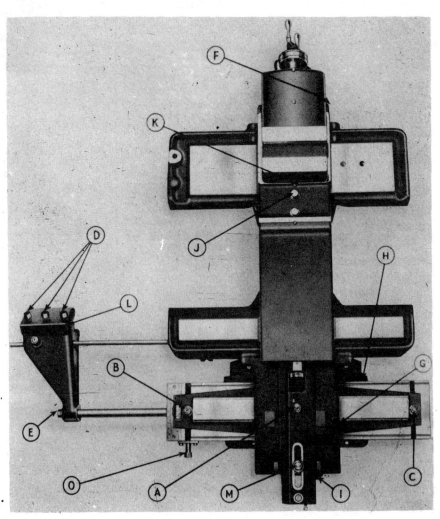

Fig. 525
Saddle with Taper Attachment

The guide bar is graduated in degrees at one end and in taper per foot at the other end.

INSTRUCTIONS FOR OPERATING TAPER ATTACHMENT

To set this attachment for any desired taper, loosen the nut "A" (see Fig. 525) holding the guide-plate to the sliding shoe, and nuts "B" and "C" at the ends of the swiveling bar. Set the swiveling bar to the desired taper indicated on the scale, using the micrometer adjusting screw, and tighten the nuts mentioned above. Except when setting the taper, these nuts should always be tight even when the taper attachment is not being used. Clamp screw "M" and lock nut "L" must be loose when using taper attachment and tightened only when attachment is not in use. To engage or disengage the taper attachment, tighten or loosen the nuts marked "D" which clamp the locking arm to the bed. When the taper attachment is being used, nut "E" holding the bar to the locking-arm should be tight. Be sure to have the point of the tool on the center-line.

If the taper attachment is used consistently on one job, it would be advisable to shift the sliding bar occasionally to equally distribute any wear on the swiveling bar. The taper gib "G" in the sliding shoe can be adjusted to compensate for any wear which might have occurred.

The compound rest and the taper attachment slide should move freely, but there should be no looseness or play. If chatter or non-uniform taper occurs this is the result of looseness in these slides and can be corrected by adjusting the gib "F" on the compound rest base and gibs "G", "H" and "I" on the taper attachment.

Should any looseness or back-lash develop in the cross feed screw nuts it can be removed by loosening screw "J" and tightening screw "K", then retightening screw "J".

The relieving or backing-off attachment shown in Fig. 526 is used for backing off cutters, taps, etc. It requires very little time to attach to a lathe and the attachment can be removed when the required tools are relieved or backed off.

The milling attachment shown in Fig. 527 illustrates how mill work can be done on a lathe and shows the operations in milling a

Milling a Standard Keyway.   Squaring the End of a Shaft.

Fig. 527

(a)   (b)   (c)

Fig. 526. (a) Form relieving; (b) internal relieving; (c) helical (or spiral) flute relieving.

dovetail, milling a square end on a shaft, and cutting a keyway.

Fig. 528 shows the steady rest used for turning long shafts of small diameter

Fig. 528

and for boring and threading spindles. It is necessary to support long work while machining it, to prevent taper caused by springing of the work.

The follower steady rest shown in Fig. 529 is attached to the saddle of the lathe for machining work of small diameter that is likely to spring if it has no support. The adjustable jaws of the follower

rest bear directly on the finished diameter of the work, following the cutting edge of the tool on the opposite side of the work. As the tool feeds along the work the follower rest, being attached to the saddle, travels with it.

Fig. 529

11. Name the two most commonly used lathe chucks and give the use of each.
A. The four jaw independent lathe chuck, Fig. 530, is used for holding most of the work that should be held in a chuck. The hardened steel jaws are reversible and will hold work of different sizes and shapes. Each jaw is independent of the other.

The three jaw universal chuck in Fig. 531 holds cylindrical work. All three jaws move universally to bring the work on center. This chuck is easier to operate than the four jaw chuck but it can be used only on round or hexagon stock.

12. Name two plates that are used extensively on the lathe.
A. The driving plate in Fig. 532 is used

Fig. 530   Fig. 531   Fig. 532   Fig. 533

Fig. 534   Fig. 535   Fig. 536

for driving work held between centers. The face plate, Fig. 533, is used to hold work while it is being machined.

13. Name two lathe dogs.

A. Fig. 534 shows a bent tail lathe dog used for driving cylindrical pieces. The clamp dog in Fig. 535 is used for work with flat sides.

14. What attachment should be used for holding cylindrical shafts of small diameter?

A. The draw-in collet chuck attachment in Fig. 536 is used on the lathe for the production of small metal parts and in the tool room for fine accurate work. Long rods are passed through the draw bar and held in the collet chuck while they are being machined. Short work may also be held in this chuck.

tools are held in tool holders, some of which are shown in Fig. 538.

16. Tell how the tool holder should be placed in the tool post for turning work.

A. As a general rule place the tool holder in the tool post at approximately 90° with the center line or a little in the direction of the dead center when feeding toward the headstock. The point of the cutting tool should be on the center line between the dead center and the live center, and should not extend far from the tool holder. Neither should the tool holder extend far from the tool post if chattering is to be avoided. A very good illustration is shown in Fig. 539.

CAUTION. When taking a heavy cut do not have the tool pointing in the direction of the live center. If the tool is pointed in that direction and runs into a hard spot the material will have a tendency to push the

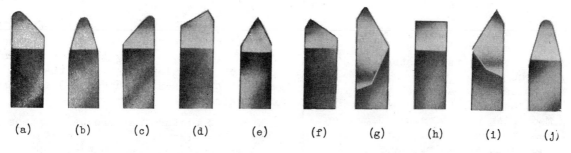

(a)   (b)   (c)   (d)   (e)   (f)   (g)   (h)   (i)   (j)

Fig. 537.   (a) Left hand turning tool; (b) round nose turning tool; (c) right hand turning tool; (d) left hand corner tool; (e) threading tool; (f) right hand corner tool; (g) left hand side tool; (h) square nose tool; (i) right hand side tool; (j) brass tool.

15. What are some of the most important cutting tools and how are they held?

A. Fig. 537 shows some of the different types of tools used in lathe work. These

tool away, forcing the tool to dig into the material and perhaps scrapping the job. This caution is unnecessary in taking a light cut.

Bear in mind that when you force the

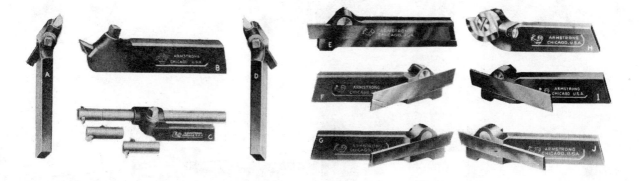

Fig. 538.   (a) Left hand turning tool; (b) straight-shank turning tool; (c) boring tool; (d) right hand turning tool; (e) straight shank cut-off tool; (f) left hand off-set cut-off tool; (g) left hand off-set side tool; (h) threading tool; (i) right hand off-set cut-off tool; (j) right hand off-set side tool.

tool into the work .001" on the periphery you are taking .002" off the diameter.

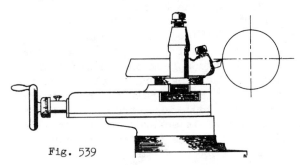

Fig. 539

17. Describe the difference between the dead center and the live center.

A. The dead center is supported in the tailstock and does not revolve. The live center is contained in the spindle and revolves with the work.

CAUTION. Keep the dead center well lubricated with oil.

18. When is the combined drill and countersink used?

A. To machine a piece of stock on centers, each end must be countersunk. This is accomplished with a combined drill and countersink (shown in Fig. 540). The included

Fig. 540

angle of the combined drill and countersink is 60°. This angle is the same as that of the dead and live centers.

19. Why should a countersunk hole be used for a center hole instead of a large prick punch mark?

A. If a large prick punch mark is used instead of a hole made by a combined drill and countersink, the work will ride on the point of the steel center and not on the sides as it should. The work will soon heat and run untrue, due to improper location and lack of lubrication. The hole produced by the combined drill and countersink should contain center compound to lubricate the point of the dead center.

20. What extra care should be taken in inserting a centered piece of work between centers?

A. Do not force the dead center against the work close to the countersunk hole, as that causes the hole to be burred or nicked and the work will not run true.

21. How is the work mounted on centers?

A. When mounting work between centers be careful to see that the centers are in good condition. The dead center in the tailstock will be the first to show wear. As the live center in the spindle revolves with the work, the dead center must be lubricated. Incorrect mounting on centers is shown in Fig. 541. A lathe dog that will not clear

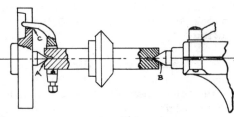

Fig. 541

the bottom of the slot in the driving plate has been used (shown at C). A and B show incorrect bearings which cause the work to revolve eccentrically.

Fig. 542 shows correct mounting between centers. The lathe dog is fastened to

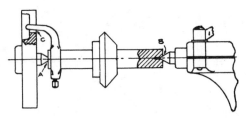

Fig. 542

the work and the tail should clear the bottom of the slot. The work is held firmly but not too tightly on the live and dead centers. Adjusting the centers to the work can be learned in a short time. Too great a pressure of the dead center against the work will cause a squeaking noise and the center and work will heat up so as to cause wear from the friction.

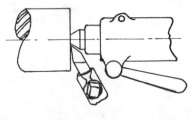

Fig. 543

22. How should the facing or squaring tool be set and used?

A. Fig. 543 shows a cylindrical piece of work being faced off to true the ends. The tool is pointing toward the live center and is not at right angles with the center line. Only light cuts are taken to true the

work. Use a steady rest and chuck if you
wish to remove a large amount of stock on the
end of a shaft.

23. What is the first thing to do when given
    a piece of work, and a drawing which
    gives several length dimensions and the
    overall length?
    A. Add the several lengths and see if
    they coincide with the overall length given.
    Also determine whether the work has suffi-
    cient stock to finish to the O D shown on
    the drawing.

24. Give several reasons why the end of a
    piece of work may not be 90° with the
    center line when faced with a side tool.
    A. The lathe centers may be out of line,
    the tool may be springing away from the work,
    the cutting edge of the tool may have worn
    off, or the gibs of the compound rest may be
    loose.

25. When using a driving plate and a steady

rest, what is the best method for holding a shaft on center so that the end of
the shaft can be bored?
    A. The best way to do this job is to tie
it to the driving plate as shown in Fig.
544. The plate is unscrewed about three or
four revolutions from the shoulder of the
spindle. Then the work is placed on the
center and tied securely to the driving
plate with a rawhide strap. Finally the
driving plate is screwed to the shoulder of
the spindle. This tightens the strap on the
work and holds it firmly.

26. Name some operations performed on the
    lathe.
    A. Straight (cylindrical) turning, taper
turning (by using the compound tool rest or
taper attachment), facing (cutting at right
angles to the axis of the work), either

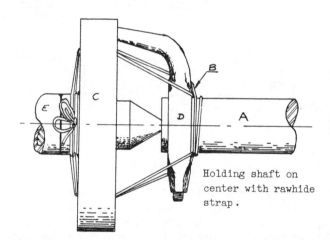

Holding shaft on
center with rawhide
strap.

Fig. 544

Fig. 545. Turning the O D.

Fig. 546. Machining Radius with
Boring Tool.

Fig. 547. Internal Threading
using Boring Bar.

Fig. 548. Knurling.

straight or taper boring, knurling, and external and internal thread cutting. A thread may be chased with a tap or die or cut with a threading tool.

Some lathe operations are shown in Figs. 545 to 551.

Fig. 549. Cutting a National Standard Thread.

27. What is the difference between an arbor and a mandrel?

A. An arbor is a shaft used for holding and driving cutting tools. Usually one end of the arbor is tapered to fit the tapered hole in the spindle of the machine.

A mandrel is a shaft or bar for holding work to be machined. Plain mandrels are usually tapered .001 to .005 of an inch. Expanding mandrels are made in many types and sizes.

28. How much grinding stock should be left on a shaft 1" in diameter and 48" long if it is to be ground soft?

A. Leave .015 of an inch.

Fig. 551. Cutting a square thread on work which is being supported by a follower rest.

29. How much grinding stock should be allowed on a shaft 1" in diameter and 48" long if it is to be hardened and ground?

A. Leave .035 of an inch.

30. How much grinding stock should be left

Fig. 550. Boring With A Boring Bar.

on the O D and I D of a bushing with 2" bore and 2½" O D if it is to be ground soft?

A. Leave .015 of an inch on each diameter.

31. How much grinding stock should be allowed on the bushing in question 30 if it is to be made of high speed steel, hardened and ground?

A. Leave .025 of an inch.

32. How may cast iron be machined rapidly and efficiently?

A. Use soda water with a slow speed and heavy feed.

33. How should a tool be ground for finishing cast iron?

A. Grind the tool with a flat top, rounded nose, and with small side clearance.

34. When a drawing calls for grinding on a diameter and against a shoulder, what kind of an under-cut should be given?

A. The under-cut should be made with a narrow, round nose tool fed in at an angle of 45°.

35. What kind of a tool is used to under-cut corners on a job that is to be hardened and ground?

A. Use a round nose tool, as a square nose tool will have a tendency to cause the steel to crack or break when it is hardened.

36. Why is it a good policy to use a coarse feed when machining stock that is to be ground?

A. Much time will be saved by taking a coarse feed.

37. Should a deep or shallow cut be made in cutting cast iron? Explain.

A. Cast iron has a very hard scale on the outside, caused by the iron becoming chilled in casting. In cutting this iron the tool must be immediately forced under this scale or the cutting edge of the tool will be ruined. Never take a shallow cut on cast iron unless the size of the stock is almost down to the size of the finished job.

38. State the usual cutting speeds for the most commonly used metals.

A. The peripheral speeds in feet per minute (fpm) recommended for cutting metals on the lathe when using high speed steel tool bits are shown in the chart below.

MATERIAL	FPM	
Chrome, Non-Shrink, H S Steel.....	40 -	50
Cast Iron.........................	55 -	65
"RR" (Tool) Steel.................	60 -	65
"AAA" Steel......................	80 -	85
Cold Rolled Steel, Machine Steel..	110 -	120
Brass............................	115 -	125
Bronze, Copper, Aluminum.........	200 -	250

39. What precaution should be taken in fit-ting a plug gage into work that has just been bored?

A. Before fitting a plug gage into work be sure that the temperature of the work is almost the same as that of the gage or the plug gage may freeze in the work.

40. Name four methods of producing tapers on the lathe.

A. For short tapers the compound rest is generally swiveled to the required angle and the cutting tool is fed by hand. The taper attachment, Fig. 525, may be used to produce

Fig. 552

internal and external tapers. The tool is fed automatically. The third method used to produce a taper is to off-set the tailstock. The tailstock, Fig. 521, has set screws on the side which can be adjusted. The fourth method, producing a taper with a square nose tool, is illustrated in Fig. 552.

CAUTION. In producing tapers the

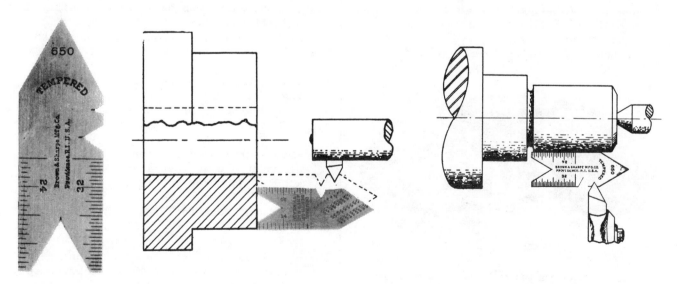

Fig. 553   Fig. 554   Fig. 555

point of the tool must be on the center line
of the work.

41. What method should be used to produce a
    short taper having an included angle of
    60°, such as a lathe center?
    A. This taper may be produced by using
either the compound rest or the square nose
tool. The method which uses the compound
rest is considered the better of the two.

42. What method should be used in producing
    a long shaft having a taper of .015" per
    foot?
    A. Use the taper attachment for this type
of job.

43. Name several ways of checking tapers.
    A. A taper may be checked by a microm-
eter, by measuring the large and small diameters, by taper gages, by fitting to the
spindle or sleeve for which it is intended,
and by using the sine bar and Johansson Gage
Blocks.

44. What is the center gage and for what is
    it used?
    A. The center gage is a small flat piece
of steel which is graduated in fractions of
an inch. Included angles of 60° are cut in
it, as shown in Fig. 553. It is used in setting American National and Sharp V threading
tools, as shown in Figs. 554 and 555.

45. Explain clearly what is meant by the
    pitch and lead of a screw thread.
    A. The pitch is the distance from a point
on one thread to the corresponding point on
the next, measured parallel with the axis of
the work. The lead is the distance a nut
would advance on a screw thread in each revolution. The lead of a single thread screw is
equal to the pitch, the lead of a double
thread screw is equal to twice the pitch,
etc.

46. In setting a threading tool for cutting a
    tapered thread, as on a pipe, should the
    center line of the tool be set at 90°
    with the center line of the work or at
    90° with the line of taper?
    A. Set the center line of the tool at 90°
with the center line of the work.
         NOTE. The taper attachment of the
lathe should be used in cutting pipe threads
or on any other tapered thread job.

47. When a lathe has been set for cutting
    threads and a trial cut taken on the job,
    how is it checked for the correct number

of threads per inch?
    A. Use a rule and count the number of
threads per inch, as shown in Fig. 556, or

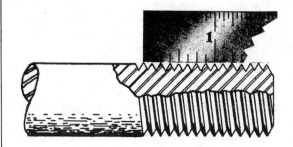

Fig. 556

use a screw pitch gage, as shown in Fig. 557.

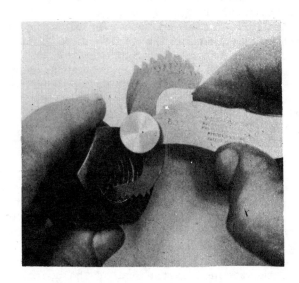

Fig. 557

Major Diameter	Threads Per Inch	Major Diameter	Threads Per Inch
$\frac{1}{4}$	20	$\frac{9}{16}$	12
$\frac{5}{16}$	18	$\frac{5}{8}$	11
$\frac{3}{8}$	16	$\frac{3}{4}$	10
$\frac{7}{16}$	14	$\frac{7}{8}$	9
$\frac{1}{2}$	13	1	8

48. State the number of threads per inch for
    standard American National Coarse
    threads.
    A. The table above gives the number of
threads per inch for the standard American
National Coarse thread series (up to one
inch in diameter).

49. Explain the difference in cutting a
    right-hand and a left-hand thread.
    A. To cut a right-hand thread the com-
pound rest is set over toward the right on
the saddle at the proper angle. The feed of
the carriage is toward the headstock. In
cutting a left-hand thread the compound rest
is moved toward the left on the saddle and
the feed of the carriage is toward the tailstock.

50. Describe three methods of cutting a dou-
    ble thread.
    A. The following examples will show how
to cut a double thread having 1/4" lead and
1/8" pitch:
        Swivel the center line of the com-
pound rest to a line parallel with the dead
and live centers. In cutting the first
groove, use a roughing tool, which is narrower than the finishing tool. By dividing one
inch by the lead, which is 1/4 of an inch,
four single threads per inch will be obtained. Set the lathe to cut four threads
per inch. Force the tool into the work the
required depth by using the cross slide feed.
By using the feed screw on the compound rest,
on which a graduated collar is attached, move
the roughing tool over the length of the
pitch and proceed to rough out as before.
        Another way to do this job is to use
a face plate having two equally spaced slots
in which to insert the tail of the lathe dog.
Be careful not to move the dog after the job
is started.
        Still another way to do this job is
to mark two teeth equally spaced on the gear
that is attached on the end of the spindle,
and after roughing out the first groove, disengage the marked gear on the spindle and
turn both spindle and gear one-half a revolution then mesh. You are then ready to rough
out the second groove.

51. Describe knurling.
    A. Knurling is a process of making a
series of indentations or depressions on the
surface of the work, and is done by a knurling tool held in the tool post. Fig. 558
shows a good example of knurling.
        A slow speed should be used to knurl
a piece of work. Force the knurling tool
slowly into the work to a depth of approximately 1/64 of an inch. Set the feed so that
the tool will feed across the surface of the
work. Reverse the feed of the carriage at

the end of the cut, and repeat the operation
until the knurling has been completed. The

Fig. 558. (Left) coarse, 14 pitch; (center) medium,
21 pitch; (right) fine, 33 pitch.

knurls should be lubricated while cutting.
An illustration of a knurling job is shown
in Fig. 548.

52. What are some of the safety rules which
    should be observed in the lathe depart-
    ment?
    A. The following safety rules apply par-
ticularly to the lathe department. They
should be observed by every one working on a
lathe.
1. Have the first piece of every job inspected.
2. Do not touch any chips with your hand.
   Use a wooden paddle.
3. Do not use a Westcott wrench on any nut or
   screw. Use an open end wrench, a compound
   wrench, and the standard tool post wrench.
4. Do not leave the chuck wrench in the chuck
   at any time. Oil the spindle nose before
   putting on chuck or plate.
5. Be neat. Do not allow oil or chips to
   collect around your lathe.
6. Keep tools off ways and top of carriage.
7. Cover all boring and internal threading
   tools with a towel while using gages or
   calipers to check bore.
8. Goggles must be worn on the following
   types of work:
        Turning of square, hexagon, or ir-
        regular stock of any kind;
        Turning of all broach jobs;
        On bronze or brass at all times;
        On all tough stock where the chips
        have a tendency to fly.

Chapter 20

TURRET LATHES

HORIZONTAL UNIVERSAL TURRET LATHE

The screw machine was an early development from the lathe and was primarily designed for the rapid production of screws, as the name implies. This machine gained wide prominence later by the addition of certain tools and attachments which could be used for the completion of small cylindrical duplicate parts.

Fig. 559

"B" HEADSTOCK (No. 4 and No. 5 Universal)
  4. Surface Speed Preselector
  5. Preselector Drum Housing
  6. Spindle Control Lever
  7. Preselector Handwheel

"C" CHANGE GEAR BOX
  (For halving or doubling all feeds)
  8. Feed Shaft

"D" UNIVERSAL CROSS SLIDE CARRIAGE UNIT
  9. Cross Slide
  10. Carriage
  11. Square Turret
  12. Square Turret Indexing and Binder Handle
  13. Longitudinal Feed Handwheel
  14. Cross Feed Handwheel

15. Carriage Clamp
16. Cross Feed Engagement Lever
17. Longitudinal Feed Engagement Lever
18. Carriage Apron
19. Feed Shift Levers
20. Feed Reverse Lever
21. Longitudinal Stop Rod
22. Master Stop Screw
23. Longitudinal Feed Stop Roll

"E" HEXAGON TURRET UNIT
  24. Hexagon Turret
  25. Circumference Binder Ring
  26. Turret Slide
  27. Saddle
  28. Slide Binder Lever

29. Turnstile
30. Saddle Apron
31. Feed Shift Levers
32. Feed Engagement Lever
33. Stop Roll

"F" COLLET CHUCK
  34. Bar Chuck Lever
  35. Bar Chuck Wedge
  36. Finger Holder

"G" RATCHET BAR FEED
  37. Bar Feed Head
  38. Ratchet Pawl Lever
  39. Bar Feed Tube Clamp
  40. Bar Feed Tube

A turret, usually of hexagon shape, and a saddle were placed on the machine. The turret held different tools, such as drills, reamers, counterbores, and cutters. The machine was then called a turret lathe by several manufacturers; a horizontal type is shown in Fig. 559. The horizontal turret lathe will be found in most tool rooms, as it can produce small duplicate cylindrical parts in less time than the engine lathe.

The universal carriage (Fig. 560) has both power cross feed and longitudinal feed and may be used for turning, necking, facing, etc., while the hexagon turret is turning and boring. A graduated dial with observation clips gauges the depth of a cut when using the cross feed.

Fig. 560. Universal Cross Slide

The chasing attachment (Fig. 561) cuts threads by means of a leader and fol-

lower. The leader is clamped on the carriage feed rod and the follower nut is bolted to the chasing lever. Each leader cuts only one thread.

The taper attachment (Fig. 562) is used to turn or bore tapers up to certain sizes and lengths. The tool block of this attachment slides on a base which is fitted to the carriage.

Fig. 562. Taper Attachment for Carriage

The attachment at the extreme right of Fig. 563 offers a means of leading-on taps and dies to the full length of the turret slide stroke. Each leader and double half nut will cut any one pitch 4 to 36 threads per inch when using standard gears in the head end gear box. The control interlock prevents the leading-on and feeding mechanisms from engaging at the same time. The turret slide stop screws

Fig. 561. Chasing Attachment for Carriage

Fig. 563. Leading-on Attachment for Ram Type Turret Lathes

can be set to disengage the single clutch at the end of the thread.

FULL UNIVERSAL THREADING TURRET LATHE

Tool room threading performance, and high speed production advantages are combined in a thread cutting lead screw machine (Fig. 564). The single thread fine pitch lead screw is used for thread cutting only. The remote control lever, located at the operating position, gives forward and reverse motion to the carriage. It is not necessary to disengage the double half nuts or reverse the spindle for the threading operation. This remote control lever is also used in cutting left-hand threads. An automatic knock-off for right-hand threads permits threading to accurate shoulder lengths. The change from threading to regular feeding is accomplished by shifting a single lever on the head end lead screw gear box. A safety interlock prevents the screw cutting and feeding mechanisms from engaging at the same time. The threading range is 2 to 56 threads per inch in lengths to 20 inches.

Operating pre-selector (Fig. 565):

A) Set the pre-selector when setting up tools. First, rotate the drum until the desired feet per minute which show on the dial, correspond to the work diameter to be cut. Then, number that cut, using master numeral (1) at the top of the drum.

B) Set the indicator for the second cut, by selecting the desired surface feet per minute for the second work diameter. Number this cut with master numeral (2). Continue similarly for additional operations.

C) Cuts are now shown in order by the master numerals 1, 2, 3, etc. Now, use these numerals in their turn from one cut to the next. In this way the correct feet per

Full Universal Threading Turret Lathe.

Fig. 564

Pre-Selector.

Fig. 565

minute will be set for each cut, regardless of the number of speed changes.

D) In actual use it is possible to pre-select the next speed by turning to the next numeral while the machine is cutting. Then, when the cut is finished it is only necessary to move the lever, bringing the next speed into play, without stopping the spindle. Forward--reverse--stop--start with the one lever.

Pre-selector is set for surface speeds as shown in Fig. 566.
The operator can select the correct spindle r.p.m. to produce the proper surface feet per minute for any diameter to be cut by turning the handwheel.

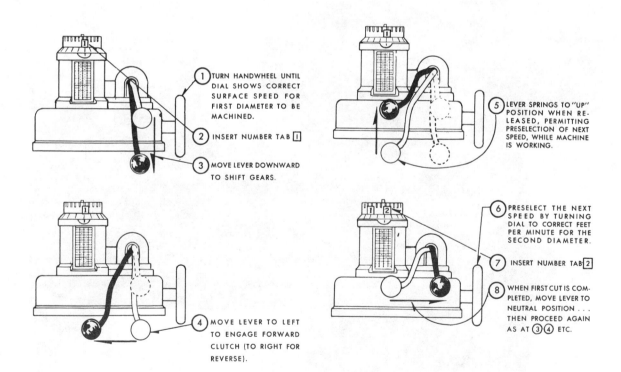

Setting Pre-Selector for Surface Speeds.

Fig. 566

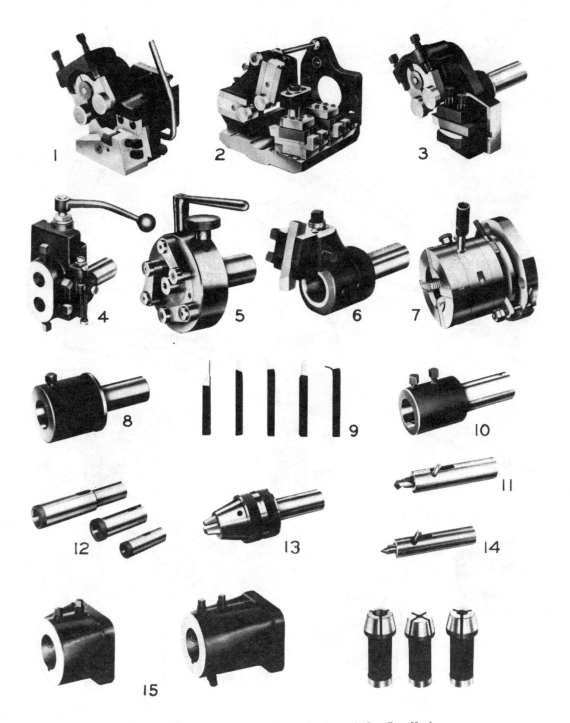

Fig. 567. Universal Tooling Equipment for Bar Work

1. Single Cutter Turner
2. Multiple Cutter Turner
3. Combination End Facer and Turner
4. Quick Acting Slide Tool
5. Center Drilling Tool
6. Adjustable Knee Tool
7. Die Head
8. Clutch Tap and Die Holder
9. Forged Cutters for Square Turret
10. Floating Tool Holder
11. Combination Stock Stop and Starting Drill
12. Taper Drill Sockets
13. Drill Chuck
14. Combination Stock Stop and Center
15. Flanged Tool Holders (Short and Long)

Fig. 568.   Universal Chucking Tools

1. Three-Jaw Geared Scroll Chuck
2. Reversible Adjustable Angle Cutter Holder
3. Reversible Straight and Angle Cutter Holder
4. Flanged Tool Holders (Short and Long)
5. Stationary Overhead Pilot Bar and Pilot Sleeves
6. Piloted Boring Bar
7. Spindle Pilot Bushing

8. Slide Tool
9. Forged Cutters for Square Turret
10. Adjustable Single Turning Head
11. Angular Cutter Stub Boring Bar
12. Straight Cutter Stub Boring Bar (Double Ended)
13. Straight Shank Taper Socket and Holder with Taper Shell Sockets

A permanent set-up of the Universal Tooling Equipment can be arranged so that the large and heavy tools of the flanged type are permanently mounted in their logical order on the machine. If necessary for certain jobs, the tool stations can be back-indexed, or skip-indexed, to suit the requirements of the job, but the flanged tools themselves are not changed from one turret face to the other. Ordinarily, the extra machine-handling time required to skip-indexed is less than the time would be required to remove these tools from the machine or to change their position on the turret.

The lighter tools are of the shank type, which can be quickly mounted in the turret or holder.

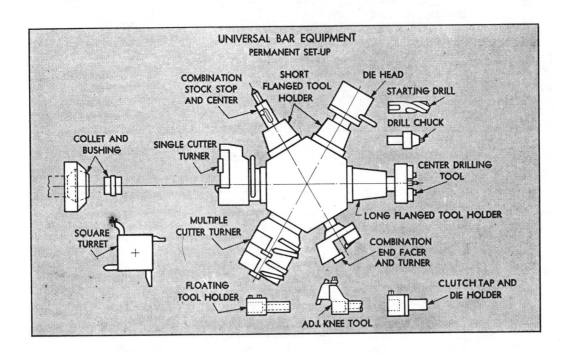

Fig. 569. Permanent Set-Up for Bar Work

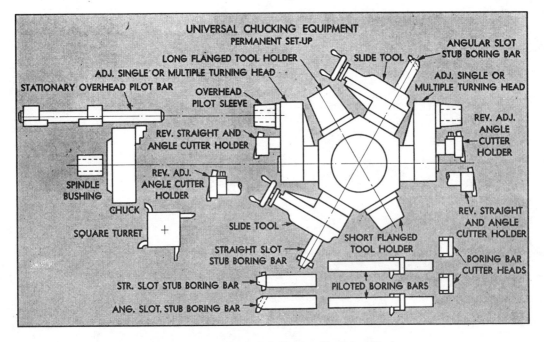

Fig. 570. Permanent Set-Up for Chucking Work

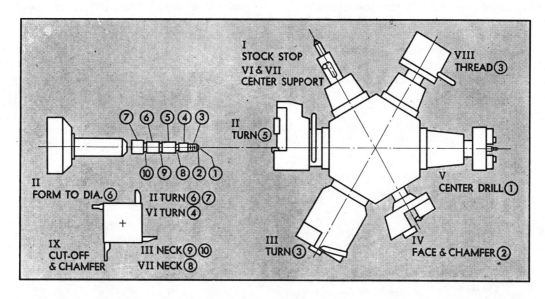

Fig. 571. Tooling Set-Up for Shaft with Twenty Pieces in Lot

The complete tooling layout is shown in Fig. 571. All the tool stations are used.

The cutters are high speed steel and of standard stock shapes. The order of operation is shown in Fig. 573.

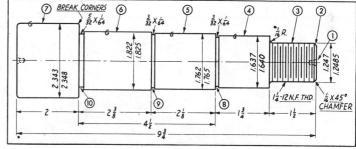

Fig. 572. Typical Bar Job--a Shaft

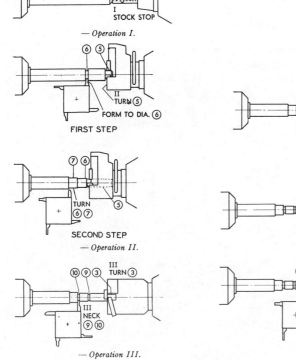

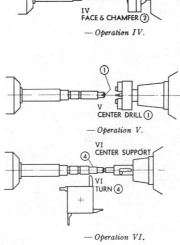

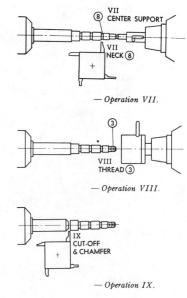

Fig. 573. Order of Operation

THE STATIONARY OVERHEAD PILOT BAR

The Stationary Overhead Pilot Bar, Fig. 574, which fastens to the head of the machine, can be heavier than the pilot bars which are attached to the turning heads because this bar does not add any weight to the hexagon turret and its tools. This type of pilot bar can be adjusted endwise in the machine for various lengths of work, and clamped in place. The purpose of the overhead pilot bar is to tie the turret of the machine to the headstock to provide greater rigidity, and to allow deeper cuts to be taken with heavier feeds.

When the accuracy requirements of the job are studied, it is found that the diameters do not have to be turned to very close limits since they are to be ground after hardening. However, it is necessary to keep the diameters closely concentric in order to allow for "clean-up" in the grinding operations. The thread diameter, which is not ground, must run true with the other diameters.

VERTICAL TURRET LATHE

The vertical lathe in Fig. 577 is a machine having a horizontal rotary table with adjustable chuck jaws and a vertical turret. It is adapted for turning, boring, facing, drilling, reaming, and similar operations. A thread cutting attachment is also used on some of these machines.

Fig. 574. Combined and Multiple Cuts

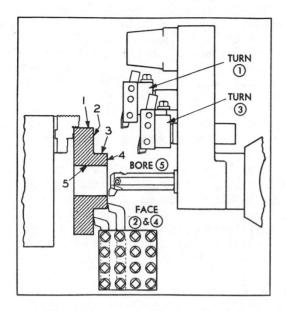

Fig. 575. Combined and Multiple Cuts Being
Taken on a Simple Gear Blank Job in Quantity Lots.

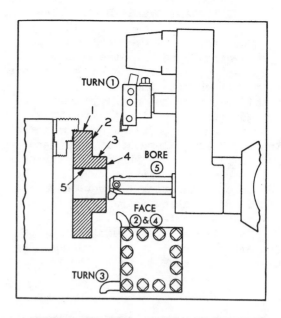

Fig. 576. Combined and Multiple Cuts Being
Taken on a Simple Gear Blank Job in Small Lots.

Fig. 577. Vertical Turret Lathe

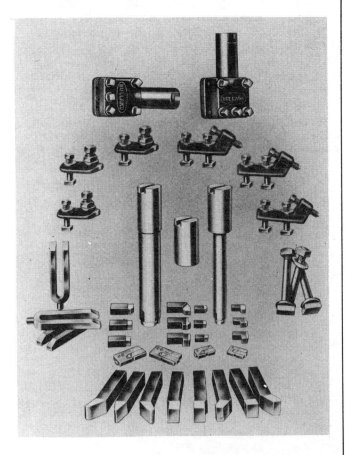

Fig. 578
Standard Tool Equipment for Vertical Turret Lathe

Fig. 580
Drilling and Machining Operation

The illustration (Fig. 580) shows machining from solid stock. Main Head operations are drilling, rough bore, finish bore, first and second ream with a chamfer or burring operation. Simultaneously the Side Head takes a rough and finish face, chamfer, rough and finish turn. The sequence of operations and the simultaneous use of Side and Main Heads provide a method for cutting waste time between cuts.

Fig. 579. Machining a Cylinder on the Vertical Lathe

In as much as the angles shown in Fig. 581 require accuracy and fine finish, a rough cut is first taken with the Side Head Ram, using a single point cutter. To do this, the Main Head is indexed to present the cam located in a holder of the turret. The Side Head then feeds against this cam forcing the special tool holder downward at the cam angle, thus taking a rough angular facing cut. To obtain the fine finish and accuracy, the Main Head is then indexed to present the sweep tools for the two angles.

After finish turning the O.D., (Fig. 582, the next operation of the Side Head is grooving. This may be done in one or two passes according to the depth of groove and the degree of finish required. To save time while this operation is going on, it is customary practice to finish turn or finish bore the inside diameters with the Main Head. If necessary, finish facing across the top or chamfering may also be done with the Main Head.

Fig. 581
Angular Facing Operation

Fig. 582
Grooving Operation

Chapter 21

MILLING MACHINES

The milling machine is a machine that removes metal from the work with a revolving milling cutter as the work is fed against it. The milling cutter is mounted on an arbor where it is held in place by spacers or bushings. It is made from high speed steel and can be had in different sizes and shapes. Regular or irregular shaped work may be produced on this machine, designs varying according to the particular class of work wanted. Milling machines may be grouped into various classes according to this variation in general appearance design, as the column and knee type, the manufacturing types, and the planer type.

The important parts of the milling machine (see Fig. 583) are:

1. Starting levers
2. Spindle
3. Column
4. Knee
5. Table
6. Elevating screw
7. Dividing head
8. Speed levers
9. Feed levers
10. Table movement levers
11. Foot stock
12. Arbor yoke

Fig. 583. Universal Milling Machine--Column and Knee Type

156

The most common type of milling machine used in the tool room is the "column and knee" type. This machine is so-named because of its design. The main casting consists of a high column, to which is fastened a bracket or "knee," which supports the table. The knee is adjustable on the column, so that the table can be raised or lowered to accommodate jobs of various sizes. Vertical cuts may be taken by feeding the table up or down. The table may be moved in the horizontal plane in two directions, either at right angles to the axis of the spindle (longitudinal feed) or parallel to the axis of the spindle (transverse feed).

The Universal milling machine (Fig. 583) is a development that embodies all the principal features of the other types of milling machines. It is designed to handle practically all classes of milling machine work. The table has the same movements as the plain milling machine, but in addition it can be swiveled on the saddle so that it moves at an angle to the spindle in the horizontal plane. When fitted with an attachment known as an index head, angular, spiral, and helical cuts may be made. The universal type of milling machine is used to cut helical gears, twist drills, milling cutters, and various kinds of straight and taper work. The universal mill-

Fig. 584
Plain Milling Machine

Fig. 585
Vertical Spindle Milling Machine

The Plain milling machine shown in Fig. 584 is one in which the longitudinal travel of the table is fixed at right angles to the spindle. In this machine the table has three movements: longitudinal (at right angles to the spindle), transverse (parallel to the spindle), and vertical (up and down). It is the practice to take heavy cuts at fast speeds and coarse feeds in the classes of work for which the medium and larger sizes of plain milling machines are adapted. The rigid construction of the machine enables this to be done successfully, and it is this ability that is the chief value of the plain machine.

ing machine is regarded by many as one of the most important machines in the tool room.

The Vertical Spindle milling machine shown in Fig. 585 embodies the principles of the drilling machine. The spindle and table are similarly located, and the cutter is mounted in the spindle. The spindle has a vertical movement, and the table has vertical, longitudinal, and transverse movements. The vertical spindle milling machine is used for face milling, profiling, die sinking, and for various odd-shaped jobs. Owing to the position of the spindle, this type of machine can be used advantageously in boring holes.

The Manufacturing milling machine shown in Fig. 586 is used primarily for the production of small parts of typewriters, sewing machines, and similar machine and tool parts. The spindle is supported in bearings located in the adjustable head, which can be raised or lowered. The capacity of the machine is limited for work of great height because the table cannot be raised or lowered, the only adjustment being in the head. Furthermore, there is no transverse table feed, the only transverse movement being obtained by a slight adjust-

ment of the spindle.

The Planer milling machine shown in Fig. 587 is designed for the heaviest classes of slab and gang milling. It bears a marked resemblance to the planer, from which it derives its name. The spindle is mounted in bearings carried in a vertically adjustable slide similar to that of a planer, and is in a corresponding position. The class of work produced on the planer mill is identical to that of the column and knee type machine, therefore the same principles are involved.

The "Precision" Horizontal Boring, Drilling and Milling Machine, Fig. 588, is used for performing any or all of the stated operations on a piece of work at a single

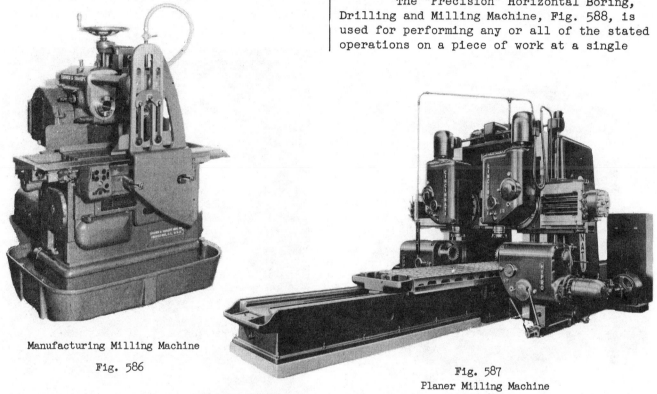

Manufacturing Milling Machine

Fig. 586

Fig. 587
Planer Milling Machine

Fig. 588. Horizontal Boring Machine

Fig. 589. Thread Miller

setting. This saves time in resetting the work and insures accuracy in relation of the machined surfaces to each other. The Thread Mill, Fig. 589, machines the different kinds of standard threads, using a special cutter.

Fig. 590 shows the Spline Mill used to mill slots and keyways with either open or closed ends.

In order to increase the range of work that can be done` on a milling machine, many mechanisms known as attachments are used. Some of the most commonly used attachments are discussed in the following paragraphs.

The Compound Vertical Milling Attachment, Figs. 591(a) and 591(b), is particularly applicable to a large variety of milling, because it can be set in two planes. It is especially advantageous when it is desired to set the spindle at an angle to the table, as in milling angular strips, table ways, etc., for with the spindle in this position, the full length of the table travel is available, and an ordinary end mill, instead of an angular cutter, can be used for milling the angle.

The Universal Milling Attachment, Fig. 592, as its name implies, is fully universal in regard to setting the spindle. Its range of work is very much the same as' is covered by the preceding attachments, and, in addition, it can do many unusual jobs because its spindle may be set at any angle in both planes. The outer end of this attachment is so made that an arbor yoke can be used to provide additional stability.

Rotary Attachments are used on a

variety of circular milling such as circular T-slots, segment outlines, etc., and on a great deal of tool and die-making jobs that require splining, slotting, or irregular form milling. Fig. 593 shows a hand feed unit, and Fig. 594 a unit equipped with a power feed. Both attachments are graduated on the circumferences of the tables, and both have adjustable dials on the worm shaft.

The Slotting Attachment, Fig. 595, is

Fig. 590. Spline Mill

(a)   (b)

Fig. 591

Fig. 592

Fig. 593

Fig. 594

Fig. 595

Fig. 596

lengths. The attachment can be set at any angle between 0 and 90°, either side of the center line, the position being indicated by graduations on the circumference of the head. The tool is held in place by a clamp bolt, and a stop that swings over the top of tool shank making it impossible for the tool to be pushed up.

The Rack Milling Attachment, Fig. 596, is used

largely used in tool making, such as in forming box tools for screw machines, making templates, splining keyways, and work of a similar character. The working parts consist of a tool slide that is driven from the machine spindle by an adjustable crank that allows the stroke to be set for different

for cutting teeth in racks. It can also be used in connection with the Universal Spiral Index Centers for cutting worms, on Universal Milling Machines, and for other miscellaneous operations. The cutter is mounted on the end of a spindle that extends through the attachment case parallel to the table T-slots. This

spindle is driven from the machine spindle by a train of hardened steel bevel and spur gears.

The High Speed Milling Attachment, Fig. 597, consists of a pair of gears for increasing the speed, and an auxiliary spindle that drives the cutter. It is used in order to obtain the correct speed for small milling cutters which should be run more rapidly than the fastest spindle speed when cutting keyways and slots, etc.

Fig. 597

The Tilting Table, Fig. 598, is designed primarily for use in connection with index centers when fluting taper reamers, taps, etc. In addition to this work, many other kinds of taper pieces can be accurately reproduced.

Fig. 598

Micrometer spindle and micrometer table setting attachments are used to facilitate the performing of extremely accurate boring and milling jobs. They are of particular advantage in making jigs, fixtures, dies, and other tool room work where accurate adjustments are required. Measuring bars used in connection with micrometer heads and dial gages are employed. These bars may be used in combinations to obtain various adjustments.

Fig. 599

Fig. 599 shows a micrometer spindle setting attachment.

The Index Head, Fig. 600, is used for obtaining equally spaced divisions on the periphery of work such as gears, drills, reamers, cutters, etc., and also for helical, spiral, and general taper work.

Triple Index Centers, Fig. 601, are used for manufacturing purposes where economy

Fig. 600

Fig. 601

and rapid production are important factors. All the spindles are operated simultaneously by moving the index crank. The ratio of the worm to the worm-wheel is 40 to 1, which is the same as the usual index head. When using the three spindles, a job 4 inches in diameter can be swung, and when using the two outside spindles, a job 8 inches in diameter can be swung.

Fig. 602

The Plain Vise, Fig. 602, is used for light milling operations. The bed and slides are made of cast iron, and the jaws are made of tool steel, hardened and ground. It is fastened to the table by means of a screw that passes through the bed and threads into a nut inserted in the table T-slot.

Fig. 603 shows the Flanged Vise, which differs very little from the plain vise, except that a slotted flange is provided at each end so that a T-slot bolt and nut can be used to fasten the vise to the surface of the table.

The upper part of the Swivel Vise in Fig. 604 is the same as the flanged vise, but the base is held to the table with a swivel, allowing the vise to be turned or swiveled at any angle. The entire circumference of the base is graduated in degrees.

The Toolmakers' Universal Vise in Fig. 605 is of great advantage to the toolmaker, as a job can be set at any angle in the horizontal or vertical plane.

The Spring Chucks for milling machines shown in Fig. 606 are convenient for holding wire, small rods, straight shank drills, mills, etc. The collet holder is made of steel, is ground to fit a standard tapered hole, and has a hole through its entire length. The spring collet is held in place by a cap nut that forces it against the tapered seat and closes the chuck concentrically.

The Collets shown in Fig. 607 are used for holding cutters and tools, depending

Fig. 606

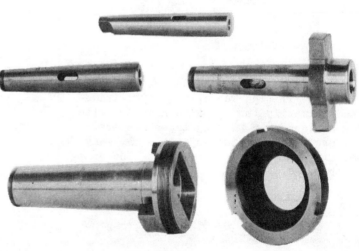

Fig. 607

Fig. 603   Fig. 604   Fig. 605

upon the nature of the work. Notice that one has a tang, one has a tapped hole to receive a draw bar, one has a tenon to help in driving, and one is used in connection with a split collet and has a cap which is used to close it to hold the cutter.

helical end-mills, right-hand cutters usually have left-hand helices and left-hand cutters have right-hand helices, so that there will not be a tendency for a cutter to pull out of the spindle when cutting.

Fig. 608. Old Style Spindle Arbor

Fig. 609. Standardized Spindle Arbor

Milling machine spindles were formerly made to accommodate arbors of the type shown in Fig. 608. Since 1927, when the National Machine Tool Builders' Association adopted a standard taper spindle end having a taper of 3½" per foot, arbors like the one shown in Fig. 609 have been used.

A milling machine operator should be familiar with the different types of milling machine cutters. Figures 610 to 625, on the following page, show some of the more common types of cutters.

Fig. 628. Straddle Tooth Sprocket Cutter

Fig. 629. Woodruff Key Seat Cutter

OPERATIONS

Some end-mills are right-hand (RH) and some are left-hand (LH). To distinguish between them, hold the shank of the cutter in your hand, with the shank end toward you. If it cuts when it revolves to the right (clockwise), it is a right-hand cutter, and if it cuts when it revolves to the left (counterclockwise), it is a left-hand cutter. In

Fig. 626. (Top) Convex and (Bottom) Concave Cutters

Fig. 630 illustrates a method of milling square or hexagon heads on bolts, using a

Fig. 627. Corner Rounding Cutters

Fig. 630

Fig. 610.
Coarse-Tooth Milling Cutter

Fig. 614.
Helical Plain Milling Cutter

Fig. 622.
Staggered Tooth Side
Milling Cutter

Fig. 611.
Angular Cutter

Fig. 615.   Taper Shank Cutter

Fig. 616.   Double-End Cutter

Fig. 617.   Two Lipped Taper Shank

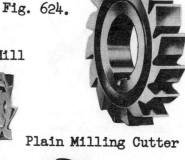

Fig. 623.   Face Milling
Cutter with Inserted Teeth

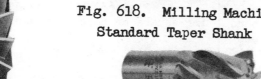

Fig. 618.   Milling Machine
Standard Taper Shank

Fig. 612.
Double Angle Cutter

Fig. 619.   Straight Shank Spiral End Mill

Fig. 624.

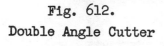

Fig. 620.   T-Slot Cutter

Plain Milling Cutter

Fig. 613.   Shell End Mill

Fig. 621.
Metal Slitting Saw

Fig. 625.
Involute Spur Gear Cutter

chuck on the dividing head spindle for holding the work. This figure also furnishes a
good example of the use of a pair of side
milling cutters as "straddle mills."   Two
sides are finished at a cut, completing a
square bolt head with two cuts or a hexagon
head with three cuts. In indexing the work,
the worm of the dividing head is thrown out
of mesh and the divisions are obtained from
the rapid index plate on the spindle nose.

　　　Fig. 630a illustrates face milling,
using an inserted tooth milling cutter. The
cutter is mounted directly on the nose of
the spindle and the longitudinal feed is
used.

Fig. 631

Fig. 630a

　　　The helical milling cutter in Fig.
631 is placed as near the spindle as possible. By using plen-
ty of cutting lubricant this
heavy plain slabbing cut can be
taken with a good finish.

　　　The operation shown in
Fig. 632 illustrates an excellent example of the use of the
circular milling attachment in
connection with the vertical
spindle milling machine, for
cutting circular T-slots in the
saddle of a universal milling
machine.

　　　The first or plain slot
is cut out on a boring mill or
can be milled at the same setting shown in Fig. 632, using
the two-lipped end mill, which
is then replaced by a T-slot cutter.

　　　Fig. 633 shows the method of setting a vise at 90° with
an arbor. In checking, the
point of the indicator is placed
against the solid jaw and the
longitudinal feed is used.

Fig. 632

Fig. 633

Fig. 634

The Gang Milling operation shown in Fig. 634 consists of cutting two grooves each 1.170" wide and $\frac{5}{16}$" deep in three steel forgings at one traverse of the table.

The Triple Index Centers shown in Fig. 601 are employed.

In heavy milling, the arbor must be rigidly supported by arbor yokes to prevent chattering.

A coolant should be used in milling almost all metals except cast iron.

SETTING CUTTER CENTRAL WITH WORK

In Fig. 636(a), the keyway is improperly cut. Notice that it is off-center. This would make it difficult to fit a key into the job when assembled with mating part.

Cutters are sometimes set central with a shaft by using the sight method, in which the cutter is forced into the work a few thousandths and fed across the work. A small oval, which should not exceed the width of the cutter by more than 1/16, is thus

SETTING CUTTER CENTRAL WITH WORK

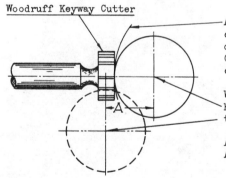

Woodruff Keyway Cutter

A small piece of paper is placed between cutter and work so that cutter will not burr or mark the side of the work.
CAUTION: Cutter must not be revolving when checking this dimension.

Work is moved from here to this position, for cutting keyway.

A = amount to move cross feed, or
A = $\frac{1}{2}$ diameter of work + $\frac{1}{2}$ thickness of cutter + thickness of paper.

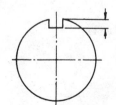

Depth of cut is always taken from the side and never from the center of the cut.

Look up the depth of cut in the American Machinists' Handbook.

Fig. 635

formed. The cutter is then set so that an equal portion of the oval is visible on each side of the cutter (see Fig. 636(b)). This may be measured with a rule if desired. The cutting edge of the cutter may also be lined up with the center line of the work by sight, as shown in Fig. 636(c).

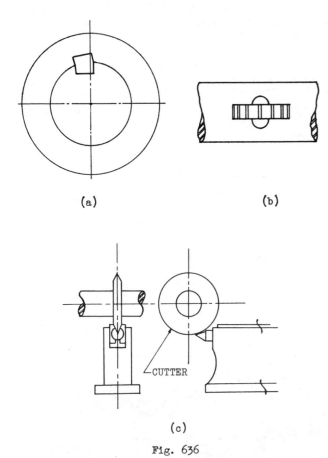

(a)   (b)

CUTTER

(c)

Fig. 636

DIVIDING HEAD

Principle parts of the dividing head:

1. Lead screw crank
2. 2nd gear on stud
3. 1st gear on stud
4. Gear on spindle
5. Lever for locking rapid index plate
6. Angular graduations
7. Angular locating point
8. Rapid index plate
9. Pin for locking index plate
10. Index plate
11. Sector arm
12. Index crank
13. Table lock
14. Table
15. Gear on worm
16. Idler gear
17. Graduated lead screw collar
18. Position of lead screw gear

The index head (Fig. 638) is a device used to divide the periphery of a piece of work into any number of equal parts, and to hold the work in the required position while the cuts are being made. The most essential parts of the index head are the worm and worm wheel, index plates, sector arms, and change gears.

The worm wheel has 40 teeth and the worm has a single thread. The worm wheel is keyed to the spindle, and when the worm, which is turned by the index crank, is turned 40 times, the spindle is revolved once. Fractional parts of a turn are obtained by means of the index plates which are furnished with each head. The sector arms are used to mark off the number of holes on the index plate which are required to make a fractional part of a turn of the index crank, without counting them each time the index crank is moved. The

Fig. 638

Fig. 637

change gears control the movement of the index plates and the spindle in differential
indexing and helical milling. Rapid, plain,
and differential indexing are the three
methods most commonly used.

RAPID INDEXING. In rapid indexing,
the worm and worm wheel are disengaged and
the spindle is moved by hand. The required
number of divisions on the work is made by
means of the rapid index plate (see Fig. 638),
located on the nose of the spindle. This
plate usually has 24 equally spaced holes,
and only the numbers which will divide evenly into 24 can be indexed (that is, 2, 3, 4,
6, 8, 12, and 24). An index pin placed in
one of the holes of the rapid index plate locates the spindle in the proper position, and
a clamping device locks it while the cut is
being made. This method is used when a large
number of duplicate parts are being milled.

RULE. Divide 24 by the number of divisions required, and the result equals the
number of holes to move in the rapid index
plate. This may be written as the following
formula (N = number of divisions required).

Number of Holes to Move = $\dfrac{24}{N}$

EXAMPLE. Index for a
hexagon head screw.

$24 \div 6 = 4$

To index for a hexagon
head screw, machine one side
and move four holes in the
rapid index plate for each of
the other sides.

CAUTION. After you have
finished rapid indexing, see
that the worm and worm wheel are
fully engaged, so that the machine is again set for plain indexing.

PLAIN INDEXING. Plain
indexing is a method of indexing
for numbers beyond the range of
rapid indexing. In this operation, the index head spindle is
moved by turning an index crank
attached to a worm which meshes
with the worm wheel. The worm
wheel has 40 teeth and the worm
has a single thread. For each
turn of the index crank, the
worm wheel moves one tooth, or

1/40 of a revolution. To cause the spindle
to make one turn, the index crank must make
40 turns, or in other words, the ratio between
the revolutions of the index crank and those
of the spindle is 40 to 1. A wide range of
divisions may be indexed by using this method.

The number of turns or fractional
parts of a turn of the index crank necessary
to cut any required number of divisions may
be easily determined by the following rule.

RULE. Divide 40 by the number of divisions required, and the result will equal
the number of turns of the index crank. Expressed as a formula (where N = number of divisions required and T = turns of the index
crank), this is:

$$\frac{40}{N} = T.$$

EXAMPLES. Index for five divisions

$$\frac{40}{N} = \frac{40}{5} = 8T.$$

Index for eight divisions

$$\frac{40}{N} = \frac{40}{8} = 5T$$

Fig. 639

Index for ten divisions

$$\frac{40}{N} = \frac{40}{10} = 4T.$$

Fig. 639 shows an index head set up for a plain indexing job.

When the number of divisions required does not divide evenly into 40, the index crank must be moved a fractional part of a turn. This is done by using index plates. The plates furnished with a Brown & Sharpe index head have circles containing the number of holes listed below. LEARN THE NUMBER OF HOLES IN EACH OF THESE CIRCLES.

Plate #1 - 15, 16, 17, 18, 19, 20
Plate #2 - 21, 23, 27, 29, 31, 33
Plate #3 - 37, 39, 41, 43, 47, 49

As another example, let us index for 18 divisions.

$$\frac{40}{N} = \frac{40}{18} = 2\frac{4}{18}T.$$

The whole number indicates the complete turns of the index crank. The denominator of the fraction represents the index circle to use and the numerator represents the number of holes to move in that circle. Since there is an index circle which contains 18 holes, we index for 18 equally spaced divisions by moving the index crank two complete turns and four holes in the 18 hole circle.

When the denominator is smaller or larger than the number of holes contained in any of the index circles, it can be increased or reduced by multiplying or dividing both terms of the fraction by a number which will give a fraction whose denominator is the same as the number of holes in one of the index plates.

For example, assume that it is necessary to index for 1-1/3 turns of the index crank. Select an index circle on one of the index plates in which the number of holes is a multiple of 3, and it will be found that either the 15, 18, 21, 27, 33, or 39 hole circle may be used. Divide the number of holes in the selected circle by 3 and obtain the common multiple for both terms of the fraction. Assuming that the 27 hole circle was selected, the common multiple would be obtained by dividing 27 by 3, giving a result of 9. Multiplying each term of the fraction by 9, we have

$$\frac{1}{3} \times \frac{9}{9} = \frac{9}{27}$$

The denominator indicates the number of holes in the circle and the numerator indicates the number of holes to move the index crank pin that circle. When the index crank pin is moved 9 holes in the 27 hole circle, 1/3 of a turn has been made.

When the number of divisions to be indexed exceeds 40, both terms of the fraction may be divided by a common divisor to obtain an index circle which is available. If 160 divisions are required, for example, the fraction to be used is

$$\frac{40}{N} \text{ or } \frac{40}{160}$$

Since there is no 160 hole circle, this fraction can be reduced to 1/4 by dividing each term of the fraction by 40. The only index circle in the standard index plates which is a multiple of 4 is the 16 hole circle, and by multiplying each term of the fraction by 4, we obtain the following result:

$$\frac{1}{4} \times \frac{4}{4} = \frac{4}{16}$$

EXAMPLES

Index for 6 divisions

$$\frac{40}{N} = \frac{40}{6} = 6\frac{2}{3}T$$

$$\frac{2}{3} \times \frac{5}{5} = \frac{10}{15} \text{ or}$$

6 turns 10 holes 15 circle

Index for 9 divisions

$$\frac{40}{N} \times \frac{40}{9} = 4\frac{4}{9}T$$

$$\frac{4}{9} \times \frac{2}{2} = \frac{8}{18} \text{ or}$$

4 turns 8 holes 18 circle

Index for 65 divisions

$$\frac{40}{N} = \frac{40}{65} = \frac{8}{13}$$

$$\frac{8}{13} \times \frac{3}{3} = \frac{24}{39} \text{ or}$$

24 holes 39 circle

Index for 136 divisions

$$\frac{40}{N} = \frac{40}{136} = \frac{5}{17}$$

$$\frac{5}{17} \times \frac{1}{1} = \frac{5}{17} \text{ or}$$

5 holes 17 circle

CAUTION. In setting the sector arms to space off the proper number of holes in the index circle, DO NOT count the hole the index crank pin is in.

DIFFERENTIAL INDEXING. The differential method of indexing is used in indexing for numbers beyond the range of plain indexing. This is accomplished by connecting the index plate to the spindle by means of a gear train, so that the index plate can rotate in relationship to the movement of the spindle. By a proper arrangement of the gearing, the index plate can be made to move fast or slow, and in the same direction (positive) or in the opposite direction (negative) to the index crank. This causes the movement of the index plate to be either greater or less than the actual movement of the index crank. Before differential indexing is attempted, gearing and the forming of gear ratios should be understood (see GEARING section).

The standard change gears (12 in all) that are furnished with each index head have the following number of teeth: 24 (two gears), 28, 32, 40, 44, 48, 56, 64, 72, 86, and 100. Special gears having 46, 47, 52, 58, 68, 70, 76, and 84 teeth may also be furnished. LEARN THE NUMBER OF TEETH IN EACH OF THESE GEARS.

When the required number of divisions cannot be indexed by plain indexing, an approximate number of divisions which can be indexed by the plain indexing method is selected. The difference between the movement of the spindle thus secured and the necessary movement is corrected by the use of change gears. The proper gearing is found in the following manner.

1st Step.   Plain Indexing $-\dfrac{40}{n} = T$

(n = selected number)

Select some number either greater or less than the required number for which plain indexing can be used.

2nd Step.   Gear Ratio - $(n - N) \times \dfrac{40}{n}$

The gear ratio is found by using the formula $(n - N) \times \dfrac{40}{n}$, in which n = approximate number and N = required number.

3rd Step.   Select Gears.

The gears may be either simple or compound. State idlers to be used.

Simple Gearing

$$\frac{3}{8} \times \frac{8}{8} = \frac{24}{64}$$

Compound Gearing

$$\frac{16}{33} = \frac{\overset{24}{\cancel{2}} \times \overset{64}{\cancel{16}}}{\underset{72}{\cancel{3}} \times \underset{44}{\cancel{11}}} = \frac{24}{72} \times \frac{64}{44}$$

In differential indexing, the numerators of the fractions indicate the driving gears and the denominators indicate the driven gears. Idler gears control the direction of rotation of the index plate and are arranged as follows.

Simple Gearing   - 1 idler for positive motion of index plate
                   2 idlers for negative motion of index plate

Compound Gearing - 1 idler for negative motion of index plate
                   No idlers for positive motion of index plate

In case #1 an approximate number greater than the required number was selected. Note that when the approximate number is greater than the required number, the index plate must turn in the positive direction. Using simple gearing, this requires one idler gear.

In case #2 an approximate number smaller than the required number was used. Notice that when the approximate number is less than the required number, the index plate must turn in the negative direction. Using simple gearing, two idler gears are required.

In case #3 an approximate number smaller than the required number was used. Notice that in this case compound gearing is necessary, and that since the approximate number is less than the required number, the index plate must turn in the negative direction. This requires one idler gear.

These examples show that the approximate number may be greater or less than the required number and that the speed and the direction of rotation of the index plate can be controlled by the change gears. The difference between the approximate number and the required number is limited only by the index hole circles and change gears which are available.

To index for 57 divisions:

	CASE #1	CASE #2	CASE #3
STEP #1 PLAIN INDEXING	$\frac{40}{N} = \frac{40}{57}$ $\frac{40}{60} = \frac{2}{3}$ $\frac{2}{3} \times \frac{7}{7} = \frac{14}{21}$ 14 Holes 21 Circle	$\frac{40}{N} = \frac{40}{57}$ $\frac{40}{56} = \frac{5}{7}$ $\frac{5}{7} \times \frac{3}{3} = \frac{15}{21}$ 15 Holes 21 Circle	$\frac{40}{N} = \frac{40}{57}$ $\frac{40}{54} = \frac{20}{27}$ 20 Holes 27 Circle
STEP #2 GEAR RATIO	$(n - N) \times \frac{40}{n} =$ $(60 - 57) \times \frac{40}{60} =$ $\cancel{3} \times \frac{2}{\cancel{3}} = \frac{2}{1}$	$(n - N) \times \frac{40}{n} =$ $(56 - 57) \times \frac{40}{56} =$ $-1 \times \frac{5}{7} = -\frac{5}{7}$	$(n - N) \times \frac{40}{n} =$ $(54 - 57 \times \frac{40}{54} =$ $-\cancel{3} \times \frac{20}{\cancel{27}} \quad -\frac{20}{9}$ 9
STEP #3 SELECT GEARS	$\frac{2}{1} \times \frac{24}{24} = \frac{48}{24}$ Driver Driven 1 Idler	$-\frac{5}{7} \times \frac{8}{8} = \frac{-40}{56}$ Driver Driven 2 Idlers	$\quad\quad\quad\quad 64 \quad 40$ $-\frac{20}{9} = \frac{-4 \times 5}{\cancel{3} \times \cancel{3}}$ Driver Driven $\quad\quad\quad\quad 48 \quad 24$ 1 Idler

Fig. 640 shows an index head geared for differential indexing (simple gearing) while (Fig. 641) shows an index head geared for differential indexing (compound gearing).

Fig. 641

Fig. 640

A chart for plain and differential indexing, similar to the one shown below, is available in most milling machine departments.

INDEX TABLE 51 to 92.

NUMBER OF DIVISIONS	INDEX CIRCLE	NO. OF TURNS OF INDEX	GRADUATION	GEAR ON WORM	1ST GEAR ON STUD	2ND GEAR ON STUD	GEAR ON SPINDLE	IDLERS NO. 1 HOLE	IDLERS NO. 2 HOLE
69	20	12/20	118	40			56	24	44
70	49	28/49	112						
71	21	12/21	113	72			40	24	
72	27	15/27	110	72			40	24	
	18	10/18	109						
73	27	15/27	110						
73	18	10/18	109	28			48	24	44
73	49	28/49	112	28			48	24	44
74	21	12/21	113						
75	37	20/37	107						
76	15	8/15	105	32			48	44	
77	19	10/19	103						
78	20	10/20	98	48			24	44	
79	39	20/39	101						
80	20	10/20	98	48			24	44	
81	20	10/20	98	32			48	24	44
82	41	20/41	96						
83	26	10/26	98						
84	21	10/21	94	40			24	44	
85	17	8/17	92						
86	43	20/43	91						
87	15	7/15	92	72			32	44	
88	33	15/33	89	72			32	44	
89	27	12/27	88						
	18	8/18	87						
90	27	12/27	88						
90	18	8/18	87	24			48	24	44
91	39	18/39	91	24			48	24	44
92	23	10/23	86						

NUMBER OF DIVISIONS	INDEX CIRCLE	NO. OF TURNS OF INDEX	GRADUATION	GEAR ON WORM	1ST GEAR ON STUD	2ND GEAR ON STUD	GEAR ON SPINDLE	IDLERS NO. 1 HOLE	IDLERS NO. 2 HOLE
51	17	14/17	33*	24			48	24	44
52	39	30/39	152	56	40	24	72		
53	49	35/49	140	56	40	24	72		
54	21	15/21	142						
55	27	20/27	147						
56	33	24/33	144						
56	49	35/49	140	56			40	24	44
57	21	15/21	142	56			40	24	44
57	49	35/49	140						
58	21	15/21	142						
58	29	20/29	136	48			32	44	
59	39	26/39	132	48			32	44	
59	33	22/33	132	48			32	44	
60	18	12/18	132						
60	39	26/39	132						
61	33	22/33	132	48			32	24	44
61	18	12/18	132	48			32	24	44
62	31	20/31	127	48			32	24	44
63	39	26/39	132	24			48	24	44
63	33	22/33	132	24			48	24	44
64	18	12/18	132	24			48	24	44
64	16	10/16	123						
65	39	24/39	121						
66	33	20/33	120						
67	49	28/49	112	28			48	44	
67	21	12/21	113	28			48	44	
68	17	10/17	116						

PLAIN & DIFFERENTIAL INDEXING

INDEX TABLE 2 to 50

Gear on Spindle 64 T.
Gear on Worm 40 T.
Idler 24 T.
2nd Gear on Stud 32 T.
1st Gear on Stud 56 T.
GEARED FOR 107
No. 1 Hole
No. 2 Hole

NUMBER OF DIVISIONS	INDEX CIRCLE	NO. OF TURNS OF INDEX	GRADUATION
40	Any	1	
41	41	40/41	3*
42	21	20/21	9*
43	43	40/43	12*
44	33	30/33	17*
45	27	24/27	21*
46	18	16/18	21*
47	23	20/23	172
48	47	40/47	168
49	18	15/18	165
49	49	40/49	161
50	20	16/20	158

GRADUATIONS IN TABLE INDICATE SETTING FOR SECTOR ARMS WHEN INDEX CRANK MOVES THROUGH ARC "A," EXCEPT IN CASES MARKED * WHEN THE INDEX CRANK MOVES THROUGH ARC "B."

NUMBER OF DIVISIONS	INDEX CIRCLE	NO. OF TURNS OF INDEX	GRADUATION
26	39	1 21/39	106
27	27	1 13/27	95
28	49	1 21/49	83
29	21	1 9/21	85
29	29	1 11/29	75
30	39	1 13/39	65
31	33	1 11/33	65
32	18	1 6/18	65
33	31	1 9/31	56
34	20	1 5/20	48
35	33	1 7/33	41
36	17	1 3/17	33
36	49	1 7/49	26
37	21	1 3/21	28
38	27	1 3/27	21
38	18	1 2/18	21
39	37	1 3/37	15
39	19	1 1/19	9
39	39	1 1/39	3

NUMBER OF DIVISIONS	INDEX CIRCLE	NO. OF TURNS OF INDEX	GRADUATION
13	39	3 3/39	14
14	49	2 42/49	169
14	21	2 18/21	170
15	39	2 26/39	132
15	33	2 22/33	132
16	18	2 12/18	132
17	20	2 10/20	98
18	17	2 6/17	69
18	27	2 6/27	43
19	18	2 4/18	43
20	19	2 2/19	19
21	Any	2	18*
22	21	1 19/21	161
23	33	1 27/33	147
24	23	1 17/23	132
24	39	1 26/39	132
24	33	1 22/33	132
25	18	1 12/18	132
25	20	1 12/20	118

NUMBER OF DIVISIONS	INDEX CIRCLE	NO. OF TURNS OF INDEX	GRADUATION
2	Any	20	
3	39	13 13/39	65
3	33	13 11/33	65
3	18	13 6/18	65
4	Any	10	
5	Any	8	
6	39	6 26/39	132
6	33	6 22/33	132
6	18	6 12/18	132
7	49	5 35/49	140
7	21	5 15/21	142
8	Any	5	
9	27	4 12/27	88
9	18	4 8/18	87
10	Any	4	
11	33	3 33/33	126
12	39	3 13/39	65
12	33	3 11/33	65
12	18	3 6/18	65

INDEX TABLES

EXERCISES

1. Briefly describe the index head and explain its purpose.
2. Describe the following parts of an index head and explain the purpose of each:
   (a) Worm and worm wheel, (b) Index plates, (c) Sector arms, and (d) change gears.
3. Tell what the usual index head ratio is, and explain what it means.
4. Name and explain briefly three kinds of indexing in common use.
5. Explain how the sector arms should be set for indexing twelve divisions.
6. Explain what must be done in changing from rapid indexing to plain indexing.
7. Explain what must be done to change the index head from plain indexing to differential indexing.
8. List the number of holes in each of the standard index plates, and the number of teeth contained in the standard change gears furnished with a Brown & Sharpe index head.
9. Explain the meaning of the following terms: (a) Gears, (b) Gear ratio, (c) Gear train, (d) Idler gear, (e) Simple gear ratio, and (f) Compound gear ratio.
10. Index for the following equally spaced divisions: 8, 12, 24, 37, 43, 56, 61, 96, 129, 173.

GRADUATING

Flat rules and verniers may be graduated (divided into regular intervals) on a milling machine, by using a pointed tool and an index head. The tool is held stationary in a fly cutter holder (Fig. 642). This is mounted in the spindle of the machine, or it may be fastened to the spindle of a vertical milling machine or rack cutting attachment. The work is clamped to the table parallel to the "T" slots. The index head spindle is geared to the table feed screw with gears having a 1-1 ratio. The table is moved longitudinally by turning the index crank. Fractional parts of a turn are obtained by means of the index plates, the same as in plain indexing. The lines are cut by moving the table transversely under the point of the tool. The movement of the table is controlled with the hand feeds.

Fig. 642

Fig. 643 shows the milling machine set-up for graduating. <u>Notice how the gears are arranged.</u> When the index crank is turned one revolution, the spindle turns 1/40 of a revolution, and through the 1-1 ratio causes the lead screw to move 1/40 of a revolution. Since the usual lead of the lead screw is .250, one turn of the index crank will move the table.

$$\frac{1}{40} \text{ of } \frac{.250}{1} = .00625.$$

When the table is to be moved any required distance, divide the required distance by the distance advanced in one turn of the index crank, and the result will be the number of turns of the index crank that are necessary. This may be written as a formula, as follows:

$$T = \frac{W}{.00625}$$

where T = number of turns of index crank, W = width of divisions, and .00625 = the distance that the table moves in one turn of the index crank.

SHOW HOW TO INDEX FOR LINES 1/32 (.03125) APART

$$\frac{W}{.00625} = \frac{.03125}{.00625} = 5$$

Turn the index crank five turns in any circle.

For instructions in combining fractions and continued fractions see Practical Shop Mathematics, Volume II--Advanced, by Wolfe and Phelps, McGraw-Hill Book Co.

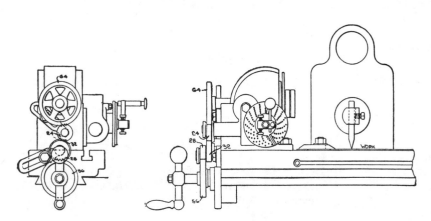

Fig. 643

SHOW HOW TO INDEX FOR LINES .0481 APART

$$\frac{W}{.00625} = \frac{.04810}{.00625} = 7\frac{435}{625} \text{ turns of the index crank}$$

$7\frac{435}{625}$ may be changed to $7\frac{16}{23}$ by continued fractions, as follows:

435	625	1
380	435	2
55	190	3
50	165	2
5	25	5
	25	
	0	

		1	2	3	2	5	
1	0	1	2	7	16	87	x 5 = 435
0	1	1	3	10	23	125	x 5 = 625
		1	2	3	2	5	

To index for 7-16/23 turns, turn the index crank 7 turns and 16 holes in the 23 hole circle.

TO FIND THE ERROR IN EACH DIVISION

The error in each division is equal to the difference between the distance the table should move and the distance the selected indexing actually moves it.

$$\frac{87}{125} \times \frac{.00625}{1} = .0043500 \text{ distance the table should move}$$

$$\frac{16}{23} \times \frac{.00625}{1} = .0043478 \text{ distance the table actually moves}$$
$$\underline{} \quad .0000022 \text{ error in each division}$$

TO FIND THE ERROR IN EACH INCH

The error in each inch is equal to the number of divisions in one inch multiplied by the error in each division.

$1.000 \div W$ = number of divisions in 1"

$1.000 \div .0481$ = 20.79 divisions in 1"

Error in each division X divisions in 1" = error in 1"

$.0000022$ X 20.79 = 0.000045 error in 1"

USE OF ANGULAR INDEXING TABLE

60 Seconds (") = 1'   90° (Degrees) = 1 Right Angle

60 Minutes (') = 1°   360° (Degrees) = 1 Circle

The index head requires 40 turns of the index crank to move the spindle through one revolution, or 360°. Therefore one turn of the index crank will move the spindle 1/40 x 360° = 9°.

$$9° = 9 \times 60' \text{ or } 540'$$
$$9° = 9 \times 60' \times 60" \text{ or } 32,400"$$

In angles given only in degrees, divide the required number of degrees by 9 to obtain the required turns of the index crank. In angles involving degrees and minutes, reduce the angles to minutes and divide by 540. In angles involving degrees, minutes, and seconds, reduce the angles to seconds and divide by 32,400. Carry the division to the fourth decimal place, and the result will be the number of turns of the index crank necessary to index the angle.

To find the index circle which must be used for the decimal part of a turn, find the required decimal or the nearest decimal to it in the table on page 175. The index circle is listed under "C" and the number of holes to move in that circle is listed under "H". The following examples show application of the table.

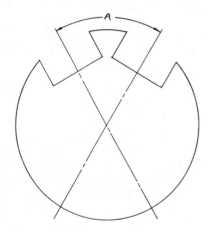

Fig. 644

1. Angle A = 24°45'

$$24° = 24 \times 60' = 1440'$$
$$45' = \underline{45'}$$
$$1485'$$

$$1485 \div 540 = 2.7500$$

In the table opposite .7500, 12 is under "H" and 16 is under "C". TOTAL INDEXING = 2 Turns, 12 Holes, 16 Circle.

ANGULAR INDEXING TABLE

Value	H	C	Value	H	C	Value	H	C	Value	H	C	Value	H	C	Value	H	C	Value	H	C	Value	H	C
.0204	1	49	.1395	6	43	.2609	6	23	.3846	15	39	.5106	24	47	.6279	27	43	.7500	12	16	.8718	34	39
.0213	1	47	.1429	3	21	.2632	5	19	.3871	12	31	.5116	22	43	.6296	17	27	.7500	15	20	.8723	41	47
.0233	1	43	.1429	7	49	.2653	13	49	.3878	19	49	.5122	21	41	.6316	12	19	.7551	37	49	.8750	14	16
.0244	1	41	.1463	6	41	.2667	4	15	.3888	7	18	.5128	20	39	.6326	31	49	.7561	31	41	.8776	43	49
.0256	1	39	.1481	4	27	.2683	11	41	.3902	16	41	.5135	19	37	.6341	26	41	.7568	28	37	.8780	36	41
.0270	1	37	.1489	7	47	.2703	10	37	.3913	9	23	.5151	17	33	.6364	21	33	.7576	25	33	.8788	29	33
.0303	1	33	.1500	3	20	.2727	9	33	.3939	13	33	.5161	16	31	.6383	30	47	.7586	22	29	.8824	15	17
.0323	1	31	.1515	5	33	.2759	8	29	.3953	17	43	.5172	15	29	.6410	25	39	.7619	16	21	.8837	38	43
.0345	1	29	.1538	6	39	.2766	13	47	.4000	6	15	.5185	14	27	.6452	20	31	.7647	13	17	.8888	16	18
.0370	1	27	.1579	3	19	.2777	5	18	.4000	8	20	.5217	12	23	.6471	11	17	.7674	33	43	.8888	24	27
.0408	2	49	.1613	5	31	.2791	12	43	.4043	19	47	.5238	11	21	.6486	24	37	.7692	30	39	.8919	33	37
.0426	2	47	.1622	6	37	.2821	11	39	.4054	15	37	.5263	10	19	.6500	13	20	.7742	24	31	.8936	42	47
.0435	1	23	.1628	7	43	.2857	14	49	.4074	11	27	.5294	9	17	.6512	28	43	.7755	38	49	.8947	17	19
.0465	2	43	.1633	8	49	.2857	6	21	.4082	20	49	.5306	26	49	.6522	15	23	.7760	36	47	.8966	26	29
.0476	1	21	.1666	3	18	.2903	9	31	.4103	16	39	.5319	25	47	.6531	32	49	.7777	21	27	.8974	35	39
.0488	2	41	.1702	8	47	.2927	12	41	.4118	7	17	.5333	8	15	.6552	19	29	.7777	14	18	.8980	44	49
.0500	1	20	.1707	7	41	.2941	5	17	.4138	12	29	.5349	23	43	.6585	27	41	.7805	32	41	.9000	18	20
.0513	2	39	.1724	5	29	.2963	8	27	.4146	17	41	.5366	22	41	.6596	31	47	.7826	18	23	.9024	37	41
.0526	1	19	.1739	4	23	.2973	11	37	.4186	18	43	.5385	21	39	.6666	10	15	.7838	29	37	.9032	28	31
.0541	2	37	.1765	3	17	.2979	14	47	.4194	13	31	.5405	20	37	.6666	12	18	.7872	37	47	.9048	19	21
.0555	1	18	.1795	7	39	.3000	6	20	.4211	8	19	.5454	18	33	.6666	14	21	.7879	26	33	.9070	39	43
.0588	1	17	.1818	6	33	.3023	13	43	.4242	14	33	.5484	17	31	.6666	18	27	.7895	15	19	.9090	30	33
.0606	2	33	.1837	9	49	.3030	10	33	.4255	20	47	.5500	11	20	.6666	22	33	.7907	34	43	.9130	21	23
.0612	3	49	.1852	5	27	.3043	7	23	.4286	9	21	.5510	27	49	.6666	26	39	.7931	23	29	.9149	43	47
.0625	1	16	.1860	8	43	.3061	15	49	.4286	21	49	.5517	16	29	.6735	33	49	.7949	31	39	.9184	45	49
.0638	3	47	.1875	3	16	.3077	12	39	.4324	16	37	.5532	26	47	.6744	29	43	.7959	39	49	.9189	34	37
.0645	2	31	.1892	7	37	.3103	9	29	.4348	10	23	.5555	10	18	.6757	25	37	.8000	12	15	.9231	36	39
.0666	1	15	.1905	4	21	.3125	5	16	.4359	17	39	.5555	15	27	.6774	21	31	.8000	16	20	.9259	25	27
.0690	2	29	.1915	9	47	.3158	6	19	.4375	7	16	.5581	24	43	.6809	32	47	.8049	33	41	.9268	38	41
.0698	3	43	.1935	6	31	.3171	13	41	.4390	18	41	.5610	23	41	.6829	28	41	.8065	25	31	.9302	40	43
.0732	3	41	.1951	8	41	.3191	15	47	.4419	19	43	.5625	9	16	.6842	13	19	.8085	38	47	.9310	27	29
.0741	2	27	.2000	3	15	.3226	10	31	.4444	8	18	.5641	22	39	.6875	11	16	.8095	17	21	.9333	14	15
.0769	3	39	.2000	4	20	.3243	12	37	.4444	12	27	.5652	13	23	.6897	20	29	.8108	30	37	.9355	29	31
.0811	3	37	.2041	10	49	.3256	14	43	.4468	21	47	.5676	21	37	.6923	27	39	.8125	13	16	.9362	44	47
.0816	4	49	.2051	8	39	.3265	16	49	.4483	13	29	.5714	12	21	.6939	34	49	.8140	35	43	.9375	15	16
.0851	4	47	.2069	6	29	.3333	5	15	.4490	22	49	.5714	28	49	.6957	16	23	.8148	22	27	.9388	46	49
.0870	2	23	.2093	9	43	.3333	6	18	.4500	9	20	.5745	27	47	.6969	23	33	.8163	40	49	.9394	31	33
.0909	3	33	.2105	4	19	.3333	7	21	.4516	14	31	.5757	19	33	.6977	30	43	.8181	27	33	.9412	16	17
.0930	4	43	.2121	7	33	.3333	9	27	.4545	15	33	.5789	11	19	.7000	14	20	.8205	32	39	.9444	17	18
.0952	2	21	.2128	10	47	.3333	11	33	.4595	17	37	.5806	18	31	.7021	33	47	.8235	14	17	.9459	35	37
.0968	3	31	.2162	8	37	.3333	13	39	.4615	18	39	.5814	25	43	.7027	26	37	.8261	19	23	.9474	18	19
.0976	4	41	.2174	5	23	.3404	16	47	.4634	19	41	.5854	24	41	.7037	19	27	.8276	24	29	.9487	37	39
.1000	2	20	.2195	9	41	.3415	14	41	.4651	20	43	.5862	17	29	.7059	12	17	.8293	34	41	.9500	19	20
.1020	5	49	.2222	6	27	.3448	10	29	.4667	7	15	.5882	10	17	.7073	29	41	.8298	39	47	.9512	39	41
.1026	4	39	.2222	4	18	.3469	17	49	.4681	22	47	.5897	23	39	.7097	22	31	.8333	15	18	.9524	20	21
.1034	3	29	.2245	11	49	.3478	8	23	.4694	23	49	.5918	29	49	.7143	15	21	.8367	41	49	.9535	41	43
.1053	2	19	.2258	7	31	.3488	15	43	.4706	8	17	.5926	16	27	.7143	35	49	.8372	36	43	.9565	22	23
.1064	5	47	.2308	9	39	.3500	7	20	.4737	9	19	.5946	22	37	.7179	28	39	.8378	31	37	.9574	45	47
.1081	4	37	.2326	10	43	.3514	13	37	.4762	10	21	.5957	28	47	.7209	31	43	.8387	26	31	.9592	47	49
.1111	2	18	.2340	11	47	.3529	6	17	.4783	11	23	.6000	9	15	.7222	13	18	.8421	16	19	.9630	26	27
.1111	3	27	.2353	4	17	.3548	11	31	.4815	13	27	.6000	12	20	.7234	34	47	.8462	33	39	.9655	28	29
.1163	5	43	.2381	5	21	.3590	14	39	.4828	14	29	.6047	26	43	.7241	21	29	.8485	28	33	.9677	30	31
.1176	2	17	.2414	7	29	.3617	17	47	.4839	15	31	.6060	20	33	.7273	24	33	.8500	17	20	.9697	32	33
.1212	4	33	.2424	8	33	.3636	12	33	.4848	16	33	.6087	14	23	.7297	27	37	.8511	40	47	.9730	36	37
.1220	5	41	.2432	9	37	.3659	15	41	.4865	18	37	.6098	25	41	.7317	30	41	.8519	23	27	.9744	38	39
.1224	6	49	.2439	10	41	.3673	18	49	.4872	19	39	.6111	11	18	.7333	11	15	.8537	35	41	.9756	40	41
.1250	2	16	.2449	12	49	.3684	7	19	.4878	20	41	.6122	30	49	.7347	36	49	.8571	18	21	.9767	42	43
.1277	6	47	.2500	4	16	.3704	10	27	.4884	21	43	.6129	19	31	.7368	14	19	.8571	42	49	.9787	46	47
.1282	5	39	.2500	5	20	.3721	16	43	.4894	23	47	.6154	24	39	.7391	17	23	.8605	37	43	.9796	48	49
.1290	4	31	.2553	12	47	.3750	6	16	.4898	24	49	.6170	29	47	.7407	20	27	.8621	25	29			
.1304	3	23	.2558	11	43	.3784	14	37	.5000	8	16	.6190	13	21	.7419	23	31	.8649	32	37			
.1333	2	15	.2564	10	39	.3793	11	29	.5000	9	18	.6207	18	29	.7436	29	39	.8666	13	15			
.1351	5	37	.2581	8	31	.3810	8	21	.5000	10	20	.6216	23	37	.7442	32	43	.8696	20	23			
.1379	4	29	.2593	7	27	.3830	18	47	.5102	25	49	.6250	10	16	.7447	35	47	.8710	27	31			

Examples showing the use of this table are given on pages 174 and 176.

2. Angle A = 16°26'

$$18° = 18 \times 60' = 1080'$$
$$26' = \underline{26'}$$
$$1106'$$

$$1106 \div 540 = 2.0481$$

Required decimal = .0481
Nearest decimal = $\underline{.0476}$
$$.0005$$

$$540 \times .0005 = .27'$$
$$.27 \times 60'' = 16.2'' \text{ ERROR}$$

In the table opposite .0476, 1 is under "H" and 21 is under "C". TOTAL INDEXING = 2 Turns, 1 Hole, 21 Circle.

3. Angle A = 24°54'23"

$$24° = 24 \times 60' \times 60'' = 86,400''$$
$$54' = 54 \times 60'' = 3,240''$$
$$23'' = \underline{23''}$$
$$89,663''$$

$$89,663 \div 32,400 = 2.7674$$

In the table opposite .7674, 33 is under "H" and 43 is under "C". TOTAL INDEXING = 2 Turns, 33 Holes, 43 Circle.

4. Angle A = 39°51'21"

$$39° = 39 \times 60' \times 60'' = 140,400''$$
$$51° = 51 \times 60'' = 3,060''$$
$$21'' = \underline{21''}$$
$$143,481''$$

$$143,481 \div 32,400 = 4.4284$$

Nearest decimal = .4286
Required decimal = $\underline{.4284}$
$$.0002$$

$$32,400'' \times .0002 = 6.48'' \text{ ERROR}$$

In the table opposite .4286, 9 is under "H" and 21 is under "C". TOTAL INDEXING = 4 Turns, 9 Holes, 21 Circle.

ANGULAR INDEXING. Angular indexing is the type of indexing used when the measurement on the job to be indexed is given as an angle. If the work requires a greater degree of accuracy than can be obtained by use of the Angular Indexing Table, differential indexing may be used.

EXAMPLES

Index for 25° (N = degrees in given angle)

$$\frac{N°}{9°} = T \qquad \frac{N°}{9°} = \frac{25°}{9°} = 2\frac{7}{9}T \qquad 2\frac{7}{9} \times \frac{2}{2} = 2\frac{14}{18}T$$

Index 2 turns and 14 holes in 18 hole circle.

Index for 12°12'. (N' = minutes in given angles)

$$\frac{N'}{540'} = T \qquad 12° = 12 \times 60' = 720'$$
$$\phantom{\frac{N'}{540'} = T \qquad} 12' = \underline{12'}$$
$$\phantom{\frac{N'}{540'} = T \qquad 12° = 12 \times 60' = }732'$$

$$\frac{732}{540} = 1\frac{192}{540} = 1\frac{1}{3} \text{ approx.} \qquad 1\frac{1}{8} \times \frac{5}{5} = 1\frac{5}{15}T$$

Index 1 turn and 5 holes in 15 hole circle.

Gearing.

$$1\frac{1}{3} = \frac{4}{3}$$

$$\frac{\cancel{732}}{\cancel{540}} \times \frac{\cancel{6}}{\cancel{4}} \times \frac{40}{1} = \frac{122}{3} = 40\frac{2}{3} \text{ turns}$$

$$40\frac{2}{3} - 40 = \frac{2}{3} \text{ Gear Ratio}$$

$$\frac{2}{3} \times \frac{24}{24} = \frac{48 \text{ Driving Gear}}{72 \text{ Driven Gear}} \; - 1 \text{ Idler}$$

Index for 29°25'16". Check error in arc. (·N" = seconds in given angle)

$$\frac{N''}{32400''} = T \qquad 29° = 29 \times 60' \times 60'' = 104400$$
$$\phantom{\frac{N''}{32400''} = T \qquad} 25' = 25' \times 60'' = 1500$$
$$\phantom{\frac{N''}{32400''} = T \qquad} 16'' = \underline{16}$$
$$\phantom{\frac{N''}{32400''} = T \qquad 29 \times 60' \times 60'' = 10}105916$$

$$\frac{105916}{32400} = 3\frac{3}{16} \text{ turns approx., or 3 turns and}$$

3 holes in 16 hole circle.

Gearing.

$$3\frac{3}{16} = \frac{51}{16} \qquad \frac{105916}{32400} \times \frac{16}{51} \times \frac{40}{1} = 41\frac{473}{20655}$$

$$41\frac{473}{20655} - 40 = 1\frac{473}{20655} \qquad 1\frac{473}{20655} = \frac{21128}{20655}$$

20655	21128	1
20339	20655	43
316	473	1
314	316	2
2	157	78
2	156	2
	1	

	1	43	1	2	78	2
0	1	1	44	(45)	134	10497	21128
1	0	1	43	(44)	131	10262	20655
	1	43	1	2	78	2

$$\frac{45}{44} = \frac{\cancel{8} \times \cancel{8}}{\cancel{4} \times \cancel{11}} = \frac{72}{64} \times \frac{40}{44} \quad \frac{\text{Driving Gear}}{\text{Driven Gear}}$$

Check for Error.

$$41\frac{473}{20655} = 41\frac{1}{44} \text{ approx.}$$

$$\frac{1805}{44} \times \frac{1}{40} \times \frac{51}{16} \times \frac{32400"}{1} = 105915.55"$$

$$105916" - 105915.55" = .45" \text{ error.}$$

HELICAL AND SPIRAL MILLING. When the spindle of an index head is geared to the lead screw of a milling machine so that the work revolves on its axis as the table moves along the ways, a helical or spiral cut is produced. When the cut is made on cylindrical work it is called a helical cut and when made on a tapered piece it is called a spiral cut. Helical milling cutters, helical gears, twist drills, counterbores, and similar work are produced in this way.

Before a helical cut can be made, the lead of the helix, the angle of the helix, and the diameter of the work must be known. The lead is equal to the distance the table advances when the work makes one revolution. Any change in the gearing connecting the index head spindle and the lead screw will change the lead of the helix. The helix angle is the angle the cut makes with the axis of the work, and changes with any change in the lead or in the diameter of the work. THE TABLE MUST BE SET AT THE HELIX ANGLE.

The index head spindle is geared to the lead screw of the table by means of a train of change gears, as shown in Fig. 645. These gears are called the gear on the screw (D), the first gear on the stud (C), the second gear on the stud (B), and the gear on the worm (A). The gear on the screw and the first gear on the stud are the driving gears and the second gear on the stud and the gear on the worm are the driven gears. This may be expressed as a ratio:

$$\frac{\text{Driven Gears}}{\text{Driving Gears}} = \frac{\text{2nd} \times \text{Worm}}{\text{1st} \times \text{Screw}} = \frac{A \times B}{C \times D}$$

By using different combinations of change gears, the distance that the table moves while the spindle revolves once may be changed. In other words, the lead that is cut depends directly on the gears that are used. Usually (though not always) the gear

ratio is such that the work is advanced more than one inch while it makes one revolution. Therefore, the lead is expressed in inches per revolution rather than in revolutions per inch, as in threads. For example, a helix is said to have an eight inch lead rather than that its pitch is one-eighth turn per inch.

Fig. 645

The table feed screw usually has four threads per inch and a lead of one-fourth of an inch. Motion is transferred from the lead screw to the spindle through the worm and worm wheel, which have a 40 to 1 ratio. When the spindle makes one revolution, the table moves ten inches along the ways if even gearing (1 to 1 ratio) is used. One revolution of spindle X Index head ratio X Gear ratio X Lead of lead screw = Lead of machine, or

$$1 \times \frac{40}{1} \times \frac{1}{1} \times \frac{1"}{4} = \text{Lead of machine.}$$

The standard lead of a milling machine is 10" and all change gears are figured on this basis. Any change in gear ratio makes a corresponding change in the lead.

EXAMPLES

(1) Using 1 to 4 ratio

$$1 \times \frac{\cancel{40}}{1} \times \frac{1}{\cancel{4}} \quad \frac{1"}{\cancel{4}} = \frac{5}{2} = 2.500" \text{ Lead}$$

(2) Using 2 to 1 ratio

$$1 \times \frac{\overset{10}{\cancel{40}}}{1} \quad \frac{2}{1} \quad \frac{1''}{\cancel{4}} = 20'' \text{ Lead.}$$

The compound ratio of the driven to the driving gears equals the ratio of the lead of the required helix to the lead of the machine. Expressing this in fraction form:

$$\frac{\text{Driven Gears}}{\text{Driving Gears}} = \frac{\text{Lead of Required Helix}}{\text{Lead of Machine}}$$

Or, since the product of each class of gears determines the ratio, and the lead of the machine is 10 inches, the

$$\frac{\text{Driven Gears}}{\text{Driving Gears}} = \frac{\text{Lead of Required Helix}}{10}$$

The compound ratio of the driven to the driving gears may always be represented by a fraction whose numerator is the lead to be cut and whose denominator is 10. That is, if the required lead is 20, the ratio is 20:10. To express this in units instead of tens, divide both terms of the ratio by 10. This is often a convenient way to think of the ratio, a lead of 40 giving a ratio of 4:1, a lead of 25 a ratio of 2.5:1, etc.

To illustrate the usual calculations, assume that a helix of 12 inch lead is to be cut. The compound ratio of the driven to the driving gears equals the desired lead divided by 10, or it may be represented by the fraction 12/10. Resolving this into two factors to represent the two pairs of change gears,

$$\frac{12}{10} = \frac{3}{2} \times \frac{4}{5}$$

Both terms of the first factor are multiplied by a number (24 in this case) that will make the resulting numerator and denominator correspond with the number of teeth of two of the change gears furnished with the machine (such multiplications not affecting the value of a fraction).

$$\frac{3}{2} \times \frac{24}{24} = \frac{72}{48}$$

Treating the second factor similarly,

$$\frac{4}{5} \times \frac{8}{8} = \frac{32}{40}$$

Selecting 72, 32, 48, and 40 teeth gears,

$$\frac{12}{10} = \left(\frac{72 \times 32}{48 \times 40}\right)$$

The numerators of the fractions represent the driven gears, and the denominators the driving gears. The 72 teeth gear is the worm gear, the 40 is 1st on the stud, the 32 is 2nd on the stud, and the 48 is the screw gear. The two driven gears or the two driving gears may be transposed without changing the helix. That is, the 72 teeth gear could be used as the 2nd on the stud and the 32 teeth gear could be used as the worm gear, if desired. A third combination could also be made.

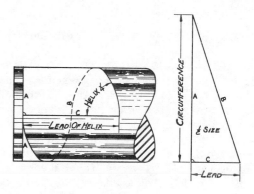

Fig. 646

Determine the gears to be used in cutting a lead of 27 inches.

$$\frac{27}{10} = \frac{3}{2} \times \frac{9}{5} = \left(\frac{3}{2} \times \frac{16}{16}\right) \times \left(\frac{9}{5} \times \frac{8}{8}\right) = \frac{48}{32} \times \frac{72}{40}$$

Determine the lead that would be cut by the gears, with 48, 72, 32, and 40 teeth, the first two being used as the driven gears.

$$\text{Helix to be cut} = \frac{10 \times 48 \times 72}{32 \times 40} = 27 \text{ inches to one revolution}$$

The milling machine table must always be set to the angle of the job.

The angle of the helix depends upon the lead of the helix and the diameter to be milled. In the sketch of Fig. 646, let "a" equal the circumference and "c" the lead of helix. The greater the lead of the helix for a given diameter, the smaller the helix angle, and the greater the diameter for a given lead, the greater the helix angle. Any change in the diameter of the work or in the lead will make a corresponding change in the helix angle.

$$\frac{\text{Cir.}}{\text{Lead}} = \frac{SO}{SA} = \text{Tangent of Helix Angle}$$

$$\frac{\text{Lead}}{10} = \text{Gear Ratio}$$

Lead = Cir. x Cot of Helix Angle

$$\frac{Lead}{10} = Gear\ Ratio$$

Gear Ratio x 10" = Lead Cut

Required lead - lead gears will cut = Error
    in Lead

Find gearing required to mill the
flutes on a 3" diameter cutter (Fig. 645)
when the helix angle is 35°8'.

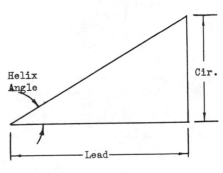

Fig. 647

Lead = Cir. (π · D) x Cot of Helix Angle

Lead = 3.1416 x 3 x 1.4211 = 13.3935

Gear Ratio $= \dfrac{Lead}{10} = \dfrac{13.3935}{10} \times \dfrac{10000}{10000} = \dfrac{133935}{100000}$

100000	133935	1
67870	100000	2
32130	33935	1
30685	32130	17
1445	1805	1
1440	1445	4
5	360	72
	360	
	0	

	1	2	1	17	1	4	72	
0	1	1	3	4	71	(75)	371	26785	X 5 = 133935
1	0	1	2	3	53	(56)	277	20000	X 5 = 100000
	1	2	1	17	1	4	72	

$$\frac{75}{56} = \frac{\overset{48}{\cancel{6}} \times \overset{100}{\cancel{25}}}{\underset{64}{\cancel{4}} \times \underset{56}{\cancel{14}}} = \frac{48}{64} \times \frac{100}{56} = \frac{Driven\ Gears}{Driving\ Gears}$$

$$\frac{75}{56} \times \frac{10}{1} = \frac{750}{56} = 13.3928\ Lead$$

13.3935 Required Lead
13.3928 Lead Cut
  .0007 Error in Lead

Set table at the helix angle.

Find helix angle and gearing required
for a lead of 3.140" on 1½" dia.

$$\frac{Cir.}{Lead} = Tangent\ of\ Helix\ Angle$$

$$\frac{3.1416 \times 1.5}{3.140} = \frac{4.7124}{3.140} = 1.50076$$

1.50076 is the tangent of 56°19'24"

Gear Ratio $= \dfrac{Lead}{10} = \dfrac{3.140}{10} = \dfrac{314}{1000}$

314	1000	3
290	942	5
24	58	2
20	48	2
4	10	2
4	8	2
0	2	

	3	5	2	2	2	2	
1	0	1	5	11	(27)	65	157	X 2 = 314
0	1	3	16	35	(86)	207	500	X 2 = 1000
	3	5	2	2	2	2	

$$\frac{27}{86} = \frac{\overset{24}{\cancel{6}} \times \overset{72}{\cancel{18}}}{\underset{64}{\cancel{2}} \times \underset{86}{\cancel{43}}} = \frac{24}{64} \times \frac{72}{86}\ \begin{matrix}Driven\ Gears\\Driving\ Gears\end{matrix}$$

$$\frac{27}{86} \times \frac{\overset{5}{\cancel{10}}}{1} = \frac{135}{43} = 3.13953\ lead\ cut$$

Required lead - lead cut = error
    3140" - 3.13953" = .00007".

Fig. 648 shows how the indexing head
and the rack cutting attachment are set up to
mill a helical groove or thread in a worm.
The worm shown has a triple thread, 2" pitch
diameter, .500 pitch, 1.500 lead, and a helix
angle of 76°34'30". The gear ratio and the
table setting for this worm are found as follows.

$$\frac{Lead}{10} = Gear\ Ratio \qquad\qquad \frac{1.5}{10} \times \frac{10}{10} = \frac{15}{100}$$

$$\frac{15}{100} = \frac{\overset{24}{\cancel{6}} \times \overset{40}{\cancel{20}}}{\underset{64}{\cancel{4}} \times \underset{100}{\cancel{25}}} = \frac{24}{64} \times \frac{40}{100}\ \begin{matrix}Driven\ Gears\\Driving\ Gears\end{matrix}$$

When using the rack cutting attachment, the cutter is held at a 90° angle with the work and the table is set at the complement of the helix angle.

$$90° = 89°59'60''$$
$$\underline{76°34'30''}$$
$$13°25'30''$$

SHORT LEAD MILLING. When very small leads are required, the dividing head worm and worm wheel may be disengaged and the gearing connected directly from the dividing head spindle to the table lead screw. With even gearing, when the dividing head spindle revolves once, the lead screw (which has four threads per inch) makes one revolution and

Fig. 648

the table is moved a distance equal to the lead, or .250". The rack cutting attachment shown in Fig. 648 is used with this method.

EXAMPLE. Find the gears to cut a lead of .3492".

$$\frac{\text{Lead}}{.250} = \text{Gear Ratio} \qquad \frac{.3492}{.250} \times \frac{10000}{10000} = \frac{3492}{2500}$$

2500	3492	1
1984	2500	2
516	992	1
476	516	1
40	476	11
36	440	1
4	36	9
	36	

		1	2	1	1	11	1	9	
0	1	1	3	4	7	81	88	873	x 4 = 3492
1	0	1	2	3	5	58	63	625	x 4 = 2500
		1	2	1	1	11	1	9	

$$\frac{88}{63} = \frac{\cancel{8} \times \cancel{11}}{\cancel{9} \times \cancel{7}} = \frac{64}{72} \times \frac{44}{28} \quad \begin{array}{l}\text{Driven Gears}\\\text{Driving Gears}\end{array}$$

$$\frac{88}{63} \times \frac{.250}{1} = .349206 \text{ lead cut}$$

Required lead - lead cut = error

.349206 - .3492 = .000006 error

The regular means of indexing cannot be used in short lead milling. Have the number of teeth in the gear on the spindle some multiple of the number of divisions required. The gears may then be swung out of mesh and

Fig. 649

Fig. 650

the gear on the spindle advanced the number of teeth necessary to index the work one division.

Sometimes it is necessary to mill a few teeth on a cylindrical shaft or plunger. If a rack cutting attachment is not available, the work may be done as shown in Fig. 649. The shaft is supported on a parallel and clamped in a vise, and the teeth are indexed by means of the graduated dial on the cross feed screw, the movement being equal to the linear pitch, or 3.1416 divided by the diametral pitch. Before indexing, care should be taken to remove backlash from the screw.

Fig. 650 shows a cylindrical cam being milled with an end mill, producing a helical slot with parallel sides. The dividing head centers are brought to a level with the center of the machine spindle. The table is set at right angles to the spindle and the angle of the helix is obtained by the combination of change gears used. Either right-hand or left-hand helices may be cut in this way by leaving out or adding an extra idler gear. When this method is used for cylindrical cam milling, the gears are calculated and placed the same as for helical milling, as shown in Fig. 645.

The Cam Cutting Attachment in Fig. 651 is used for cutting either face, peripheral, or cylindrical cams from a flat cam former (shown by letter "A"). The cam former is made from a disk about one-half inch thick, on which the required outline is laid out. The disk is machined or filed to the required shape. The table of the machine remains clamped in one position during the cutting, and the necessary rotative and longitudinal

movements are contained in the mechanism itself. The rotative movement is obtained by a worm driving a wheel fixed to the spindle of the attachment. The cam former is secured to the face of the worm wheel, and as the wheel revolves, the cam former depresses the sliding rack which in turn drives a pinion geared to another rack in the sliding bed of the attachment. This gives the necessary longitudinal movement on the face of the worm wheel.

Fig. 652 illustrates the cutting of a face cam with the cam cutting attachment. The necessary rotative movement is obtained by hand feed, and the necessary longitudinal movement is produced by the cam former and the

Fig. 652

mechanism of the attachment, as described in the discussion of Fig. 651. A peripheral cam may be milled in the same manner.

The face cam illustrated in Fig. 653 is machined by another method. The work is held in the horizontal plane, in the dividing head, and an end mill is used in the vertical spindle attachment.

In Fig. 653 we have a cam with two lobes (a lobe is a projecting part of a cam wheel), one (A) having a rise of 2.493" in $169°12'$ and the other (B) having a rise of 2.442" in $104°24'$.

The lead in $360°$ of lobe A

$$= \frac{360°}{169°12'} \times 2.493 = 5.304"$$

The lead in $360°$ of lobe B

$$= \frac{360°}{104°24'} \times 2.442 = 8.421"$$

Fig. 651

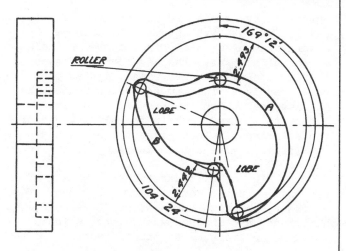

Fig. 653

To gear up the dividing head to cut lobe A,

$$\frac{\text{Lead of Lobe}}{\text{Lead of Machine}} = \frac{\text{Driven}}{\text{Drivers}} = \frac{\text{Second} \times \text{Worm}}{\text{First} \times \text{Screw}}$$

$$\frac{5.304}{10} = \frac{35}{66} \text{ approx. (by continued fractions)}$$

$$\frac{35}{66} = \frac{\overset{28}{\cancel{7}} \times \overset{40}{\cancel{5}}}{\underset{44}{11} \times \underset{40}{6}} = \frac{28}{44} \times \frac{40}{48}$$

To gear up the dividing head to cut lobe B,

$$\frac{8.421}{10} = \frac{16}{19} \text{ approx.} = \frac{16 \times 4}{19 \times 4} = \frac{64}{76}$$

(use 2 idlers)

The path of the roller should first be rough drilled. The parts of the cam, other than lobes A and B, can be scribed, drilled, and then milled to the scribed lines.

A method often followed in cutting peripheral cams, especially those for use on automatic screw machines, is that of using the dividing head and a vertical spindle milling attachment. This is illustrated in Fig. 654. The dividing head is geared to the table feed screw, the same as in cutting an ordinary helix, and the cam blank is fastened to the end of the dividing head. An end mill is used in the vertical spindle milling attachment, which is set to mill the periphery of the cam at right angles to its sides. In other words, the axes of the dividing head spindle and attachment spindle must always be parallel to mill cams by this method. The cutting is done by the teeth on the periphery of the end mill. The principle of this method may be explained in the following way.

Fig. 654

Suppose the dividing head is elevated to 90°, or at right angles to the surface of the table (see Fig. 655), and is geared for any given lead. It is apparent that as the table advances and the blank is turned, the distance between the axes of the dividing head spindle and the attachment spindle becomes less. In other words, the cut becomes deeper and the radius of the cam is shortened, producing a spiral lobe with a lead which is the same as that for which the machine is geared.

Now suppose the same gearing is retained and the dividing head is set at zero, or parallel to the surface of the table, as shown in Fig. 656. It is apparent, also, that the axes of the dividing head spindle and the attachment spindle are parallel to each other. Therefore, as the table advances and the blank is

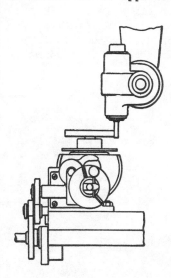

Fig. 655

turned, the distance between the axes of the
dividing head spindle and the attachment
spindle remains the same. As a result, the
periphery of the blank, if milled, is concentric, or the lead is zero.

To calculate the leads, gears, and
angles to incline the dividing head and vertical spindle attachment for a cam having a
.470" rise in 85°, .750" rise in 75°, and
.358" rise in 58° (see Fig. 658), proceed as
follows:

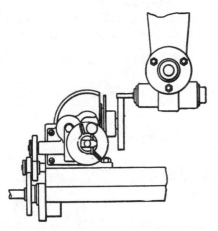

Fig. 656

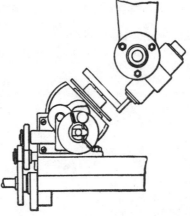

Fig. 657

Fig. 658

$$.470" \text{ rise in } 85° = \frac{360}{85} \times .470$$

$$= 1.99$$

$$.750" \text{ rise in } 75° = \frac{360}{75} \times .750 = 3.600$$

$$.358" \text{ rise in } 58° = \frac{360}{58} \times .358 = 2.222$$

If, then, the dividing head is elevated to any angle between zero and 90°, as
shown in Fig. 657, the amount of lead given
to the cam will be between that for which the
machine is geared and zero. Hence it is
clear that cams with a very large range of
different leads can be obtained with one set
of change gears, and the problem of milling
the lobes of a cam is reduced to a question
of finding the angle at which to set the head
to obtain any given lead.

To cut the smallest possible lead
with the dividing head geared to the lead
screw, place a 24-tooth gear on worm, an 86-
tooth gear first on stud, a 24-tooth gear
second on stud, and a 100-tooth gear on worm.
Calculate the lead as follows:

$$\frac{24}{86} \times \frac{24}{100} \times \frac{40}{4} = .66976 \text{ or } .67$$

$\frac{40}{4}$ is number of turns of index crank to one spindle
$\frac{40}{4}$ is number of threads per inch on lead screw

To find the angle to set the dividing
head and vertical spindle attachment, divide
the lead of the cam by the lead of the machine. The lead of the machine must always
be greater than the lead of the cam.

To find the lead of the cam, that is,
the theoretical continuous rise in one complete revolution, if the rise in 27° is
.127", calculate as follows: 360° ÷ angle in
which rise occurs × rise = rise in 360°.

$$\frac{360}{27} \times .127 = 1.693 \text{ rise in } 360°$$

The machine must be geared with a
greater lead than that of the cam having the
greatest lead. As an example, use the lead
3.657" and gear the machine as follows:

$\frac{3.657}{10.000}$ by continued fractions equals

$$\text{approximately } \frac{64}{175} = \frac{\overset{32}{\cancel{4}}}{7} \times \frac{\overset{64}{\cancel{16}}}{\underset{56}{\cancel{25}} \underset{100}{}}$$

Place the 32-tooth gear on
the worm, the 56-tooth gear the
first on the stud, the 64-tooth gear
the second on the stud, and place the
100-tooth gear on the lead screw.

The sine of the angle at which to incline the dividing head is found by dividing
the lead of the cam by the lead of the machine.

Lead of cam having .470" rise in 85° = $\frac{1.99}{3.657}$

$$= .54415 \text{ sine} = 32°58'$$

Lead of cam having .750" rise in 75° = $\frac{3.600}{3.657}$

$$= .9844 \text{ sine} = 79°52'$$

Lead of cam having .358" rise in 58° = $\dfrac{2.222}{3.657}$

$$= .6076 \text{ sine} = 37°25'$$

Note in the above work that the machine is geared the same for all leads, and that the dividing head and vertical spindle attachment are inclined at different angles to mill the different leads.

A cylindrical cam is milled and the gears calculated in the same manner as a helical groove is milled, an end mill being used instead of a milling cutter.

All tool rooms have, or can obtain, a chart showing the many different leads and gears used to cut these leads.

CUTTER SPEEDS

The cutting speed of a cutter is the rate at which it engages the work (usually expressed in surface feet per minute). No definite rule can be made for the speed of a milling cutter, because too many factors, such as the depth of cut, amount of feed, material cut, type of job, condition of machine and cutter, and finish required, must be considered. With H.S.S. cutters, the following feeds are usually satisfactory:

High Speed Steel (H.S.S.) - 40 - 50 F.P.M.
Chrome Non-Shrink (C.N.S.) - 40 - 50   "
Tool Steel (R   RR)   - 60 - 80   "

Machine Steel (M.S.)   - 80 - 100 F.P.M.
Cast Iron (C.I.)   - 80 - 100   "
Brass   - 150 - 200   "

A cutter should never be run at a speed which would cause excessive heat and dull or burn the cutting edge.

The following formulas may be used in figuring the speed of a milling cutter.

(a) $\text{R.P.M.} = \dfrac{\text{Cutting Speed} \times 12}{3.1416 \times \text{dia.}}$

Example. What is the speed of the spindle for a $1\frac{1}{4}$" cutter running 40 F.P.M.?

$$\text{R.P.M.} = \frac{40 \times 12}{3.1416 \times 1.25} = 122+$$

(b) $\text{Cutting Speed} = \dfrac{3.1416 \times \text{dia.} \times \text{R.P.M.}}{12}$

Example. Find the cutting speed of a $1\frac{1}{2}$" end mill running 382 R.P.M.

$$\text{Cutting Speed} = \frac{3.1416 \times 1.5 \times 382}{12} = 150+$$

The following table may be used to find directly the R.P.M. of cutters of different diameters (for the more common surface speeds):

TABLE OF CUTTING SPEEDS

Feet per Minute	40	45	50	55	60	65	70	75	80	90	100	110	120	130	140	150
Diam., Inches	REVOLUTIONS PER MINUTE															
1/4	611	688	764	840	917	993	1070	1146	1222	1375	1528	1681	1833	1986	2139	2292
5/16	489	550	611	672	733	794	856	917	978	1100	1222	1345	1467	1589	1711	1833
3/8	407	458	509	560	611	662	713	764	815	917	1019	1120	1222	1324	1426	1528
7/16	349	393	437	480	524	568	611	655	698	786	873	960	1048	1135	1222	1310
1/2	306	344	382	420	458	497	535	573	611	688	764	840	917	993	1070	1146
5/8	244	275	306	336	367	397	428	458	489	550	611	672	733	794	856	917
3/4	204	229	255	280	306	331	357	382	407	458	509	560	611	662	713	764
7/8	175	196	218	240	262	284	306	327	349	393	437	480	524	568	611	655
1	153	172	191	210	229	248	267	287	306	344	382	420	458	497	535	573
1-1/8	136	153	170	187	204	221	238	255	272	306	340	373	407	441	475	509
1-1/4	122	138	153	168	183	199	214	229	244	275	306	336	367	397	428	458
1-3/8	111	125	139	153	167	181	194	208	222	250	278	306	333	361	389	417
1-1/2	102	115	127	140	153	166	178	191	204	229	255	280	306	331	357	382
1-5/8	94.0	106	118	129	141	153	165	176	188	212	235	259	282	306	329	353
1-3/4	87.3	98.2	109	120	131	142	153	164	175	196	218	240	262	284	306	327
1-7/8	81.5	91.7	102	112	122	132	143	153	163	183	204	224	244	265	285	306
2	76.4	85.9	95.5	105	115	124	134	143	153	172	191	210	229	248	267	287
2-1/4	67.9	76.4	84.9	93.4	102	110	119	127	136	153	170	187	204	221	238	255
2-1/2	61.1	68.8	76.4	84.0	91.7	99.3	107	115	122	138	153	168	183	199	214	229
2-3/4	55.6	62.5	69.5	76.4	83.3	90.3	97.2	104	111	125	139	153	167	181	194	208
3	50.9	57.3	63.7	70.0	76.4	82.8	89.1	95.5	102	115	127	140	153	166	178	191
3-1/4	47.0	52.9	58.8	64.6	70.5	76.4	82.3	88.2	94.0	106	118	129	141	153	165	176
3-1/2	43.7	49.1	54.6	60.0	65.5	70.9	76.4	81.9	87.3	98.2	109	120	131	142	153	164
3-3/4	40.7	45.8	50.9	56.0	61.1	66.2	71.3	76.4	81.5	91.7	102	112	122	132	143	153
4	38.2	43.0	47.7	52.5	57.3	62.1	66.8	71.6	76.4	85.9	95.5	105	115	124	134	143
4-1/2	34.0	38.2	42.4	46.7	50.9	55.2	59.4	63.6	67.9	76.4	84.9	93.4	102	110	119	127
5	30.6	34.4	38.2	42.0	45.8	49.7	53.5	57.3	61.1	68.8	76.4	84.0	91.7	99.3	107	115
5-1/2	27.8	31.3	34.7	38.2	41.7	45.1	48.6	52.1	55.6	62.5	69.5	76.4	83.3	90.3	97.2	104
6	25.5	28.6	31.8	35.0	38.2	41.4	44.6	47.8	50.9	57.3	63.7	70.0	76.4	82.8	89.1	95.5

Chapter 22

GAGES AND GAGE BLOCKS

1. What is a gage?

A. A gage is a device for determining whether or not one or more of the dimensions of a manufactured part are within specified limits. A master gage is one whose gaging dimensions represent as exactly as possible the physical dimensions of the component. An inspection gage is a gage for the use of the manufacturer or purchaser in accepting the product, and a working gage is one used by the manufacturer to check the work as it is produced.

2. Name several types of gages.

A. Gages are divided into the following types according to the purposes for which they are used: ring gage, plug gage, receiving gage, indicating gage, snap gage, and caliper gage.

Fig. 659. Ring Gage

3. What is a ring gage?

A. A ring gage, Fig. 659, is one whose inside measuring surfaces are circular in form. The measuring surfaces may be cylindrical or conical.

4. What is a plug gage?

A. A plug gage is one whose outside measuring surfaces are arranged to verify the specified uniformity of holes. A plug gage may be straight or tapered and of any cross-sectional shape. Figs. 660, 661 and 662 show different types of plug gages.

Fig. 660. Cylindrical Plug Gage

Fig. 661. Spline Plug Gage

5. What is a snap gage, and what is a caliper gage?

A. A snap gage, Fig. 663, is a

Fig. 662. Taper Plug Gage

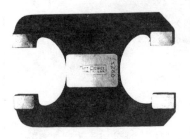

Fig. 663

fixed gage arranged with inside measuring surfaces for calipering diameters, lengths, thicknesses, etc., and a caliper gage, Fig. 664, is one which, for internal members, is similar to a snap page, and for external members, is similar to a plug gage.

Fig. 664

6. What is an indicating gage?

Fig. 665

A. An indicating gage, Fig. 666, is one that exhibits visually the variations in the uniformity of dimensions or contour, the amount of the variation being indicated by lever on graduated scale, dial, flush pin, plunger, gages, etc.

7. What is a receiving gage?

A. A receiving gage, Fig. 665, is one whose inside measuring surfaces are arranged to verify the specified uniformity of size and contour of manufactured material.

Fig. 666

8. Explain what a sine bar is and tell how it
   is used.

A. The sine bar is used either for measuring angles accurately or for locating work
to a given angle within very close limits. It
consists of a bar in which 2 hardened and
grouped plugs of the same diameter are set.
The center distance between the plugs is usually 5 or 10 inches. The edges of the bar

Fig. 667

must be parallel with the line of the plug
centers. The sine bar is always used in conjunction with a true surface, such as a surface plate from which measurements are taken.
Fig. 667 shows a 10" straight sine bar and
Fig. 669 shows how it is used. In Fig. 668,
the difference in the height of the measuring
plugs, A-B, divided by the length of the sine
bar is equal to the sine of the required angle.

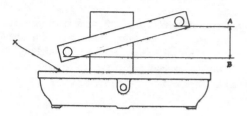

Fig. 668

For example, in setting a 5" sine bar,
as shown in Fig. 668, distance A-B was determined by means of a Vernier Height Gage to be
2.68525". This number, divided by 5, (because the 5" sine bar was used) gave a result
of 0.53705", the sine of the angle required.
A table of natural sines will show the value
of this angle to be 32°29'. If a 10" sine
bar were used, distance A-B would have been
5.3705", or 10 times the natural sine of an
angle of 32°29'. An error in setting the 10"
sine bar would cause but half the inaccuracy
in measurement of the angle that would be
caused by the same error in setting the 5"
bar. In setting up for a given angle, the
method outlined above is reversed. By referring to a table of natural sines, the sine of
an angle of 32°29' is found to
be 0.53705". Then by setting
the 5" sine bar in a manner
such that the perpendicular distance A-B is five times the
natural sine, or 2.68525", the
correct angle is at once established between the edge of the
sine bar and the surface plate
on which it is mounted.

9. Tell what the amplifying
   comparator, Fig. 671, is
   and explain its use.

A. The amplifying comparator
is a precision measuring instrument which amplifies any error
in the work 10 times.

A combination of Jo-
blocks, or a gage of pre-
determined size is placed under

Fig. 669. Using Straight Sine Bar

Fig. 670. Using Sine Bar Fixture

the indicating point and the dial indicator
hand set at zero. The Jo-blocks or gage is
then removed and the job to be checked is
placed under the point. Any deviation shown
on the dial graduations indicates ten times
the amount of error between the work and the
pre-determined size. Gages should be checked
for accuracy at 68°F.

10. In fitting a shaft to a hole, which is
    usually made standard, the hole or shaft?
    A. Usually the hole is made standard and
any allowance for fit is made on the shaft.

    When a micrometer has been used with
the spindle locked in one place for some
time checking hardened work such as reamers,
cutters, or similar jobs, the end of the
spindle and the anvil will be worn slightly
concave, Fig. 673. The spindle and the anvil should be checked occasionally with a
straight edge, Fig. 674, to see whether they
are worn or not. If these surfaces are worn,
they should be lapped flat and parallel with

Fig. 671

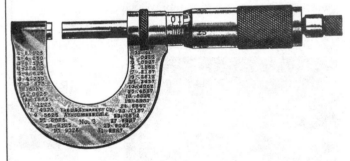

Fig. 673

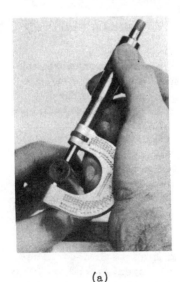

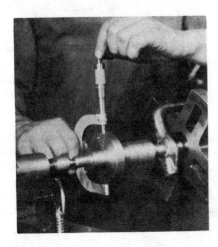

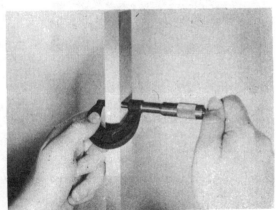

(a)   (b)   (c)

Fig. 672. Suggestions for Holding a Micrometer

each other. When adjustments are made they should be checked both with flat Jo-block, Fig. 675, and with round jaws, Fig. 676, to be certain that the reading is correct.

Fig. 674

Fig. 676

Fig. 675

PRECISION MEASUREMENT

It is said that all important discoveries have been made through the medium of fine measurement of Time, Mass, or Length, and of these three, precision measurement of length is the most difficult and has become more and more important until today it is one of the greatest problems before the mechanical world.

plugs in a bronze bar one inch square and thirty-six inches long.

We have no standard yard in the United States. All measurements of length are referred to our copy of the International Prototype Meter, which is kept at the Bureau of Standards, Washington, D.C.

The International Prototype Meter is defined as the length at 0 degrees centigrade (or 32 degrees Fahrenheit) between two lines on a platinum-iridium alloy bar, kept in the International Bureau of Weights and Measures near Paris, France.

In the United States, the yard has been defined by Executive Order, dated April 15, 1893, as 3600/3937 of a meter, or expressed as a decimal 0.9144 meters.

An inch in the United States is defined as 1/36 of a yard or 2.54 centimeters.

Johansson Gage Blocks are the standard of precision measurement for the world. They have no counterpart in either science or industry. They measure accurately in millionths of an inch, an accomplishment considered impossible before their introduction.

A few years ago, the mechanics who worked in machine shops were "all around" machinists; that is, they were proficient in the operation of all the machine equipment as well as being able to perform bench and assembly operations.

Fig. 677. 1" Johansson Gage Block

Fig. 678. The "Home" of Johansson Gage Blocks, Ford Motor Company, Dearborn, Michigan.

The units of length are the Inch and the Millimeter, fractional parts of the yard and the meter.

An inch is a relative quantity and is an English unit of length that has been definitely established at 1/36 of the "Imperial Standard Yard," which Great Britain defines as the distance at 62 degrees Fahrenheit between the central traverse lines in two gold

Today the "all around" machinist has given way (except on tool and gage work) to the highly specialized operator who is instructed and trained to perform a certain operation on a particular piece of apparatus. The tools and gages furnished make it possible for him to make all measurements in accordance with pre-determined specifications which are more accurate than were thought possible by the "all around" mechanic of yesteryear.

The gradually increasing quantity production of recent years is on such a huge scale that the "cut and try" method of measuring has been discarded, even by the machine shop that builds only one complete apparatus.

The ideal measuring tool for making possible precision and interchangeable manufacture must meet the following requirements:

First--Must give external and internal measurements in one ten-thousandth part of an inch.

Second--Must be of such design to be used directly on the work, to eliminate the possibility of error in transferring accurate measurements.

Fig. 679. Measuring OD with Accessory Set

Third--Must be positive, and dependable in a sense, foolproof.

fied size, hardened, ground, stabilized, and finished to an accuracy within a few millionths part of an inch from specified size. They embody in their commercial manufacture the solving of four universally recognized metallurgical and mechanical problems, namely: Flat Surfaces in Steel; Parallel Surfaces in Steel; Accuracy as to Dimension in Steel; and Effective Heat Treatment and Seasoning of Steel.

Fig. 683

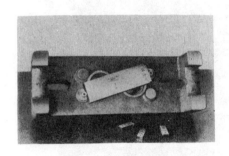

Fig. 680. Checking Distance Between Test Plugs

Fig. 681. Measuring ID with Accessory Set

Fig. 682. Measuring with Accuracy.

Fourth--Must give the same result, any number of times, in the hands of different mechanics.

Fifth--Must be self-checking, so error due to wear, accident, or abuse may be readily discovered.

Sixth--Must be seasoned and stabilized to reduce to a minimum errors in accuracy due to the change that takes place in metals.

Seventh--Must have an established reputation for accuracy that is accepted by the manufacturer and customer.

The Johansson Gage Blocks (Fig. 683) are rectangular pieces of tool steel, approximately 3/8 inches by 1-3/8 inches by speci-

FLAT SURFACES IN STEEL. It is considered one of the most remarkable achievements in mechanics to make a flat surface in steel, and, by the Johansson methods, a flat surface with an extremely high finish having the appearance of burnished silver is produced, which approaches nearer the perfect plane than any other surface produced by the hand of man. These flat-lapped surfaces when thoroughly cleaned and slid one on the other with a slight inward pressure, will take hold as though magnetized. They have been known to sustain a weight of 200 pounds on a direct pull (Fig. 684), although the contacting surfaces are less than one-half square inch. Scientists have offered "atmospheric pressure,"

"molecular attraction," and "a minute film of
oil on the lapped surfaces" as an explanation
of this phenomena. Possibly it is a combination of all three.

GAGE BLOCKS
HOLDING
200 POUNDS
ON
DIRECT PULL

Fig. 684

PARALLEL SURFACES IN STEEL. The degree of parallelism attained in the manufacture of the Johansson Gage Blocks is well
shown by the fact that any block in a given
combination may be turned end for end at will
without affecting either the size or the parallelism of the two extreme surfaces of the
combination.

ACCURACY. The making of one steel
surface parallel with another is a recognized problem, but to make the parallel surface a pre-determined distance from another
surface and with an accuracy in millionths of
an inch is a most remarkable achievement.
That this has been accomplished is proven by
the way in which an equivalent combination of
Johansson Gage Blocks check against one solid
block. See Fig. 685.

EFFECTIVE SEASONING. Johansson Gage
Blocks are so seasoned by Johansson methods
that internal stresses and strains are relieved. The molecules of the steel may be
said to be at rest and because of this, the
usual warping or growing is checked.

GAGING SYSTEM. Set No. 1 will make
.0001 inch sizes from .200 inch and .001
inch sizes from .100 inch up to 12.000 inches.

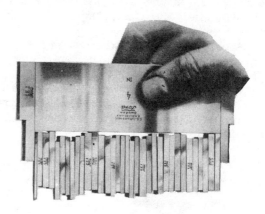

Fig. 685

A full set consists of eighty-one
blocks (Fig. 686), which have surfaces flat
and parallel within .000008 of an inch. This
set is made up of four series:

Fig. 686

Sizes Composing a Set of 81 Blocks

FIRST SERIES

.1001"	.1002"	.1003"	.1004"	.1005"	.1006"	.1007"	.1008"	.1009"

SECOND SERIES

.101"	.102"	.103"	.104"	.105"	.106"	.107"	.108"	.109"	.110"
.111"	.112"	.113"	.114"	.115"	.116"	.117"	.118"	.119"	.120"
.121"	.122"	.123"	.124"	.125"	.126"	.127"	.128"	.129"	.130"
.131"	.132"	.133"	.134"	.135"	.136"	.137"	.138"	.139"	.140"
.141"	.142"	.143"	.144"	.145"	.146"	.147"	.148"	.149"	

THIRD SERIES

.050"

.100"	.200"	.300"	.400"	.500"	.600"	.700"	.800"	.900"
.150"	.250"	.350"	.450"	.550"	.650"	.750"	.850"	.950"

FOURTH SERIES

1.000"	2.000"	3.000"	4.000"

1. The first series consists of nine blocks, ranging in size from 0.1001 of an inch to 0.1009 of an inch by steps of 0.0001 of an inch.
2. The second series consists of forty-nine blocks, ranging in size, by 0.001 of an inch, from 0.101 of an inch to 0.149 of an inch.
3. The third series consists of nineteen blocks, ranging in size, by 0.050 of an inch, from 0.050 of an inch to 0.950 of an inch.
4. The fourth series consists of four blocks, 1, 2, 3, and 4 inches in size.

The blocks of the third series can be combined with those of the fourth series to give any multiple of 0.050 between 0.050 and 10 inches. The second series is used to obtain dimensions varying by thousandths, and the first series to obtain dimensions varying by ten-thousandths.

Johansson Gage Blocks are made in three qualities--B, A, and AA. At a temperature of 68°F, these blocks have the following accuracy:

Working Set (B quality)   = 0.000008 inch
Inspection Set (A quality)   = 0.000004 "
Laboratory Set (AA quality) = 0.000002 "

The extreme accuracy of these blocks can be indicated by the following comparisons. A human hair is approximately three thousandths (.003) of an inch thick. The most accurate work in the mechanical field is that of toolmakers, who work to an accuracy of one ten-thousandth of an inch, which is thirty times finer than a human hair. To carry the illustration still further, light waves are approximately sixteen millionths of an inch long, which is 250 times finer than a human hair and 6¼ times finer than the accuracy used by a toolmaker. The accuracy of AA quality Johansson Gage Blocks, however, is two millionths (.000002) of an inch. This is

1500 times finer than a human hair;
50 times finer than a toolmaker works;

8 times finer than the length of a light wave.

JOHANSSON BLOCKS

ASSIGNMENT. To build a combination of Johansson blocks of any required size, within the range of the set.

HOW IT IS DONE

(A) TO FIND THE SIZE OF THE BLOCKS TO BE USED.

1. Acquaint yourself with the size of the blocks in the set.
2. Begin with the right-hand figure of the specified size.
3. Continue working from the right to the left.
4. Build the combination with the fewest possible number of blocks.

EXAMPLE

1.2721	1.2721	1.2721	1.2721	1.2721
.1001	.1009	.1008	.1006	.1007
.149	.1002	.1003	.1005	.1004
.123	.147	.139	.138	.141
.900	.124	.132	.133	.130
1.2721	.800	.100	.500	.600
	1.2721	.700	.300	.200
		1.2721	1.2721	1.2721

It is possible to make many combinations to obtain any required size.

(B) TO BUILD THE COMBINATION

1. Select from the set the first 2 blocks of the combination (Fig. 687(a)).
2. Wipe each of the contacting surfaces of the blocks on the palm of the hand, on the wrist, or on a piece of chamoise, and then place the contacting surfaces together (Fig. 687(b)).
3. With a slight inward pressure, slide one block on the other (Fig. 687(c)). If the contacting surfaces are clean they will cling together as though they were magnetized.
4. Continue in this manner until the required combination is completed.

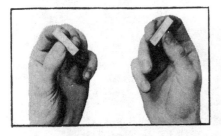  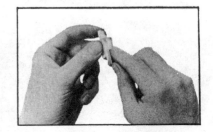

(a)   (b)   (c)

Fig. 687

YOUR JOB.   Make as many combinations of blocks as possible to form the following sizes: _____ ; _____ ; _____ .
Do not use the same block for more than one combination.

APPLICATIONS OF JOHANSSON BLOCKS

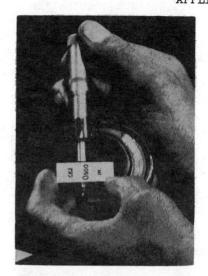

Fig. 688.   Checking Micrometer

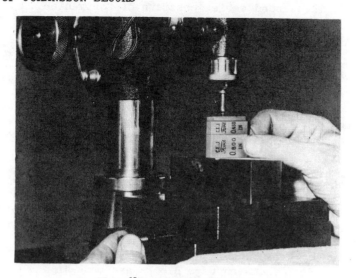

Fig. 689.   Setting Visual Gage

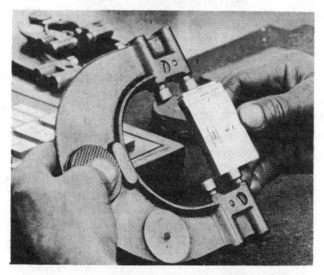

Fig. 690.   Checking Snap Gage

Fig. 691.   Using Amplifying Gage Set With Jo-Blocks

Fig. 692.   Checking Spacing of Holes in Relation Gage

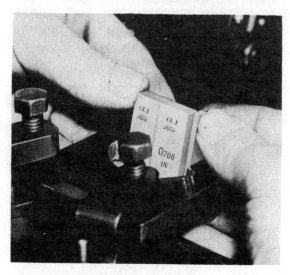

Fig. 693.   Measuring Snap Gage on a Grinder

APPLICATION OF JOHANSSON BLOCKS

Fig. 694. Gage Blocks Used to Establish Locations

Fig. 695. Checking Taper on Plug Gage

(a)

(b)

Fig. 696. (a) Checking Gage with Gage Blocks and Indicator, (b) Checking Location of Radii

(a)

(b)

Fig. 697. Setting Sine Bars with Gage Blocks

APPLICATION OF BLOCKS AND ACCESSORIES

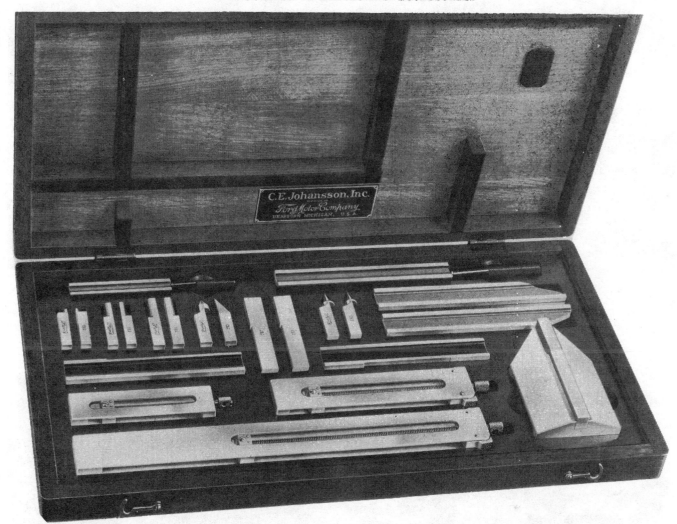

Fig. 698. Accessory Set Containing Jaws, Scriber, Center Point, Tram Points, Straight Edge, Adjustable Holder, and Foot Block.

Fig. 699. Checking Inside Diameter of a Ring Gage

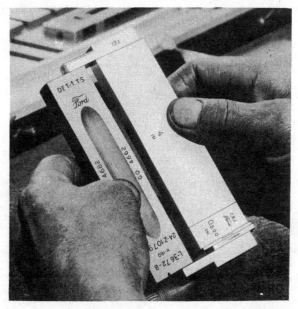

Fig. 700. Checking "GO" Size of Special Length Gage

APPLICATION OF BLOCKS AND ACCESSORIES

Fig. 701. Establishing Locations Using Accessories

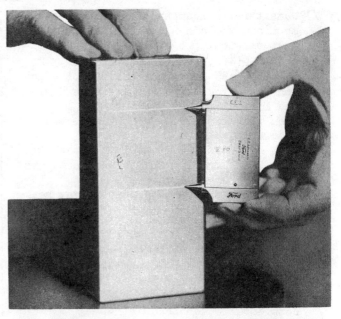

Fig. 702. Rechecking Locations Already Established

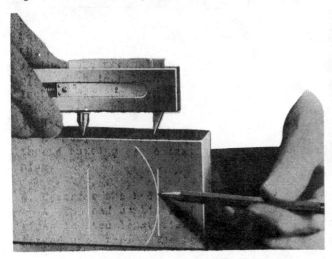

Fig. 703. Establishing a Radius Within Close Limits

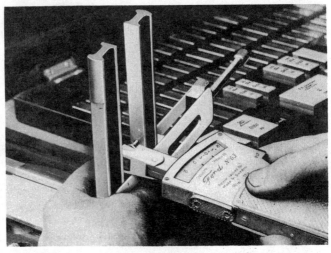

Fig. 704. Setting Internal Indicator Gage for Size

Fig. 705. Checking Outside Diameter of a Plug Gage

Fig. 706. Setting a Compound Slide by Use of Blocks

1. What is an inch?

2. Name three fundamental units of measurement and explain which one has been the most difficult to obtain.

3. State briefly what Jo-Blocks are.

4. Show how the different classes of Johansson Gage Blocks are designated and give the limits of each class.

5. From what kind of steel are the gage blocks made?

6. What are the four great problems that have been overcome in the manufacture of Johansson Gage Blocks?

7. Name five jobs where Jo-Blocks may be used.

8. Describe how Jo-Blocks may be used as (a) a height gage, (b) a compass.

9. What are the explanations for the cohesion of the blocks?

10. What would be the result of leaving the blocks "stacked" or wrung together for a considerable length of time?

11. Give at least three combinations of blocks from one set that will make .9243".

12. What rules should be followed in building combinations of gage blocks?

13. How are Jo-Blocks used to set a comparator or an amplifying gage?

14. What are precision measurements?

15. How is the accuracy of Johansson Blocks regarded in industry?

16. Why is it necessary to have some standard of measurement?

17. Name the features that an ideal measuring tool should have.

18. What is meant by "interchangeability"?

19. State what is meant by the limit system and tell why it is necessary.

20. What is the approximate Rockwell test of Jo-Blocks?

21. Does the temperature affect the size of the blocks?

22. Are Johansson Blocks used directly in connection with production?

The following problems show practical applications of the sine bar and Johansson Blocks in checking angles, and applications of plug gages in checking the diameter at the small end of a taper.

1. Figure dimensions W, X, Y, and Z used to check a #6 Jarno plug gage. The Jarno taper formulas are given below. These formulas are also given on page 85, Chap. 13, Tapers.

$$\text{Diameter at Large End} = \frac{\text{No. of Taper}}{8}$$

$$\text{Diameter at Small End} = \frac{\text{No. of Taper}}{10}$$

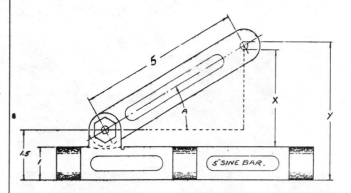

Fig. 707

A = Included angle at which the sine bar is set
X or Y = Combination of Jo-Blocks used to set sine bar at the included angle

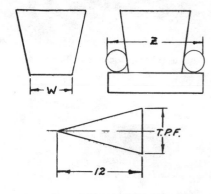

Fig. 708

W = Diameter at small end of taper
Z = Diameter over plugs at small end of taper

$$\text{Length of Taper} = \frac{\text{No. of Taper}}{2}$$

$$\text{Taper Per Foot} = .600$$

$$\text{Taper Per Inch} = .050$$

2. A #7 B & S taper plug gage is .600 in diameter at the small end, 2-7/8 long, and has a taper per foot of $\frac{1}{2}$". Find the value of W, X, Y, and Z for this gage.

3. Find the value of W, X, Y, and Z in each of the following plug gages.

Taper Per Foot	Length	Diameter at Large End
7/16	4	3.250
3/4	8	3.000
7/8	3	2.125
1/4	2	.289

Chapter 23

HEAT TREATMENT

1. What is steel?

A. Steel is an alloy of iron and carbon. Among the other elements found in steel are silicon, phosphorus, sulphur, manganese, and chromium.

2. How may steels be classified?

A. Steels may be roughly classified as straight carbon steels and alloy steels. A straight carbon steel is a steel that owes its properties chiefly to various percentages of carbon without substantial amounts of other alloying elements. When steel contains about .20% or less it may be called a low carbon steel, from .20% to about .60% a medium carbon steel, and from about .60% to 1.30% a high carbon steel. An alloy steel is a steel to which some element other than carbon has been added to improve or change the physical properties.

3. How are the different types of steel designated?

A. In Ford Motor Co. some types of steel are designated by letter. For example, "A" steel is a low carbon alloy steel; "AAA" is a medium carbon alloy steel; "RR" is a high carbon tool steel; etc. Other types are designated by name such as: chrome non-shrink, Ford hot work, insert die steel, etc. Bar stock is identified by the colors painted on the end of the bar (see steel analysis chart, page 204). Many other firms use the S A E numbering system for identifying steel.

4. What is a low carbon steel?

A. A low carbon steel is a steel that does not contain enough carbon to cause it to harden to any great extent when heated to a certain temperature and quenched in oil, water, or brine. It may be heat treated to increase its strength or case hardened to increase its resistance to wear. Some of the low carbon steels are: machine steel, cold rolled steel, and "GG" steel. Among the jobs made from low carbon steel are: clamps, pressure pads, stripper plates, bolts, nuts, washers, and similar jobs where surfaces are not subjected to continuous wear. When hard wearing surfaces and a soft core are required, low carbon steel may be

case hardened either by cyaniding or by carburizing and hardening.

5. What is a medium carbon steel?

A. A medium carbon steel is a steel that contains from about .20% to about .60% carbon. "E", "EE", "EEE", "FFF", "H", and "L" steels are medium carbon steels (see steel analysis chart, page 204). Medium carbon steels are used for a wide variety of work including nuts, bolts, stock guides, clamps, flask pins, crankshafts, crane shafts, etc. Medium carbon steels are also used extensively in production work. When heat treatment is required, these steels may be cyanided, pack hardened, or quenched directly from the furnace.

6. What are high carbon steels?

A. High carbon steels are steels which usually contain from about .70% to 1.30% carbon. Tool steel is a high carbon steel. "RR" steel, which contains .95% to 1.05% carbon, is the tool steel most commonly used in Ford Motor Co. Many of the tools and working parts of machines, guide pins, rest buttons, locating pins, dies and punches, gages, bushings, centers, etc., are made from tool steel. Tool steels contain sufficient carbon to cause them to harden; therefore, these steels can be hardened and tempered. The heat treatments used are "Q", "QQ", and "SS". (See standard heat treatment chart, page 207).

7. What are alloy steels?

A. Alloy steels are those steels which contain some alloying elements, such as: chromium, vanadium, nickel, molybdenum, tungsten, etc., in addition to carbon and iron which give them some peculiar characteristic not possessed by ordinary steel. Alloys are put into steels for the following reasons: to secure greater hardness, to secure greater toughness or strength, to enable the steel to hold its size and shape during hardening, or to enable the steel to retain its hardness at high temperatures.

Chromium is used as a hardener, nickel for strength and toughness, tungsten and molybdenum as hardeners and heat registers, manganese for strength, and vanadium as a

197

purifier and fatigue resister. "A" (low carbon alloy), "AAA" (medium carbon alloy),
chrome non-shrink, high speed, Ford hot work,
rustless 18-8, insert die steel, and vanadium
tool steel are some of the alloy steels used
by Ford Motor Co.

8. What jobs are made from "A" and "AAA"
   steel?

   A. "A" steel is used for heavy duty bushings, large worms, large worms where hard
surfaces are more important than accuracy,
and for thin sections and intricate parts.
To obtain a hard surface, "A" steel must be
carburized, hardened, and tempered. "AAA"
steel is used for acme screws, milling machine arbors, grinder spindles, lock nuts,
tap holders, small gear and worms, machine
parts, boring bars, and similar jobs (see
heat treatment chart, page 207).

9. What is high speed steel?

   A. High speed steel is an alloy steel used
to make tool bits, forged cutting tools,
milling machine cutters, forming tools, reamers, broaches, and similar jobs. It contains
a high percentage of tungsten, chromium,
vanadium, and carbon. (See analysis chart,
page 204). These elements give the steel the
ability to retain sufficient hardness at high
temperatures to cut metal.

10. Briefly explain the purpose of some of the
    operations used in heat treatment.

    A. Normalizing is a uniform heating above
the usual hardening temperatures followed by
cooling freely in air. This treatment is
used to put the steel back in a normal condition after forging or after improper heat
treatment.

    Annealing is a uniform heating above
the usual hardening temperatures followed by
cooling as desired. Annealing may be done
either to soften a piece that is too hard to
machine or to remachine a piece that has already been hardened. Annealing is also done
to relieve internal strains set up by exten-
sive machining. In annealing, the work
should be cooled slowly either in the furnace, in a box, or in air.

    Quenching is the operation of cooling
rapidly in a suitable medium such as water,
brine, oil, or air.

    Tempering (Drawing) is the operation
of reheating a piece that has been hardened
to relieve the internal strains and to increase the toughness. All tempering is done
below the critical range. The best way to
insure correct tempering of a part is to immerse it in a bath of oil or nitrate, in a

pyrometer controlled pot like that shown in
Fig. 709. Another method is to watch the

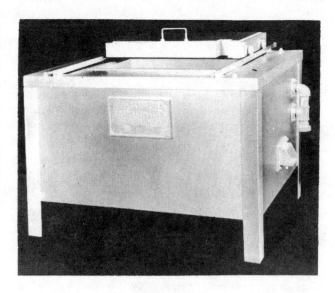

Fig. 709

color of the steel as it heats up in an open
furnace. The chart gives the approximate

Color and Temperature Chart

NO.	COLOR	TEMPERATURE
1	Straw	360 to 375° F
2	Light Straw	420
3	Medium Straw	460
4	Dark Straw	490
5	Purple	525
6	Blue	550 to 560
7	Pale Blue	600

temperatures for the various colors.

    Strain draw is a low oil draw used
on hardened steel immediately after it is
quenched to relieve strain and reduce brittleness caused by the shock of sudden
quenching. The usual temperature used to
strain draw tool steel is 375° to 400° F.

    Cyaniding and Carburizing. Low car-
bon steels (not over .20% carbon) do not become hard when heated above their critical
points and quenched, so surface hardening is
resorted to for machine and cold rolled
steels. Cyaniding is done by keeping the
piece immersed in a molten bath of sodium
cyanide from 5 to 30 minutes, depending on

size of piece and depth of penetration desired. The piece is then quenched in water,
brine, or oil, and a very hard "skin" or casing, .010" to .015" thick is formed. (See
Heat Treatment "O", page 208.) This is
called "case hardening." Another method used
is carburization. The work is placed in a
metal box containing a mixture of bone,
leather, charcoal, and carburizing materials.
The lid is sealed with fire clay and the box
is placed in a furnace for some hours at a
temperature of 1700° F. The depth to which
the carbon penetrates depends upon the length
of time the piece is left in the furnace.
When properly heat treated, the piece has a
hard outer shell and a soft core. (See Heat
Treatment "A" and "P", pages 207 and 208.)

11. What is meant by critical points?

A. The critical points or critical temperatures are the temperatures at which some
definite change takes place in the physical
properties of the steel. These points are
important because, in heat treating a piece
of steel, it must be heated to a temperature
above the upper critical point and then
quenched. When the critical points for a
certain steel are known, the heat of the furnace can be regulated by a pyrometer. Fig.
710 shows a type of pyrometer used on an
electric furnace.

Fig. 710

Not so many years ago it was the
custom for a hardener to watch the color of
the work in the furnace to determine its temperature. A cherry red, perhaps, for tool
steel and an orange, or even at times a lemon
yellow for high speed, was thought to be the

correct heat. There was a big element of
chance in this procedure and the U. S.
Bureau of Standards has demonstrated con-
clusively that at temperatures around 2000°
F. the "old timers" who depended on their
eyes were off as much as 200° F. in judging
furnace temperature. Therefore, the thermo-
couple and pyrometer have come into use for
the accurate measuring of temperatures.

12. What is meant by heat treatment of a
    metal?

A. Heat treatment is a method by which
the heat treater is able to change the physical properties of a metal. There are three
major steps in the heat treating of steel:
hardening, tempering, and annealing. The
hardening operation consists of heating the
steel above its critical range and then
quenching it in a suitable medium, such as
water, brine, or oil. Then it must be given
a tempering or drawing treatment which consists of reheating the hardened steel to a
temperature below the critical range that
will produce the physical properties desired.

The important factors in hardening
are: heating to the correct temperature,
holding at heat for correct length of time,
selecting correct quenching medium and proper method of quenching, and selecting correct drawing temperature. To harden a piece
of steel successfully it must be in the
softened or annealed state.

13. Name three types of furnaces used in
    heat treating metals.

A. Gas, oil, and electric furnaces are
the most commonly used furnaces. The heat
can be easily controlled in these furnaces,
which is an important factor. Some steels
are heated in open furnaces while others are
heated in baths of lead, cyanide of potassium, etc. Fig. 711 shows a pot furnace.
If the pot contains molten lead it is called
a "lead pot" but if it contains molten cyanide it is called a "cyanide pot." The pot
furnace can also be used for tinning baths,
for melting low fusion metals, and for other
purposes. Tools such as dies, punches,
springs, and other small steel parts may be
hardened uniformly in this furnace without
danger of oxidizing the steel. The lead pot
is especially adapted for jobs where only a
portion of the tool or part is to be hardened. Only the portion to be hardened is
immersed in the lead. The pot furnace is
rapid, convenient, and satisfactory.

The type of furnace in which the
steel is heated by gas is shown in Fig. 712.

Fig. 711

This is well adapted for most heat treat work. Fig. 713 shows the type of furnaces used for carburizing or annealing.

Fig. 712

14. What operations are involved in hardening?

A. Hardening involves both heating and cooling operations.

(a) Heating is the bringing of the steel to the desired temperature above the critical range in order to get the grain structure in the steel into the proper state for hardening.

Fig. 713

(b) Cooling is the quenching of the steel in some medium such as water, brine, caustic solution, or oil in order to preserve the structure obtained in heating. The medium must have an even temperature.

15. How are low carbon steels cyanided?

A. Low carbon steels are heated to the required temperature in a sodium cyanide bath to obtain the proper depth of case (.005" to .012"). Cyanide penetrates from .001" to .0015" for each minute of soaking. Pieces are then quenched in water or oil to produce hardness.

16. Can a cyanided low carbon steel be ground and still retain its hardness?

A. Since cyanide penetrates to a depth of not more than .015" on low carbon steel, the grind stock on a job using this steel would have to be less than .015" for the steel to retain any of its hard case.

17. What is meant by the carburizing of low carbon steels?

A. The carburizing of low carbon steels is the process of forcing carbon into steel by packing it in charcoal, bone dust, or other carburizing material and heating it to a temperature of 1700° F. in a furnace like the one shown in Fig. 713. The length of time the furnace is held at this temperature is governed by the depth of the case required. After the steel is removed from the furnace and cooled to room temperature, it can be normalized by reheating from 1560° to

1650° F. and cooling in air. It then can be hardened by inserting into a furnace or a lead pot (see Fig. 711), heating to the required temperature, and quenching.

18. For what is a carburized piece of steel recommended?

A. This steel is recommended for work requiring a hard surface and a tough core. For example, the wrist pin of the Ford car is carburized. This gives a hard surface to resist wear and a tough core to absorb the shock. Many jobs in the toolroom require this treatment. This steel can also be used for spot hardening, as in the nut shown in Fig. 714. In this nut the O D must be hard and the threads must be soft. The operations are as follows. Finish the outside

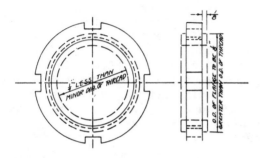

Fig. 714

diameter and the thickness to size, leaving a flange on each side 1/8" greater than the major diameter of the thread and extending 1/8" on each side (see sketch). Bore the hole for the thread 1/4" less than the minor diameter. Mill the slots, carburize, bore the hole to within 1/16" of size, face off the flanges (this removes carbon), harden, and then finish the threads.

19. In the sketch of Fig. 715 a steel block 1" x 4" x 8½", hardened and ground with two holes reamed in place, is required. What kind of steel should be used; what heat treatment is required; and how should this job be done?

A. Use machine steel with heat treatment "P" (a carburizing heat treatment). Use stock 1/4" wider than the width required and machine 1/8" off the two sides marked "A", excepting the square buttons which should be 1/4" to 5/16" larger than the diameter of the holes. The job is then ready to be carburized. After carburizing, machine off the buttons, harden, grind, drill, and ream the holes to size. Machining off the buttons after carburizing and before hardening

leaves those spots soft and immune from hardening. The holes may be drilled and reamed after the rest of the piece is hardened, thereby insuring their correct position after hardening. Bear in mind that the holes will be soft.

20. What is meant by nitriding?

A. Nitriding is a method of putting an extremely hard surface on a steel part. The process consists of exposing the steel to hot ammonia gas for some hours. The ammonia breaks down into nitrogen and hydrogen, because of the heat, and the nitrogen reacts with the steel to form the nitride case.

21. Of what importance is the element carbon in straight carbon or "simple" steel?

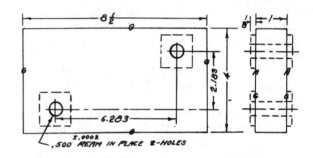

Fig. 715

A. Carbon is the element which causes straight carbon steel to harden.

22. Explain why cold rolled steel, machine steel, or type "A" steel cannot be hardened in the furnace or lead pot to a high degree of hardness.

A. These steels do not contain enough carbon to give a high degree of hardness, although they can be hardened to some extent.

23. How much carbon must be present in steel before it can be hardened noticeably?

A. Steel must contain at least .20% carbon before it can be hardened sufficiently for commercial use.

24. Give the percentage of carbon in each of the following steels: cold rolled, machine, "A", "AAA", "EE", "S", and RR (tool steel).

A. The following list gives the percentage of carbon in each of these steels.

Cold rolled	.05% to .15% carbon
Machine	.08% to .20% "
"A"	.20% to .24% "

"AAA"	.30% to .35% carbon
"EE"	.35% to .40% "
"S"	.60% to .70% "
"RR" (Tool)	.95% to 1.05% "

25. Why is charcoal kept on top of the lead in a lead pot?

A. Charcoal is kept on top of the molten lead in a lead pot to burn up the oxygen in the air, to prevent oxidation, and to keep the job clean. This prevents surface or skin softness and helps to eliminate scale.

26. What heat treatments are used for "AAA" steel? (See heat treatment chart, page 207.)

A. Heat treatments "B" and "C" are used on "AAA" steel. Heat treatment "B" is a cyanide treatment and "C" is a toughening treatment. When heat treatment "C" is called for, be sure to allow for stock from 1/16" to 1/8" to be removed after heat treatment. Heat treatment "C" is given to a piece where accuracy is the most important. It is machinable after this heat treatment.

27. What effect does heat treatment "B" and heat treatment "C" have on the hardness of "AAA" steel?

A. The Rockwell reading on annealed "AAA" steel is from 10 to 15. After heat treatment "C" is given, the reading will in-crease to from 30 to 35, and after heat treatment "B" is given, the reading should be from 48 to 52.

28. What kind of steel is usually used to make tool bits, milling cutters, reamers, drills, broaches, etc.?

A. High speed steel is usually used for making these tools.

29. Which is the most expensive: machine steel, tool steel, or high speed steel?

A. High speed steel is the most expensive because it contains expensive alloys. Tool steel rates next and machine steel is cheapest. High speed steel costs at least six or seven times as much as tool steel.

30. Why are high speed and chrome non-shrink steels preheated?

A. Some steels such as high speed and chrome non-shrink have a close, fine grain structure. They cannot stand much change in shape when cold, and will crack if put directly into a hot furnace. These steels are heated to a point slightly below the critical range and held there until thoroughly heated before being exposed to the hardening temperatures.

31. How is the hardness of steel determined?

A. The file test and the Brinell,

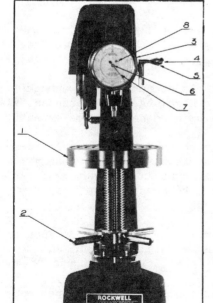

Fig. 716

Instructions for Operating
Rockwell Tester

1. Place piece to be tested upon anvil or testing table.

2. Turn wheel to elevate work into contact with test point and continue turning and forcing work against penetrator till an index shows that Minor Load is applied.

3. Turn bezel of gauge to set dial zero behind pointer.

4. Push handle back an inch to release and apply Major Load.

5. Pull handle forward, removing thereby the Major but not the Minor Load.

6. Observe when moving pointer comes to rest, then--

7. Read ROCKWELL HARDNESS Number on the dial.

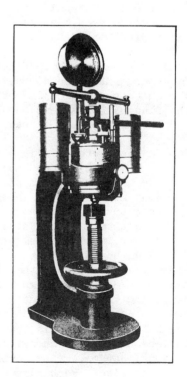

Fig. 717

Rockwell, and Scleroscope hardness testing machines are used to determine the hardness of steel.

The file test is simply trying to cut into the piece with the corner of a file. The hardness is shown by the bite the file will take. This is the oldest and still one of the most useful methods of checking hardness. While this test will not give very definite results because a new file will cut better than an old file and a fine file will make a piece appear softer than a coarse file, it will give results from quite soft to glass hardness. The greatest objection to the use of the file test is that no accurate record of results can be maintained.

In the Rockwell hardness test a 120° diamond cone for hard metals, or a 1/16" steel ball for the softer materials, is impressed into the surface to be tested by a dead weight acting through a series of levers and the depth of penetration measured. The softer the piece the deeper will be the impression under a given load. The average depth of penetration on the softest steel is only about .008 of an inch. The hardness is indicated on a dial gage graduated in the Rockwell "B" and the Rockwell "C" hardness scales. The harder the piece the higher the Rockwell number will be. For example, steel should not show a reading of more than from 30 to 35 on the Rockwell "C" scale to be machinable while a hardened high speed cutter would show a reading of from 63 to 65. When testing hard steels, the diamond point should be used and should be read on the "C"

scale; for non-ferrous metals the steel ball should be used and read on the "B" scale. Fig. 716 shows the Rockwell tester and the directions for using it. The average time for each test on a production basis is about five seconds.

In the Brinell test the hardness of the material tested is determined by the resistance it offers to the penetration of a steel ball under pressure. The Brinell hardness number is found by measuring the distance the ball is forced into the piece tested under a given pressure. The greater this distance, the softer the work, and the higher the Brinell number. The width of the indentation is measured with a microscope and the hardness number (corresponding with this width) is found by consulting a standard chart. The Brinell tester (Fig. 717) is most useful in testing soft and medium hard materials and for testing large pieces. On hard steel the impression is so small that it is difficult to read.

In the Scleroscope test a diamond pointed hammer is dropped through a guiding glass tube onto the test piece and the rebound checked on a scale. The harder the steel the higher the hammer will rebound, because the rebound is directly proportional to the resilience or springiness of the test piece. For example, if a ball is thrown on the grass it will not rebound as far as though it were thrown against the sidewalk with the same force. The Scleroscope (Fig. 718) is portable and can be used to check pieces that are too big to place on the anvil of other machines.

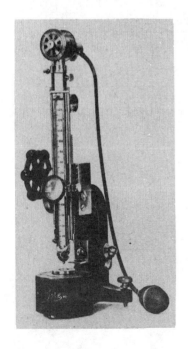

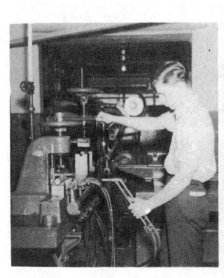

Fig. 718   Fig. 719   Fig. 720

STEEL ANALYSIS CHART

TYPE OF STEEL	CARBON	MANGANESE	CHROMIUM	OTHER ALLOYS	SILICON	PHOSPHORUS	SULPHUR	ENDS PAINTED
*A	.20 - .24	.60 - .75	.65 - .80	VANADIUM .12 - .15	.10 - .15	.03 Max	.04 Max	"
AX	.18 - .22	.65 - .75	.80 - .95		.10 - .20	.03 Max	.04 Max	White and Black
AA	.26 - .30	.65 - .80	.80 - 1.00		.10 - .20	.03 Max	.04 Max	Red and Black
*AAA	.30 - .35	.65 - .80	.90 - 1.10		.10 - .20	.03 Max	.04 Max	Red and White
AAAH	.35 - .38	.65 - .80	.90 - 1.10		.10 - .20	.03 Max	.04 Max	Red, White and Blue
AAAAL	.38 - .42	.65 - .80	.90 - 1.10		.10 - .20	.03 Max	.04 Max	Red, Green and Yellow
AAAA	.42 - .47	.70 - .90	.85 - 1.10		.10 - .20	.03 Max	.04 Max	Red and Green
AAAAA	.48 - .52	.70 - .90	.85 - 1.10		.10 - .20	.03 Max	.04 Max	Blue and White
AA Select	.28 - .32	.65 - .80	.80 - 1.00		.10 - .20	.03 Max	.04 Max	T-498 All Elements
AAA Select	.32 - .35	.65 - .80	.90 - 1.10		.10 - .20	.03 Max	.04 Max	T-12 Well Within Limits
Armature Steel	.05 Max	.30 Max			.12 - .28	.03 Max	.04 Max	
B	.95 -1.05	.20 - .30	.40 - .50		.20 - .30	.03 Max	.04 Max	
BB	.95 -1.05	.20 - .30	.90 - 1.10		.20 - .30	.03 Max	.04 Max	Blue and Red
BBB	.09 - .13	.30 - .40	1.25 - 1.50			.03 Max	.04 Max	Aluminum
Beosemer #1	.03 Max	.12 Max		COPPER OPTIONAL	.10 - .20	.09-.13	.08-.15	
C Key Stock (S.A.E. 1035)	.30 - .40	.50 - .80			.10 - .20	.05 Max	.05 Max	
C Pure Iron	.03 Max	.12 Max			.10 - .20	.01 Max	.04 Max	
*Chrome Non-Shrink	1.45 -1.60	.25 - .35	11.00 - 12.00	V .20-.25 Mo .70-.90	.20 - .40	.03 Max	.04 Max	Green With White Stripe
*CC	.44 - .50	.40 - .55	.70 - .90	TUNGSTEN 1.00 - 1.20	.15 - .25	.03 Max	.04 Max	Green
D	.45 - .52	.80 - .95	1.00 - 1.20		.10 - .20	.03 Max	.04 Max	
DD	.48 - .52	.80 - .95	1.00 - 1.20		.10 - .20	.03 Max	.04 Max	
*Die Block Ajax	.60 - .75	.30 - .40	3.25 - 3.75		.10 - .20	.03 Max	.04 Max	
*Die Block Hammer	.47 - .55	.50 - .60	.60 - .75	NICKEL 1.50 - 1.75	.10 - .20	.03 Max	.04 Max	
E	.27 - .35	.70 - .90			.07 - .15	.04 Max	.05 Max	Red
EE	.35 - .40	.70 - .90			.07 - .15	.04 Max	.05 Max	Yellow
*EEE	.40 - .45	.70 - .90			.07 - .15	.04 Max	.05 Max	Red and Yellow
Electrical	.05 Max	.30 Max			.90 -1.20	.04 Max	.03 Max	
F	.08 - .15	.80 - .99			.10 - .20	.04 Max	.08-.15	Aluminum and Black
FF	.15 - .20	.80 - .99			.10 - .20	.04 Max	.10-.15	Aluminum and Green
FFF	.34 - .40	.80 - .99			.10 - .20	.04 Max	.10-.15	Aluminum and Yellow
Ford High Speed (Taps and Drills)	.65 - .73	.25 - .35	3.75 - 4.25	V 1.00-1.25 W 17.0-18.0	.20 - .30	.03 Max	.04 Max	Red and Green With White Stripe
*Ford Hot Work	.18 - .23	.40 - .60	1.40 - 1.60	MOLYBDENUM .45 - .55	.15 - .25	.03 Max	.04 Max	(Mo. = .50 - .60)
*Ford Special High Speed	.78 - .84	.25 - .35	4.00 - 4.50	V 2.00-2.25 W 18.0-19.0	.20 - .50	.03 Max	.04 Max	Black
G	.08 - .15	.30 - .45			.07 - .15	.03 Max	.05 Max	Black
*GG	.15 - .25	.30 - .45			.07 - .15	.04 Max	.05 Max	Black
H	.27 - .37	.45 - .60			.07 - .15	.04 Max	.05 Max	Blue
Key Stock (S.A.E. 2330)	.25 - .35	.50 - .80		NICKEL 3.25 - 3.75		.04 Max	.045 Max	
L	.23 - .30	.35 - .50			.10 - .20	.04 Max	.05 Max	Green and Blue
Low Carbon, Open Hearth	.05 - .15	.35 - .50			.10 - .20	.04 Max	.05 Max	Black With Yellow Stripe
*Machine	.08 - .20	.35 - .50			.10 - .20	.04 Max	.05 Max	Black
Magnet	.82 - .90	.30 - .45	2.25 - 2.60	(High Limit Preferred)	.25 - .40	.05 Max	.04 Max	Aluminum When C is 185 or over
N	.12 - .16	.35 - .45	.30 - .40		.15 - .20	.03 Max	.04 Max	Green and Black
R	.70 - .80	.20 - .35	.10 Max	NICKEL - NONE	.15 - .25	.025 Max	.05 Max	
*RR	.95 -1.05	.20 - .35	.10 Max	NICKEL - NONE	.15 - .25	.025 Max	.05 Max	Brown
RRR	1.20 -1.30	.20 - .35			.15 - .30	.025 Max	.05 Max	
*Rustless 18-8	.05 - .10	.30 - .45	16.0 - 18.0	NICKEL 7.0 - 9.0	.15 - .30	.04 Max	.05 Max	
Rustless 18	.05 - .10	.30 - .45	16.0 - 18.0		.50 Max	.04 Max	.05 Max	
Rustless Type II	.20 - .30	.25 - .40	12.0 - 14.0		.70 -1.00	.04 Max	.05 Max	
*S	.60 - .70	.70 - .85			.15 - .20	.03 Max	.04 Max	
*SS	.70 - .85	.70 - .85			.10 - .20	.03 Max.	.04 Max	Blue and Yellow
Tap (Under 1" Diameter)	1.20 -1.30	.25 - .40	.35 - .45	V .15-.25 W 1.25-1.50	.30 - .45	.025 Max	.025 Max	
V	.35 - .45	.25 - .40	1.85 -2.50		3.60 -4.20	.02 Max	.04 Max	
Welding Wire	.10 Max	.20 Max				.03 Max	.04 Max	
*Insert Die Steel	.55 - .60	.45 - .60	.70 - .80	Mo .75-.80 Ni 2.25-2.45	.15 - .25	.025 Max	.05 Max	
*Vanadium Tool Steel	.95 -1.05	.20 - .35	.10 Max	V .40-.50 Ni None	.15 - .30	.04 Max	.05 Max	
S.A.E. #4620	.18 - .22	.30 - .60	.25 Max	Mo .20-.30 Ni 1.65-2.00		.04 Max	.04 Max	Aluminum and Red
Deep Drawing Steel	.05 - .08	.28 - .38	.04 Max	Copper .10 Max	.10 - .20	.04 Max	.04 Max	
"FH"	.20 - .25	.80 - .99				.04 Max	.10-.15	

*Most commonly used steel in the school shop.

"White if on low side of analysis and white with green stripe if on high side.

4-5-38

32. Give some suggestions that are valuable in heat treatment.

A. Work having sharp edges or different cross sections should be protected by pieces of wire and fire clay to prevent fractures.

Cold rolled steel is a low carbon steel that can be case hardened by immersing in a cyanide pot and quenching. Drill rod is made from "R" (high carbon) steel and can be hardened by heating in a lead pot or a furnace and quenching.

Steel should be heated long enough to insure a good even heat throughout. The practical rule is to keep or "soak" the piece in the furnace one hour for each square inch of cross sectional area.

To avoid warping a long slender piece when quenching, hold it vertically over the bath and plunge straight down. A warped piece may be straightened under pressure in a straightening press after heating with a blow pipe (Fig. 719).

The sand blast (Fig. 720) is used to remove scale and dirt from parts previously heat treated. This saves a great deal of the grinder's time.

33. What are some of the safety rules which must be observed in this department?

A. The following rules must be observed by students in heat treating steel:

(1) GOGGLES MUST BE WORN when working on a lead, cyanide, or nitrate pot.

(2) Do not put anything damp or wet into these pots or an explosion will occur.

(3) Hot tongs should not be left where any one can be burned by them.

(4) Never pick up anything until you know whether or not it is hot.

(5) Rubber gloves must be worn while sand blasting.

(6) Do not wear gloves while grinding.

(7) Do not use a towel to hold a piece while grinding.

The "spark test" is a method by which different steels that are not marked may be identified within certain limits by the sparks given off when they are held against a grinding wheel. In this test either a portable or a stationary grinder may be used. Only enough pressure to maintain a steady contact between the work and the wheel is necessary and less pressure is required as the wheel speed is increased. About 8,000 surface feet per minute is usually a satisfactory speed. The sparks from most steels will be practically the same whether the stock is hardened or annealed. Test pieces of known samples on which to practice or with which to compare pieces being checked should be available. The test should be made in diffused daylight if possible.

Broadly speaking, the effect of various elements is as follows: Carbon causes the sparks to burst. Manganese tends to brighten the spark and increase the spray around the periphery of the wheel. Chromium darkens the color, suppresses the stream and bursts, and causes fine carrier lines. Nickel suppresses the stream slightly and causes forked tongues. Tungsten suppresses stream and bursts, and causes fine red carrier lines. Molybdenum causes a detached spear head at the end of the ray.

The following chart shows sketches of sparks from some of the most commonly used steels. They are of necessity very general but may be used to point out certain characteristics. For example, the difference in carbon content is indicated by the difference in the number of sparks as shown in the first three sketches. If a piece of machine steel and a piece of high speed steel were sparked, the difference between them could be easily discerned. However, proficiency in identifying all steels can only be gained by practice and experience. The spark test does not analyze a piece of steel, but is simply a guide in identifying it.

SPARK TEST CHART

(a)   Low carbon steels show long club-shaped sparks, smooth light lines, light yellow in color, and no small stars. Cold rolled, machine steel, etc., are in this group.

(b)   Medium carbon steels show club-shaped lines which split into formation of a few small simple stars of light yellow color. Steels such as "E", "EE", and "EEE", are in this group. The spark from "AAA" steel, which is a medium carbon alloy, is very similar in formation but darker in color due to the chromium content.

(c)   High carbon steels such as "RR" (tool steel) show almost no club-shaped sparks, but numerous little light yellow stars. "S" and "SS" steels are included in this group.

(d)   Chrome steels show yellow spark lines with ball-shaped ends, between which occur more or less numerous and active sheafs of light depending upon the carbon content. Chrome non-shrink, the rustless steels, etc., are in this group.

(e)   High speed steels show several interrupted, dotted, brownish-red spark lines with ball-shaped, end sparks of dark blood-red color. Some of these burst into small red stars of few rays, slightly lighter in color.

Hardness Conversion Table
(Approximate)

Brinell		Rockwell			Brinell		Rockwell		
Diam. 3000 Kg., 10mm.Ball	Hardness No.	C Scale 150 Kg., 120° Cone	B Scale 100 Kg., 1/16" Ball	Shore	Diam. 3000 Kg., 10mm.Ball	Hardness No.	C Scale 150 Kg., 120° Cone	B Scale 100 Kg., 1/16" Ball	Shore
Mm									
2.20	780	68	...	96	3.45	311	32	...	43
2.25	745	67	...	94	3.50	302	31	...	42
2.30	712	65	...	92	3.55	293	30	...	41
2.35	682	63	...	89	3.60	285	29	...	40
2.40	653	62	...	86	3.65	277	28	...	38
2.45	627	60	...	84	3.70	269	27	...	37
2.50	601	58	...	81	3.75	262	26	...	36
2.55	578	56	...	78	3.80	255	25	...	35
2.60	555	55	...	75	3.85	248	24	100	34
2.65	534	53	...	73	3.90	241	23	99	33
2.70	514	51	...	71	3.95	235	22	99	32
2.75	495	50	...	68	4.00	229	21	98	32
2.80	477	48	...	66	4.05	223	20	97	31
2.85	461	47	...	64	4.10	217	18	96	30
2.90	444	46	...	62	4.15	212	17	95	30
2.95	429	44	...	60	4.20	207	16	95	29
3.00	415	43	...	58	4.30	197	14	93	28
3.05	401	42	...	56	4.40	187	12	91	27
3.10	388	41	...	54	4.50	179	10	89	25
3.15	375	39	...	52	4.60	170	8	87	24
3.20	363	38	...	51	4.70	163	6	85	23
3.25	352	37	...	49	4.80	156	4	83	23
3.30	341	36	...	48	4.90	149	2	81	22
3.35	331	35	...	46	5.00	148	0	79	21
3.40	321	34	...	45	5.10	137	−3	77	20

Heat Treatment Chart

COMMERCIAL ANNEALING   HEAT TREATMENT "D"
(1) Heat to 1425° F. (2) Cool in air. Use to restore hardened steel to machining hardness.

COMMERCIAL ANNEALING   HEAT TREATMENT "DD"
(1) Heat to 1500-1550° F. (2) Cool in mica. To eliminate scale, pack in charcoal. Use for U.S. die steel to obtain maximum softness.

CAST IRON   HEAT TREATMENT "CC"
(1) Heat in furnace 1400-1450° F. (2) Cool slowly in furnace or mica. Use for surface plates before machining.

TYPE "A" STEEL   HEAT TREATMENT "A"
(1) Carburize at 1700° F. (2) Cool in carburizing box. (3) Reheat to 1650° F. (4) Cool in air. (5) Reheat to 1560° F. (6) Quench in brine. (7) Strain draw in oil 350-375° F. Rockwell C-62-64. Use when a hard surface and a soft core is the most important requirement. Never use heat treatment "A" when minimum of distortion is desired. Use for heavy duty bushings, large worms, and large gears where a hard surface is more important than accuracy.

TYPE "A" STEEL   HEAT TREATMENT "E"
(1) Carburize at 1700° F. (2) Cool in carburizing box. (3) Reheat to 1650° F. (4) Cool in air. (5) Reheat to 1560° F. (6) Quench in oil. (7) Strain draw in oil 350-375° F. (8) Draw temper if specified. Rockwell 56-64. For thin sections and intricate parts to eliminate distortion.

TYPE "AAA" STEEL   HEAT TREATMENT "B"
(1) Heat in cyanide 1500° F. (2) Soak 5 to 10 minutes. (3) Quench in oil. (4) Strain draw in oil 350-375° F. Specify Rockwell hardness required. Use for acme screw, milling machine arbors, grinder spindles, lock nuts, tap holders, small gears and worms, etc. For large parts, where cyanide is not available, heat in furnace.

TYPE "AAA" STEEL   HEAT TREATMENT "C"
(1) Rough machine. (2) Heat to 1560° F. (3) Quench in brine or caustic solution. (4) Draw to 950-1050° F. (5) Cool in air. (6) Finish machine. Rockwell 30-34. Use for large gears, armature shafts, large boring bars, miscellaneous heavy machine parts, etc., and where accuracy and toughness are more important than hardness.

TYPE "CC" STEEL   HEAT TREATMENT "H"
(1) Heat in lead to 1500° F. (2) Quench in brine or caustic. (3) Draw in oil to 350° F. Rockwell 54-59. Use for chisels, miscellaneous hand tools, and wire machine rolls.

TYPE "E" STEEL   HEAT TREATMENT "AA"
(1) Heat to 1470-1500° F. (2) Quench in brine. (3) Strain draw in oil 350-375° F. (4) Draw temper if specified. Rockwell 43-46.

TYPE "EE" STEEL   HEAT TREATMENT "BB"
(1) Heat to 1470-1500° F. (2) Quench in brine. (3) Strain draw in oil 350-375° F. (4) Draw temper if specified. Rockwell 51-55.

TYPE "EE" AND "EEE" ST C   HEAT TREATMENT "ST"
(1) Preanneal at 1700-1750° F. (2) Cool in air. (3) Heat in cyanide to 1470° F. (4) Quench in solution. Rockwell 33-39.

TYPE "G", "GG", AND LOW CARBON STEEL
Use heat treatments "O" and "P". (See machine steel.)

TYPE "R" (TOOL) STEEL   HEAT TREATMENT "QQ"
(1) Heat to 1450-1475° F. (2) Quench in brine. (3) Strain draw in oil 350-375° F. (4) Draw temper if specified. Rockwell 55-59. For heavy duty shears, dies, and punches.

TYPE "R" (TOOL) STEEL   HEAT TREATMENT "SS"
(1) Heat in lead 1450-1475° F. (2) Quench in brine. (3) Strain draw in oil 375° F. (4) Draw temper if specified. Rockwell as specified. Use for calking tools, drift pins, machine hammers, pin drivers, picks, and screw drivers.

TYPE "RR" (TOOL) STEEL   HEAT TREATMENT "Q"
(1) Heat to 1400-1450° F. (2) Quench in brine. (3) Strain draw in oil 350-375° F. (4) Draw temper if specified. Rockwell as required. Use for chuck jaws, ball races, bushings, cams, female gages, rest buttons, arbors, locators, plug gages, snap gages, "V" blocks, etc. Use when maximum hardness is desired. Temperature to be governed by size of steel. For small parts quench in oil, if the same hardness can be obtained.

TYPE "RR" (TOOL) STEEL   HEAT TREATMENT "S"
(1) Heat in lead to 1400-1450° F. (2) Quench in brine. (3) Strain draw in oil 450° F. (4) Draw temper if specified. Use for partial hardening.

TYPE "S" (SPRING) STEEL   HEAT TREATMENT "SP"
(1) Heat to 1450° F. (2) Quench in oil. (3) Draw temper 750° F. Rockwell 41-44. Use for spring wire $\frac{1}{4}$ diameter and below.

TYPE "SS" (SPRING) STEEL   HEAT TREATMENT "SSP"
(1) Heat to 1450° F. (2) Quench in oil. (3) Draw temper 750° F. Rockwell 41-44. Use for spring wire $\frac{1}{4}$ diameter and above.

Heat Treatment Chart (Continued)

MACHINE AND COLD ROLLED STEEL   HEAT TREATMENT "O"
(1) Heat in cyanide 1500-1560° F. (soak 10 minutes). (2) Quench in brine (small parts in oil).
File hard. Use for a hard surface not subject to
continuous wear. Do not grind. Use for clamps,
locating gages, pressure and stripper pads, stock
guides, bolts, nuts, and washers.

MACHINE AND COLD ROLLED STEEL   HEAT TREATMENT "P"
(1) Carburize at 1700° F. (2) Cool in carburizing
box. (3) Reheat to 1650° F. (4) Cool in air.
(5) Reheat to 1425° F. (6) Quench in brine.
(7) Strain draw in oil 350-375° F. Rockwell 62-64.
Due to soft core, should be used for parts difficult to straighten. This heat treatment may also
be used for selective hardening where it is necessary to machine after hardening. (1) Leave stock,
(2) carburize, (3) remove case, (4) harden,
(5) machine. (See questions 17 and 18.)

HIGH SPEED STEEL   HEAT TREATMENT "G"
(1) Preheat to 1400-1450° F. (2) Superheat 2350° F.
minimum. (3) Quench in oil. (4) Draw in furnace
1050-1100° F. if not otherwise specified. Rockwell
63 minimum. Use for gear cutters, hobs, reamers.

HIGH SPEED STEEL   HEAT TREATMENT "J"
(1) Preheat to 1400-1450° F. (2) Superheat 2300-
2350° F. (3) Cool in air. (4) Draw in furnace 1050-
1100° F. if not otherwise specified. Rockwell 62-
64. Use for long keyway broaches, long and slender
tools, and to eliminate distortion. To reduce
brittleness in heat treatments "G" and "J", redraw
in nitrate 700-800° F. for 3 or 4 hours.

HIGH SPEED STEEL   HEAT TREATMENT "L"
(1) Preheat 1400-1450° F. (2) Superheat 2350° F.
(3) Quench in cyanide 1050-1100° F. (4) Draw in
furnace 1050° F. Rockwell 63 minimum. Used only
for special forming tools, threads, hobs, and similar tools, where accuracy is important.

HIGH SPEED STEEL   HEAT TREATMENT "SPG"
(1) Preheat to 1450° F. (2) Superheat to 2300° F.
(3) Cool in air. (4) Preheat to 1450° F. (5) Superheat to 2350-2400° F. (6) Quench in oil. (7) Furnace draw 1050° F. (8) Finish to print. Rockwell
C-64 minimum. Use for form tools, broaches, counterbores, spotfacers, milling cutters, cut-off
tools, reamers.

HIGH SPEED STEEL (SPECIAL)
Special heat treatment before backing off high
speed steel cutters. (1) Heat to 1650° F. (2) Draw
1200° F. (3) Quench in oil. Rockwell 34-36. Sand
blast.

SEMI-HIGH SPEED STEEL   HEAT TREATMENT "X"
(1) Preheat to 1500° F. (2) Superheat 2050-2100° F.
(3) Draw at 1175-1250° F. Rockwell 46-50. Use for
hot stock only.

CHROME NON-SHRINK STEEL   HEAT TREATMENT "M"
(1) Preheat to 1200-1250° F. (2) Superheat 1860-
1880° F. (3) Air cool. (4) Draw 980-1000° F. Rockwell 59 minimum. Use for dies.

CHROME NON-SHRINK STEEL   HEAT TREATMENT "N"
(1) Pack in charcoal. (2) Heat slowly to 1850° F.
(3) Remove from pack; air cool. (4) Draw 350° F.
Rockwell 63 minimum. Use for master gears, gauges,
etc. Parts for extreme hardness and wear.

CHROME NON-SHRINK STEEL   HEAT TREATMENT "R"
(1) Preheat 1250° F. (2) Superheat 1860-1880° F.
(3) Air cool. (4) Draw 750° F. Rockwell 56-58.
Use for small details.

CHROME NON-SHRINK ST C   HEAT TREATMENT "MM"
(1) Preheat 1250° F. (2) Superheat 1850° F. (3) Air
cool. (4) Draw 980-1020° F. (5) Air cool. Rockwell
58-62.

NON-SHRINK STEEL   HEAT TREATMENT "V"
(1) Heat to 1400-1450° F. (2) Quench in oil.
(3) Strain draw in oil 350-375° F. (4) Draw temper
if specified. Rockwell 60-62. Use for intricate
parts where minimum of distortion is desired.

NON-SHRINK STEEL   HEAT TREATMENT "Z"
(1) Heat in lead 1400-1440° F. (2) Quench in oil.
(3) Strain draw in oil 350-375° F. (4) Draw temper
if specified.

U.S. DIE STEEL   HEAT TREATMENT "F"
(1) Heat to 1560-1600° F. (2) Quench in oil. (3) Remove from oil while hot 600° F. (4) Allow oil to
burn off. Use for dies and punches for Ajax and National machines, small miscellaneous parts, and boring bars.

FORD HOT WORK STEEL   HEAT TREATMENT "FF"
(1) Heat to 1600° F. (2) Quench in oil. No draw.
Rockwell 41-45. Use for dies and punches for Ajax
and National Machines, all socket wrenches.

FORD DIE BLOCK NO. 1   HEAT TREATMENT "W"
(1) Heat in furnace to 1400-1450° F. (2) Quench in
oil. Rockwell 39-44. Use for hammer dies. Draw
800-850° F., time governed by size of die.

INSERT DIE STEEL   HEAT TREATMENT "DS"
(1) Heat to 1620° F. (2) Cool in air (fan blast).
(3) Draw to 1050-1075° F. Rockwell 42-46. Use for
hammer die inserts, hot heading dies, etc.

Chapter 24

ABRASIVES AND GRINDING WHEELS

HISTORY

AN ABRASIVE is any material having the ability to wear away a substance softer than itself. Sand and sandstone are perhaps the oldest abrasives known to mankind. Prehistoric man used sand and sandstone to form or shape edges of tools by which he earned his living and as tools became more and more important for preservation of life, he became more dependent on natural abrasives for maintenance of sharp tools.

EMERY and CORUNDUM are two "natural abrasives" which were commonly used in industry for the purpose of sharpening edged tools. They occur as a mineral deposit in the earth's crust. These abrasives, formed into wheels, were very superior to the old grindstones in that they were capable of faster cutting and could be made coarse or fine. In spite of this they could not meet the demands of industry because they contained impurities which were difficult to extract and because the percentage of the important cutting element, aluminum oxide, was not constant.

The only other element known to be harder than emery or corundum was the diamond, but its cost was prohibitive as far as industrial usage was concerned. In 1891 Dr. Edward G. Acheson set himself to the task of trying to produce artificial diamonds. His experiment consisted of combining powdered coke and corundum clay at extremely high temperatures. After the mass cooled it was found to contain brightly colored crystals which upon examination proved to possess the ability to cut glass and to have a slight cutting effect on diamonds. To this new substance Dr. Acheson gave the name "Carborundum" because it was formed from carbon and corundum; later however, this new substance was found to be composed of silicon and carbon so that the name "Silicon Carbide," having the chemical symbol SiC, was given to it.

This new material, silicon carbide, was considered as the answer for a better abrasive, but cost and limited methods of manufacture kept it from being used except as a lapping compound for finishing precious jewels; with the development of hydroelectric generators and cheap electric power, the cost of production was cut to a point where all industry could avail themselves of its use.

About the same time that Dr. Acheson was experimentally producing silicon carbide, Charles P. Jacobs, an engineer at Ampere, New Jersey was conducting similar experiments to produce a better grade of emery. His experiments consisted of trying to extract the impurities of sand, iron, and titanium oxides from clay deposits rich in aluminum oxide by use of a small electric furnace. The result was a product which consisted of about 95% pure aluminous oxide, which possessed characteristics very similar to those of silicon carbide.

CHARACTERISTICS AND MANUFACTURE

While these two artificial abrasives are very similar to each other, the properties of them differ widely. Silicon carbide is extremely hard, being rated at 9.87 on the Moh scale of hardness as compared to 10.0 for the diamond. It is easily fractured by impact and its excellence depends upon the purity of the ingredients used in making it. It has a specific gravity of 3.18, which is very low in comparison to other abrasives. Aluminum oxide is not as hard as silicon carbide but it is much tougher. On the Moh scale of hardness it is rated at 9.6 as compared to 9.87 for silicon carbide and 10.0 for the diamond.

SILICON CARBIDE is suited for grinding materials having a low tensile strength but which are very hard in nature, such as ceramics, pottery, tungsten carbide, etc.

ALUMINUM OXIDE, because of its toughness, is resistant to shock and therefore suitable for grinding materials of high tensile strength, such as tool steel, high speed steel, "AAA", etc.

Artificial abrasives have a distinct advantage over natural abrasives in that purity of product and grain size can be readily controlled. This is important because undersize grains cannot do their share of the work while oversize grains give a poor finish to the work.

Electric furnaces are used to produce

209

GRAIN SIZE CHART

Very Coarse	Coarse	Medium	Fine	Very Fine	Flour Sizes
8	12	30	70	150	280
10	14	36	80	180	320
	16	46	90	220	400
	20	60	100	240	500
	24		120		600

both abrasives. Silicon carbide is made in an open trough-like furnace, by fusing a mixture of coke, sawdust, sand, and salt. After the mass has cooled, the sides of the furnace are let down, exposing a big clinker; this clinker is broken by means of a drop weight, and the pieces put through a crushing machine. As the abrasive particles leave the last crusher, they are magnetically cleaned and washed, after which they pass onto shaker screens. The mesh of these screens vary in size from 4 to 220 meshes to the lineal inch, and by their vibrating action sort the grains according to a definite size. If the abrasive passes through a screen having 30 meshes to the lineal inch, but is retained on a screen of 36 mesh, it is called a #30 abrasive. Abrasives finer than 220 are graded for size by hydraulic or sedimentation methods. After the abrasive has been graded to size, it is dried and placed in storage bins or hoppers for future use. Above is a chart showing how the various grain sizes are classified.

GRINDING WHEELS

Grinding wheels are formed by using a suitable material to cement or bond the abrasive grains together in the desired shape. The hardness or softness of the wheel is dependent on the amount and kind of bonding material used. Since the hardness rating of the abrasive is constant it is apparent that the bond can have no effect whatever upon it. When speaking of the hardness or softness of the wheel, it is always understood to have reference to the strength of the bond.

BONDS

There are a great many different kinds of bonds but the kinds commonly used are the VITRIFIED, SILICATE, SHELLAC, RUBBER, and RESINOID. Of these bonds, the vitrified and silicate are used most. The VITRIFIED bond is used in making about 75% of all grinding wheels. The reason for this is because its strength and porosity enable it to

remove considerable stock from the job for each inch of wheel wear; it is not affected by water, acids, or ordinary temperature changes, and it is free from hard and soft spots. In the vitrified process, glass, feldspar, flint, or other ceramic substances are mixed with the abrasive and subjected to heat which causes the bond to form a glass-like structure between each abrasive particle.

In the SILICATE bond, silicate of soda is used. Besides the amount of bond used, the hardness of this wheel is governed by the amount of tamping or pressing. This produces a wheel which is milder acting than the vitrified wheel and permits the abrasive grains to be released more readily, therefore, they do not heat up so fast. Silicate wheels can be made in larger diameters than vitrified wheels and are generally used for grinding edged tools such as drills, reamers, milling cutters, etc.

RUBBER wheels are made of a mixture of abrasive, rubber, and sulphur; the mass is then pressed into shape and given a mild vulcanizing treatment. Wheels of this bond are used for high speed grinding operations and because of their high safety factor can be made very thin and can be used for cutting off steel stock.

SHELLAC bonded wheels are made by mixing the abrasive and bond in a heated machine which mixes and completely coats the abrasive with the bonding material. After the wheels are formed they are placed in an oven, covered with sand, and then baked for a short time at approximately 300° F. Wheels of this bond are used very extensively for grinding mill rolls and jobs where a high luster finish is required.

RESINOID bonded wheels are made by mixing powdered resinoid with the abrasive particles and then adding a plastic substance so that the wheels can be molded. The molding is then placed in an electric oven and heated from a few hours to three or four days depending on the size of the wheel, at approximately 300° F. Upon cooling, the wheel becomes very hard. Wheels of this bond are generally used in foundries for snagging castings or for cleaning up steel billets.

WHEEL STRUCTURE

The amount of puddling or packing that is applied in forming the grinding wheel is very influential in the way the wheel cuts, because it affects the wheel structure (Fig. 721) or grain spacing. Two

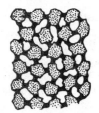

Fig. 721

Grain spacing: (a) wide; (b) medium; (c) close. This diagrammatic sketch shows how three wheels may be identical in grain size and grade (bond strength) but differ in grain spacing or STRUCTURE--and thereby differ in grinding action.

wheels of the same grade and grain size but of different grain spacing will have different cutting actions. Wheels with wide grain spacing should be used on hard dense materials and those with close spacing should be used to secure a good finish on the work. For this reason wheel life can often be increased without sacrificing grinding quality, using the same grain and grade of wheel but with a different structure.

WHEEL GRADING

As has been previously stated, the amount of bond used in making a grinding wheel determines its hardness. The degrees of hardness are specified by use of the letters of the alphabet. The Norton Company and several other companies use a lettering system by which letters at the beginning of the alphabet indicate soft wheels and letters at the end of the alphabet indicate hard wheels. The lettering system for grade of bond is shown in an accompanying chart. The Carborundum Company uses a system in the reverse order. Another chart shows the bond classifications commonly used.

In the Norton system a wheel which is marked 3860-K5BE has the following characteristics: "38" represents the kind of abrasive, which in this case is ALUNDUM or Aluminous Oxide abrasive. "60" is the grain

size, which according to the grain size chart is medium. "K" indicates the grade of the bond, which according to the chart for grade of bond is soft. "5" indicates the wheel structure; the numbers from 0 to 3 indicate close structure; 4 to 6 indicate medium structure; 7 to 12 indicate coarse structure. The last symbol in the wheel marking indicates the kind of bond. "BE" indicates the vitrified bond. "S" indicates a silicate bond; "T" is for the resinoid bond; "R" for the rubber bond; and "L" for the shellac bond.

CHART FOR GRADE OF BOND

Very Soft	Soft	Medium	Hard	Very Hard
E	H	L	P	T
F	I	M	Q	U
G	J	N	R	W
	K	O	S	Z

WHEEL SELECTION

There are several factors which affect the selection of a grinding wheel; they are the kind of material to be ground, the amount of stock to be removed, the accuracy as to size, the kind of finish required, the area of contact between the wheel and the work, and the kind of grinding machine to be used.

The nature of the material to be ground affects the selection of the wheel because, generally speaking, hard, dense materials require wheels possessing a soft bond with silicon carbide abrasive; materials that are soft and tough require a hard bond using aluminum oxide abrasive. The amount of material to be removed is important in selecting a grinding wheel because when there is a considerable amount of material to be removed the grains of a coarse grain wheel with wide spacing will be capable of taking a bigger, deeper cut without heating the work, but with slight sacrifice as to surface finish. When

COMMONLY USED GRADES AND GRAIN SIZES

	SURFACE	INTERNAL	EXTERNAL	CUTTER
GRADES	F G H I J K P	I J K L	J K L M P*	I J K L
GRAIN SIZES	36-46-60-80 120	34-46 60-120	46-60-80 120	36-46 60

*P for corners.

the amount of stock to be removed is slight, a wheel of fine grain and narrow spacing will take a smaller bite and give a good finish.

There are other factors which affect the grinding operation, such as: the speed of the wheel, the speed of the work, the condition of the grinding machine, and the knowledge and skill of the machine operator.

GRINDING THEORY

Grinding is the act of dressing, shaping, or finishing surfaces by means of a rotating abrasive wheel. In modern machine shop operation, it has been found by actual test that grinding costs will vary as much as 100% on the same work with the same kind of machine in the same factory. This is due to the difference in the handling and skill of operation of the machine by the mechanic. A good mechanic takes into consideration the factors involving the mounting, movement, size, and speed of the work and the mounting, movement, size, speed, and dressing or truing of the grinding wheel.

For precision grinding the work must be held rigidly to avoid vibration and to produce a good finish. If the work is held between centers, the center holes must be free from nicks, burrs, or dirt. The machine centers must be held securely and free from nicks. If held in a chuck or fixture, the work must be clamped so as to put the least strain on it and be solidly supported. After the work is correctly mounted the work speed must be selected so that it will move at approximately the right number of surface feet per minute to prevent distortion and excessive wear of the wheel face, and at the same time the traverse movement must be at a constant speed to prevent high and low spots on the work. The mechanism for moving the wheel must work smoothly and freely without play or bind to insure accuracy as to depth of cut.

The grinding wheel mounting is important because it must give steady and true motion to the wheel so that after it is trued up it will be free from vibration; it must give steady cutting action and be capable of producing a good finish after it is properly dressed.

GRINDING WHEEL SPEEDS

In most modern grinding machines the speed of the wheel is fixed as built by the manufacturer of the machine. Grinding wheels may be run from 5000 to 6000 surface feet per minute (S.F.P.M.). In case it is necessary to determine what the S.F.P.M. of a wheel is,

place a tachometer or speed indicator in the center of the spindle and check the revolutions for one minute; then multiply the diameter of the wheel by 3.1416 which will give the circumference of the wheel in inches. The circumference multiplied by the R.P.M. will give the distance in inches which the wheel would travel in one minute if laid on its periphery and rolled at its given R.P.M. This result divided by 12 (12 inches to the foot) gives the surface speed in feet per minute.

EXAMPLE: Determine the surface speed in feet per minute of a grinding wheel 7" diameter mounted on a surface grinder. Using a tachometer the R.P.M. is found to read 3200. Multiply the diameter of the wheel by 3.1416 by 3200 and divide by 12:

$$\frac{7 \times 3.1416 \times 3200}{12} = \frac{70371.84}{12} = 5864.3$$

Thus the surface speed in feet per minute or the S.F.P.M. equals 5864.3

By being a good observer the operator can tell many things about the work and machine. Poor or defective bearings readily show up while the wheel is being dressed, for the diamond will not show a steady red spark as it moves across the face of the wheel.

CHATTER OR VIBRATION

If the work vibrates it will be shown by lines on the work parallel to the work axis, known as chatter. This may be remedied by changing to a softer wheel, tightening the spindle bearings, checking and repairing bad belt connections, taking a lighter cut or cutting down on the work speed. If the work shows mottled surfaces it is generally an indication of vibration due to irregular motion of the grinding wheel. This irregular motion may be caused by the lead core in the wheel being improperly placed, the wheel being water logged on one side, by the structure of the wheel being of uneven density, or the face of the wheel being improperly trued.

GLAZING AND LOADING

Glazing of the wheel is a condition in which the face or cutting edge takes on a glass-like appearance and is caused by the abrasive grains wearing away faster than the bond. As long as the bond is being worn

away as fast as the abrasive particles of the wheel are being dulled the wheel will continue to have good cutting action. To remedy this condition of glazing, use a wheel having a softer bond.

Frequently the material being ground clogs up or becomes imbedded in the pores of the wheel. This condition is known as "loading" or "pinning" and is caused by the work speed being too slow, the wheel action too hard, or crowding the wheel.

The continued use of the wheel after it becomes glazed or loaded puts an added strain on the work supports and wheel spindle bearings and forces the wheel out of true, and if continued too long will produce chatter marks or surface checks on the work.

DRESSING

The cutting face of a grinding wheel must be kept in a true, clean, sharp condition if the grinding operation is to be done efficiently. This requires frequent dressing or truing of the wheel. DRESSING is the operation of cleaning or fracturing the cutting surface of a wheel for the purpose of exposing new cutting particles. TRUING a wheel is the operation of removing material from the cutting face so that the resulting surface runs concentric with the wheel spindle axis.

The operations of truing and dressing grinding wheels are usually accomplished by using a commercial diamond or piece of tungsten carbide inserted in the conical point of a piece of cold rolled steel; by a diamond dust impregnated cement, formed into a stick and encased in metal tubing; or by a piece of silicon carbide mounted as a small wheel on an axle and placed in a cast iron base. In using any of these implements the point of the dresser is brought into contact with the face of the wheel by means of a special holder and then moved mechanically or by hand across the face of the wheel at a rate of speed which will produce the desired form or surface to the cutting edge.

A clean, true wheel of the proper bond and abrasive size is a very efficient cutting tool but at best will cause the work to heat up rapidly. In the case of the lathe tool there is only one cutting point acting on the work, but even so, it is a well known fact that the cutting tool, work, and chips get quite hot. In the case of a grinding wheel there are thousands of these cutting points each doing its share of the work but all acting at the same time, so that, since the action of a lathe tool generates heat, the action of a grinding wheel would neces-

sarily develop a much greater heat. For that reason a flood of lubricant coolant at the point of contact between the wheel and work is necessary to carry off the heat and to keep the temperature of both the wheel and work as nearly constant as possible. This is especially true where a job is roughed and finish ground.

ROUGH GRINDING

The purpose of roughing the job is to remove internal strains set up by heat treatment and to remove the excess stock as rapidly as possible in preparation for finishing. Roughing is accomplished by using a slow work speed and a fast traverse, then when finishing use a high work speed and slow traverse. In both cases use a flood of good coolant.

The following are suggestions originally made by well known grinding wheel and machine manufacturers and if generally adopted should do much to eliminate grinding accidents.

CARE OF GRINDING WHEELS

1. Handle all wheels with the greatest care in storing or delivery. Wheels are frequently cracked by rough usage long before they are ever placed on a grinding machine.
2. Wheels should be stored in a dry place.
3. Before a wheel is placed on the spindle it should be sounded for cracks. A solid wheel, when tapped by a non-metallic object, gives off a dull ringing sound. A cracked wheel gives off a dull thudding sound.
4. Make sure that the grinding wheel is equipped with blotting paper gaskets on each side.
5. Never crowd a wheel on the spindle; the hole in the wheel should be .003 to .005 inch oversize to permit it to slide easily on the spindle and squarely against the flange.
6. Never mount a wheel without flanges which are properly relieved and of suitable proportions. (See Wheel Mounting, page 236.)
7. Don't screw the wheel nut too tight. The nut should be set up only tight enough so that the flanges hold the wheel firmly.
8. Keep the wheel clean and true by frequent dressing, but don't remove any more stock than is necessary to put the wheel in proper condition.
9. If a wheel vibrates excessively after it

has been properly trued, there is something wrong. Stop the machine and call
an instructor.

10. Large wheels (that is wheels over 12")
require special balancing. Don't attempt to balance them yourself.

size important?

9. How is the abrasive grain size determined?

10. How are the abrasive particles formed
into a wheel?

11. What determines the hardness or softness

Revolutions per Minute for Various Diameters of Grinding Wheels to Give Peripheral Speed in Feet per Minute as Indicated

Diameter of wheel in inches	Peripheral Speed in Feet per Minute											
	4000	4500	5000	5500	6000	6500	7000	7500	8000	8500	9000	9500
	Revolutions per Minute											
1	15279	17189	19098	21008	22918	24828	26737	28647	30558	32467	34377	36287
2	7639	8594	9549	10504	11459	12414	13368	14328	15279	16233	17188	18143
3	5093	5729	6366	7003	7639	8276	8913	9549	10186	10822	11459	12115
4	3820	4297	4775	5252	5729	6207	6685	7162	7640	8116	8595	9072
5	3056	3438	3820	4202	4584	4966	5348	5730	6112	6494	6876	7258
6	2546	2865	3183	3501	3820	4138	4456	4775	5092	5411	5729	6048
7	2183	2455	2728	3001	3274	3547	3820	4092	4366	4538	4911	5183
8	1910	2148	2387	2626	2865	3103	3342	3580	3820	4058	4297	4535
10	1528	1719	1910	2101	2292	2483	2674	2865	3056	3247	3438	3629
12	1273	1432	1591	1751	1910	2069	2228	2386	2546	2705	2864	3023
14	1091	1228	1364	1500	1637	1773	1910	2046	2182	2319	2455	2592
16	955	1074	1194	1313	1432	1552	1672	1791	1910	2029	2149	2268
18	849	955	1061	1167	1273	1379	1485	1591	1698	1803	1910	2016
20	764	859	955	1050	1146	1241	1337	1432	1528	1623	1719	1814
22	694	781	868	955	1042	1128	1215	1302	1388	1476	1562	1649
24	637	716	796	875	955	1034	1115	1194	1274	1353	1433	1512
26	588	661	734	808	881	955	1028	1176	1248	1322	1395	
26	588	661	734	808	881	955	1028	1101	1176	1248	1322	1395
28	546	614	682	750	818	887	955	1023	1092	1159	1228	1296
30	509	573	637	700	764	828	891	955	1018	1082	1146	1210
32	477	537	597	656	716	776	836	895	954	1014	1074	1134
34	449	505	562	618	674	730	786	843	898	955	1011	1067
36	424	477	530	583	637	690	742	795	848	902	954	1007

To find the R.P.M. of an 18-inch wheel having a surface speed of 6000 F.P.M., read across in the 18 line to the 6000 column which gives 1273 R.P.M.

1. What is meant by the term abrasive?
2. Name two natural abrasives.
3. Name two artificial abrasives.
4. Why are artificial abrasives used more than natural abrasives?
5. What are the characteristics of silicon carbide?
6. Name four materials on which silicon carbide should be used.
7. On what kinds of material should alumi-nous oxide be used?
8. Why is control of the abrasive grain

of a wheel?
12. Name five commonly used bonds.
13. Which bond is used most frequently? Give five reasons.
14. What are the advantages of the silicate bond?
15. Silicate bonded wheels are generally used for what purpose?
16. Rubber bonded wheels are used for what purpose?
17. How does the wheel structure affect the grinding operation?

18. How are the degrees of hardness of the wheels specified?
19. What is the difference between the Norton and Carborundum methods of rating hardness.
20. Into what five groups are the degrees of hardness divided?
21. What six factors govern the wheel selection for a given job?
22. Give four methods of overcoming "chatter."
23. What is the difference between "glazing" and "loading?"
24. What is the difference between "dressing" and "truing?"
25. Why is it necessary to keep the face of the wheel clean and true?
26. For general tool room work which abrasive is most generally used and why?
27. How might the grade and structure of the grinding wheel produce chatter?

Chapter 25

GRINDING MACHINES

DEFINITION

A grinding machine is a machine which employs a grinding wheel for producing cylindrical, conical, or plane surfaces accurately and economically and to the proper shape, size, and finish. The surplus stock is removed by feeding the work against the revolving wheel or by forcing the revolving wheel against the work.

GRINDING MACHINE TYPES

There is a great variety of grinding machines, but it is the purpose of this text to treat only of those machines which are regularly used in tool and die shops. The machines which are generally used are the cutter grinder, surface grinder, centerless grinder, external grinder, internal grinder, tool grinder, and flexible shaft or hand grinder. These machines are usually classified as to size by the largest piece of work which they can completely machine (as 6" X 18" External Grinder) or are specified by numbers such as Brown and Sharpe #13, Rivett #104 Internal, and #3 Abrasive Surface Grinder.

CUTTER GRINDER

A cutter grinder is a machine that holds the cutter while a rotating abrasive wheel is applied to the edges for the purpose of sharpening it. These machines vary in design from simple machines having a limited purpose to complex universal machines that can be adapted to any cutter grinding requirements. Fig. 723 is an illustration of a typical cutter grinder used in the Henry Ford Trade School.

The universal cutter grinder is capable of grinding cutters of various shapes by using special attachments and specially formed grinding wheels. Generally the flaring cup, plain, and dish or saucer wheels are used on cutter grinders (Fig. 722).

In selecting a grinding wheel for general purpose cutter grinding, select a soft, free cutting wheel and take very light cuts so that the temper is not drawn from the edge of the teeth. Generally speaking,

wheels of grain size from 30 to 60 and J or K bond are best adapted for high speed cutters. The shape of the wheel depends on the shape of the cutter to be sharpened.

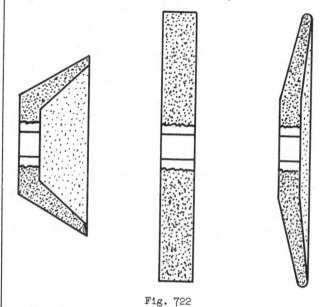

Fig. 722

SHARPENING OF CUTTERS

To sharpen the teeth of a plain milling cutter, the cutter is generally mounted on a lathe mandrel and supported between centers (as shown in Fig. 724) or the cutter is mounted on a special stub arbor and held in a universal swivel fixture. After the cutter has been mounted in the machine, a tooth rest is mounted on the table or work head and adjusted to the tooth to be first sharpened. The table and tooth rest are adjusted so that the grinding wheel follows the original land on the back of the tooth and gives the proper clearance. The cutter is then fed to the rotating grinding wheel until sparks indicate contact between it and the wheel; then the table is moved back and forth traversing the cutter until the wheel is finished cutting. Revolve the cutter backwards 180° against the spring tension of the tooth rest and without changing the depth of cut take a trial cut on this opposite tooth to check for taper. If no taper is apparent, revolve the cutter backwards and engage the next tooth to be

216

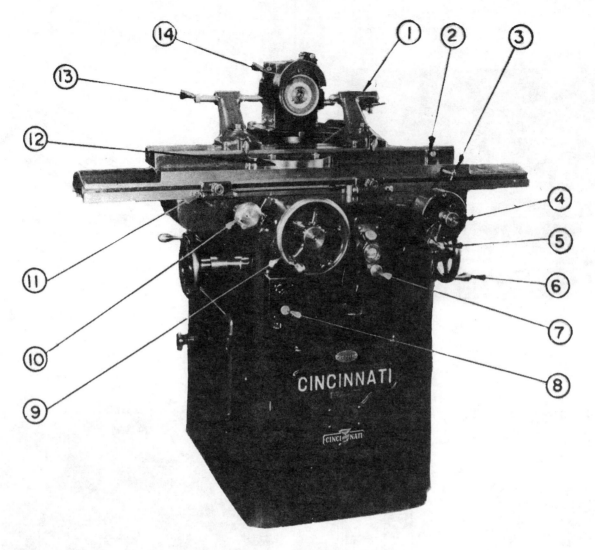

Fig. 723. Universal Cutter Grinder.

1. Quick acting tail stock.
2. Table swivel lock.
3. Fine adjustment for table swivel.
4. Slow hand movement for traversing table.
5. Crank for fast table traverse.
6. Handwheel for elevating grinding wheel head (one on each side).
7. One shot lubricating system.

8. Electrical control buttons.
9. Handwheel for adjusting table saddle (one front and rear).
10. Handwheel for fast movement of table traverse.
11. Adjustable cushioning dogs.
12. Graduated dial for swiveling table.
13. Work head center.
14. Wheel head can be raised or lowered and swiveled.

sharpened. This process is repeated until all teeth have been sharpened and ground concentric.

TOOTH RESTS

Tooth rests are of two kinds, plain and hook (Fig. 725). They consist of a piece of spring steel about .030" thick, 1/2" to 1-1/2" wide, and from 1" to 3" long brazed or riveted in a piece of round 3/8" cold rolled steel. They are held by means

of a forged clamping fixture (Fig. 726) by bolting the fixture either to the work table or the grinding wheel head.

SHARPENING SIDE TEETH OF MILLING CUTTER

The peripheral teeth of a side milling cutter are sharpened the same way as those of a plain milling cutter. To sharpen the side teeth the cutter is generally mounted on a stub expansion arbor and placed in the universal swivel fixture (Fig. 727). The

swivel fixture is then adjusted so that the side teeth will be ground about 1° out of

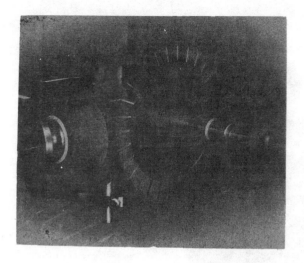

Fig. 724

parallel with the side of the cutter and have a side clearance angle of from 3° to 5°. The

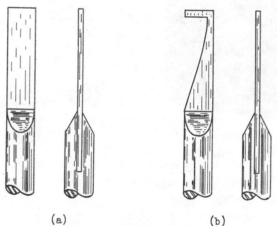

(a)   (b)

Fig. 725. (a) Plain; (b) hook.

general practice is to give the teeth of roughing cutters a side clearance of 5° and finishing cutters 3°. It is important in all cutter grinding to grind all teeth to the same height so that each tooth will do its share of cutting. For that reason after the teeth have all been sharpened a very light cut is taken on each tooth to insure them all being the same height.

SLABBING CUTTERS, that is, cutters which are of considerable length relative to their diameters, if they have straight teeth are sharpened the same way as a plain milling cutter.

HELICAL TOOTH CUTTERS are ground and

mounted the same as other cutters, but the tooth rest must be mounted in a fixed posi-

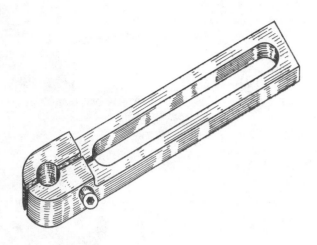

Fig. 726

tion relative to the grinding wheel. Generally the tooth rest is mounted on the

Fig. 727

wheel head (as shown in Fig. 724). This is done so that the cutter is forced to slide over the tooth rest causing the cutter to turn in such a manner that the tooth being ground will have the same helical shape at which it was originally milled. When the cutter and tooth rest are properly mounted and the grinding ready to start, the tooth should be pressed lightly against the tooth rest and held there while the table is moved longitudinally. If this is not done the tooth will leave the rest and the cutter will be damaged (see Fig. 728).

FORMED CUTTERS are sharpened by grinding the face of the teeth radially, that is by grinding the face of the teeth with a dish or saucer wheel so that the face

of the tooth comes on the radius of the cutter. (See Fig. 729.)

This practice keeps the faces of the teeth radial and maintains the correct shape of the tooth.

Fig. 728

Fig. 730

The machine is set up for grinding the formed cutter by bringing the centers in line with the face of the grinding wheel (Figs. 729, 730 and 731). If this is not done the cutter will not have the correct shape. The cutter, mounted on a mandrel is placed between the centers and the face of the tooth brought against the grinding wheel. The tooth rest is then set against the back of the tooth (Fig. 731). Move the table longitudinally and clear the wheel from the work, start the wheel and take a trial cut. To adjust the work to the wheel while grinding, revolve the cutter by moving the tooth rest toward the grinding wheel.

TO SHARPEN ANGULAR CUTTERS, the cutter is mounted on a stub expansion arbor and placed in the swivel fixture (Fig. 732).

The fixture is then swiveled to the desired clearance angle and the tooth rest set on the exact centerline of the cutter. Adjust the grinding wheel to the cutter by raising or lowering the table so it will grind the tooth supported by the tooth rest and allow the tooth immediately above to clear the grinding wheel as illustrated. The tooth rest should never be set above or below center as this will cause a change in the actual angle ground on the cutter from the

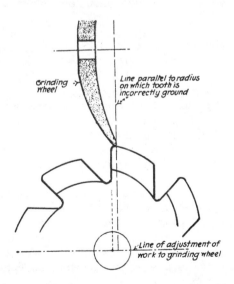

(a)

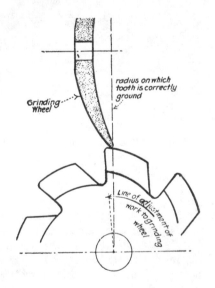

(b)

Fig. 729. (a) The Wrong Way to Grind a Gear Cutter. This brings the faces of the teeth parallel to but behind the radius and spoils the tooth outline, and therefore makes the cutter inaccurate.
(b) The Correct Way to Grind a Gear Cutter. The faces of the teeth will always be radial.

one at which the machine has been set and is often the cause of angular cutters not being accurate.

Fig. 731

END MILLS, whether they are straight or helical, are ground the same as milling

Fig. 732

cutters. The end teeth are ground the same as the side teeth of a side milling cutter.

As was previously mentioned there are three wheel shapes commonly used for cutter grinding. They are the DISK, FLARING CUP, and SAUCER wheel. The saucer wheel is used for grinding formed cutters; the disk wheel is used for grinding the clearance on cutters having narrow lands on the teeth; the flaring cup wheel is used for grinding clearance on teeth having wide lands and for gumming out slitting saws, and the spacing between the teeth of milling cutters.

In actual practice the grinding wheel revolves downward toward the cutting edge so

that the action of the wheel forces the tooth against the tooth rest. This practice results in a burr or wire edge being left on the tooth which should be oil-stoned off.

IMPORTANT FACTS TO REMEMBER WHEN SHARPENING CUTTERS:

1. Keep the cutter tooth firmly against the tooth rest.
2. Make sure the grinding wheel follows the original land on the tooth.
3. Mount the tooth rest correctly.
4. Keep the cutting surface of the grinding wheel clean.
5. Don't remove any more stock from the tooth than that required to sharpen it.
6. Be careful not to draw the temper of the tooth.

The following SAFETY RULES for cutter grinders must be observed at all times:

1. Always wear goggles on all cutter grinder work.
2. Under no circumstances is the machine to be started unless the grinding wheel is adequately guarded. Use a guard of the proper size and adjust it closely to the wheel, allowing the minimum amount of wheel exposure with which to work.
3. In mounting wheels on cutter grinders, use standard wheel bushings and safety washers. Use paper washers on large wheels.
4. When hand dressing wheels be careful to allow ample hand clearance between the wheel and table or other parts of the machine.
5. Hand dressing operations should be performed with a light pressure, especially when dressing thin wheels. A slip of the hand or a broken wheel may cause severe lacerations.
6. Any changes of guards, dogs, centers, setup, tooth rests, or any other parts of a machine are not to be made while the machine is running.
7. When grinding spotfacers, counterbores, etc., on a draw collet, use a special automatic safety guard or shut the machine down to remove the work.
8. In backing off drills, spiral reamers, etc., see that the tooth rest is properly adjusted in relation to the wheel and work to prevent slippage and consequent spinning of stock. Ask the instructor about this.
9. Care should be taken in handling sharp tools, such as reamers, drills, cutters,

and counterbores, as severe lacerations may result from stock slipping through the hands.

10. Towels are not to be used in holding small tools such as spotfacers, counterbores, and similar tools that become warm while grinding. Ask the instructor how to take care of work of this class.

11. Exhaust hoods are supplied as a safeguard for the health of grinder operators. See that they are properly adjusted at all times and that they are not abused.

QUESTIONS

1. What is a cutter grinder?
2. Name three manufacturers of universal cutter grinding machines used in the school.
3. What kinds of cutters can be sharpened on the universal cutter grinding machine?
4. What kinds of grinding wheels are suitable for general purpose cutter grinding?
5. What is a tooth rest and why is it used in cutter grinding?
6. How are plain milling cutters held for sharpening?
7. After the work and tooth rests are correctly mounted, how is the tooth sharpened?
8. Name three shapes of wheels used for cutter grinding.
9. How are the teeth of side milling cutters sharpened?
10. How are slabbing cutters having helical teeth sharpened?
11. What caution should be exercised in sharpening the teeth of helical tooth cutters?
12. How are the teeth of formed tooth cutters ground?
13. Why are formed tooth cutters ground radially on the face?
14. What is the shape of the wheel used for formed tooth grinding?
15. How are end mills generally ground?
16. Why is it important to prevent burning of the cutting edge in cutter grinding?
17. What causes burning of the cutting edge in cutter grinding?
18. Name the two types of tooth rests most generally used?

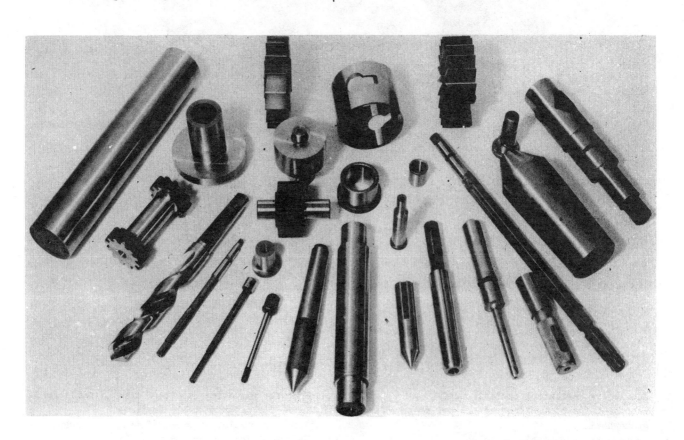

Fig. 733. Samples of external grinding as done in the Henry Ford Trade School.

CYLINDRICAL GRINDING (Shop term - External, OD; Internal, ID)

External grinding, as commonly thought of, is the act of grinding the outside diameter of a piece of work while it is revolving on its axis, for the purpose of reducing it to size, leaving a fine finish. However, external grinders are used to produce external cams, eccentrics, and special forms on the outside diameter of work. They are machines which are capable of doing many of the operations done on a lathe, but much more accurately. The big advantage lies in the fact that after a job has been heat treated a good surface finish and extreme accuracy as to size can be obtained and

where these two factors are important the extra cost involved in grinding can be overlooked. If parts can be designed so that the amount of stock to be removed is within grinding limits, then grinding is much less costly than lathe turning.

External grinders are divided into three general groups. They are the PLAIN CYLINDRICAL, UNIVERSAL, and special grinders like the CENTERLESS and CAM grinders.

The PLAIN CYLINDRICAL grinder (Fig. 734) is used to produce external cylinders, tapers, fillets, undercuts, shoulders, and may be used for form grinding by dressing the desired contour on the grinding wheel.

In any cylindrical grinder there are three movements which are very important.

Fig. 734. Landis OD Grinder

1. Handle for regulating coolant supply.
2. Handwheel for controlling in-and-out movement of grindwheel.
3. Footstock. Adjustable on table.
4. Lever for controlling movement of footstock center.
5. Knob for adjusting table for taper.
6. Trip dogs. Can be adjusted for length of work.
7. Knob for adjusting length of time of table dwell.
8. Lever for tripping table traverse.
9. Knob for regulating speed of table traverse.
10. Handwheel for traversing table by hand.
11. Lever for starting and stopping headstock spindle and table traverse.
12. Headstock unit.

They are rotation of the work on its axis, movement of the work back and forth in front of the wheel, and movement of the wheel into the work.   Trouble in cylindrical grinding, with exception of wheel content or make up, can be attributed to one of the three movements mentioned.

Rotation of the work on its axis is important because if the centers in the work are bad, or if the machine centers are of poor quality and loose, the work will be irregular in form.   The movement of the work back and forth in front of the wheel must be steady and smooth to insure a good finish and accurate sizing of the work.   Movement of the wheel spindle to revolve the wheel must be true and smooth to prevent vibration and to avoid chatter marks on the work. Movement of the wheel into the work must be without play or bind to insure accuracy as to depth of cut.

The work in external grinding usually revolves on two dead centers, one in the foot stock and one in the head stock, and is given its rotary motion by means of a drive plate, which revolves about the head stock center,

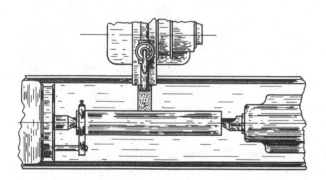

Fig. 735

driven by a motor using pulleys and a belt (Fig. 735).   The drive plate contains an adjustable arm which can be located at varying distances from the center and into which a drive pin is fitted.   This drive pin engages in the "V" slot of the grinder dog which is attached to the work and hence the revolving motion of the plate is transmitted to the work.

Besides the rotary motion of the work it is always necessary, except in the case of "plunge cut" grinding, for the work to be traversed past the grinding wheel.   The length of the table traverse should be set so as to permit the wheel to run off the end of the work about one-third of the wheel face width.   If the wheel is not permitted to over-run the end of the work, the job will be oversize at that point because the wheel will not have a chance to finish the cut.   If the wheel is permitted to completely over-run the end of the work, the job will be undersize, because the pressure required between the wheel and the work is relaxed thus permitting the work to spring toward the wheel, and at the beginning of the traverse the wheel would cut undersize.

At the end of each traverse the table stops momentarily to give the wheel a chance to grind the work to size, to permit the wheel to clear itself on the new cut and to avoid the jarring motion which would be unavoidable with an immediate reversal of the table.   If this traversing is not accomplished without jarring or without a jerking movement of the table, the work will show high and low spots due to the slight pause or dwell of the wheel on the work, which would take place at each jerk or jar of the table. The speed of the traverse depends on the width of the wheel face and the finish required on the work. It is generally such that the table will move two-thirds to three-fourths of the wheel face for each revolution of the work.

STEADY REST

Long slender work, besides being supported by the centers, should also be supported with steady rests. Fig. 736(a) shows a common type of steady rest while Fig. 736(b) shows a center rest.

When using steady rests (Fig. 737) it is very

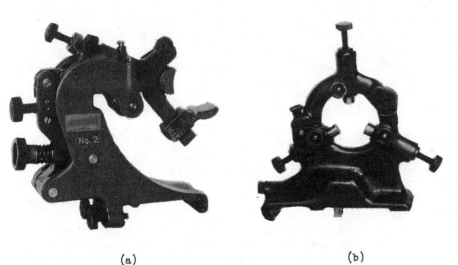

(a)   (b)

Fig. 736

important to keep the jaws properly adjusted
to the work, otherwise the work might get
caught between the lower jaw and the wheel
and be thrown from the machine or break the
wheel.

Fig. 737

FEEDS

The grinding wheel can be fed to the
work either by automatic or hand feed, and
permits feeds as small as .00005. It is not
advisable to use the hand feed except to
bring the wheel up to the work or move it
away or when taking very fine cuts. The automatic feed takes cuts from .00025 to .004
for each traverse of the table and saves
time and wear and tear on the machine by taking more uniform cuts and gives longer life
to the grinding wheel. Generally speaking,
"roughing cuts" may be from .001 to .004 at
each reversal, depending on the rigidity of
the machine and work set up and the amount of
stock to be removed. It is common practice
when the work is not to be hardened to leave
from .006 to .010 grind stock, but work
which is large, long or slender, or easily
sprung may have an allowance of from .020 to
.030 grind stock; in either case, however,
heavy infeeds require good supports for the
work. For roughing out the job use a slow
work speed, fast traverse and heavy feed. For
"finish grinding" a job use a high work
speed, slow traverse and a light feed.

The more common grades and grain
sizes of aluminous oxide wheels used for external grinding are: 46-J, 46-K, 60-K, 60-L,
80-O, 120-P.

WORK SPEEDS

To the question, "How fast should the
work speed be?" no set rule can be given. Jobs
may or may not be of the same metal, same
heat-treatment, or same diameter, any one of
which factors affects the speed. Too fast a
speed tends to wear away the wheel faster,
while too slow a speed causes the wheel to
cut harder and become dull and glazed. If the
wheel wears too fast it requires more frequent
dressing and truing and takes more time to do
a given job. Extremely hard steels may have
to be ground at 30 S.F.P.M., while very soft
steels may require up to 100 S.F.P.M. for finishing. Common work speeds are from 30 to 50
S.F.P.M. If the wheel becomes dull or glazed
it will burn the work because of the added
friction set up by forcing the wheel to cut.
If the work speed is correct and the wheel
still has a tendency to become glazed it is
an indication that the bond is too hard and
can be corrected by substituting a wheel of
the same grain but softer bond.

OPERATION ERRORS

When cracks, checks, or burns show up
on the surface of the work any one of a combination of the following conditions might be
the cause: improper speeds of work, wheel
too hard, or the wheel being glazed or loaded.
If the wheel is glazed or loaded it will cause
the work to burn and crack even though there
is a good supply of lubricant, so that best
results can be obtained by keeping the wheel
sharp and clean at all times.

Perhaps the greatest obstacle which
the operator has to overcome in the grinding
of a cylinder is that of taper in the work.
Conditions which contribute to it are poor machine or work centers, work improperly mounted, work table out of adjustment, foot stock
out of line with head stock, or steady rests
bearing too heavily or out of line with the
work.

To correct this tapering, first make
sure that the work centers are clean and true
and of sufficient depth and clearance to give
the machine centers a good bearing; then inspect the machine centers for correct taper
to fit the machine; see that they are clean
and true; mount the work and make sure that
the machine centers are engaged properly or
inspect the work table for side play and have
it corrected if necessary. If none of the
above factors are causing the taper then cor-
rection can be made by adjusting the knurled
screw located at the foot stock end of the
work table. The screw should be turned to
move the small end of the work away from the
wheel. Several adjustments may be necessary
to get the work entirely cylindrical.

PLUNGE CUT GRINDING

"Plunge cut" grinding is cylindrical grinding where the length of the work is less than the width of the wheel face so that the wheel can be fed straight into the work and completely grind the cylinder without using the table traverse. When it is necessary to grind a shouldered or stepped cylinder, plunge cut grinding is used to maintain a square corner between the cylinder and the shoulder.

The important steps in external grinding a piece of work are as follows:

1. Check the work for size to make sure grinding stock has been allowed and at the same time note any tapering of the work

2. Inspect the work centers for being clean and true. Select machine centers of suitable diameters to properly fill the work centers. The foot stock center should be cut away enough to permit the grinding wheel to clear the end of the work; a center of this type is known as a one-half full, or three-quarter full center.

3. Attach the grinding dog on the end of the job making sure that the dog does not damage such parts of the work as threads, keyways, etc., then lubricate the machine centers.

4. Set the table traverse for the length of the work, allowing for overrun of the end and the space that the grinder dog occupies.

5. If necessary, mount the steady rests and adjust the shoes to the work.

6. Dress the grinding wheel, passing the diamond across the face of the wheel quite fast, to make the wheel fast cutting. Set the work speed at the correct S.F.P.M.

7. Feed the wheel to the work by hand and take a light cut noting that the wheel starts to cut approximately at the high point of the work to conform to the check in step one.

8. Check the work for size and taper and make any table adjustments necessary to insure that the work will be straight.

9. Rough grind the job to rough size. If several pieces are to be done, set the stop on the feed ratchet and proceed as before, roughing the balance of the pieces.

10. After the pieces have been roughed out in this manner, place the dog on the rough ground end of the work and grind the unfinished end. If this end is shorter than the width of the wheel face it may be plunge cut ground. Make sure that the grinding wheel is kept sharp and clean by frequent dressing.

11. To finish grind, set the machine for fast work speed and slow traverse and dress the wheel by passing the diamond slowly across the wheel face.

12. Insert the piece to be finish ground and take a light trial cut. Check it for size and make any corrections necessary for removal of taper and if steady rests are used keep them adjusted to the work.

13. After the first piece has been ground to finish size reset the stop on the feed ratchet so that the infeed will produce the required size and then set the shoes on the steady rests for the finished diameter.

The above outline pertains to grinding a plain cylinder. If the work to be ground has shoulders, keyways, or slots, some deviations from the outline must be made. If the work to be ground has a keyway, open at each end, or splines and steady rests are to be used, it should be filled with key stock or other suitable material to prevent the steady rest shoes from catching on the work.

Sometimes it is desirable to finish grind a cylindrical piece in one operation. When a condition of this kind arises, an angle iron bracket or other suitable projection can be sweated to the end of the work (Fig. 738) to act as a driver, but it must be fixed so as not to interfere with the overrun of the grinding wheel or work center.

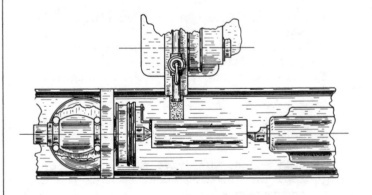

Fig. 738

SHOULDER WORK

If the work must be ground to a shoulder, it is advisable to locate the grinding wheel up against the shoulder and by plunge

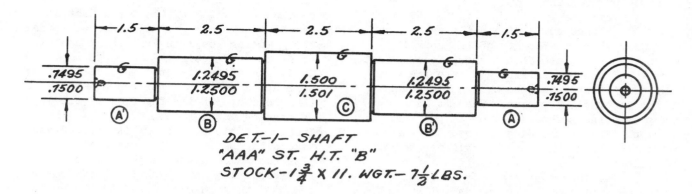

DET.-1- SHAFT
"AAA" ST. H.T. "B"
STOCK-1¾ X 11. WGT.- 7½ LBS.

Fig. 739

cut grinding, grind the work to the correct diameter. This method leaves the finished diameter with a square corner at the shoulder rather than a filleted one which would be left if the work was traversed up to the shoulder. After the job has been sized square with the shoulder, the balance of it can be done by traversing the table.

THE FOLLOWING OUTLINE IS A TYPICAL ANALYSIS OF THE PROCEDURE IN EXTERNAL GRINDING:

Type of job	Shaft as per sketch. (Fig. 739)
Type of machine	External grinder
Type of steel	"AAA"
Heat treatment	"B"
Kind of grinding wheel.	60-L
Operations required . .	Rough and finish grind as per sketch.

PROCEDURE

1. Check all diameters for sufficient grind stock.
2. Check work centers for being free from dirt and nicks.
3. Dress grinding wheel for roughing cut.
4. Mount a grinder dog of correct size on the work.
5. Set foot stock to function for correct length of work.
6. Mount the work in the machine and properly adjust the drive pin to the dog.
7. Feed the grinding wheel to diameter "A" and take a clean-up cut, seeing that the wheel closely follows the work.
8. Check diameter "A" for straightness, make any necessary table adjustments, take a trial cut and recheck the work for straightness.
9. With the wheel cutting straight, plunge cut diameter "A" at the shoulder and

rough grind, leaving .003 to .005 for finishing. Repeat this operation on A', B, B', and C in the order stated, then have the job inspected.
10. Dress the grinding wheel for finishing and with the machine cutting straight, finish grind A, A', B, B', and C in the order stated.
11. Have job inspected on all dimensions.

GAGING METHODS

CYLINDRICAL WORK having dimensions which must be held to close limits requires the use of gages which are calibrated to a finer degree of accuracy than ordinary micrometers. For this purpose an "amplifying comparator" is used. These amplifiers are of several types and calibrations, but the kind commonly used has a ratio of 10 to 1. This means that one division on the scale is .0001". Fig. 740 shows a common type of electric amplifier and the standards by which it is set. To use the amplifier, Jo-blocks are stacked to the size of the dimension to be checked. The blocks are then placed on the amplifier block and the amplifier adjusted so that when the gaging point contacts the blocks, the indicator pointer will have a deflection of not more than 10 divisions and so the pointer rests on zero. The Jo-blocks are then removed and the work placed under the contact point with a sliding or rolling motion. Any variation between the Jo-block size and the work size is then visually recorded in tenths of thousandths on the dial or scale.

TAPER WORK

ACCURATE TAPER WORK may be produced on the cylindrical grinders by one of two methods, either by swiveling the work table or the headstock. For slight tapers the

Fig. 740

table may be set by means of the swivel adjustments and table graduations, but inasmuch as this can only be approximately accurate, a standard taper ring gage, female taper gage, or sine bar is necessary to check the taper. The grinding of tapers is very much the same as grinding a cylinder, the exception being that the swivel table or wheel stand slide is set to produce the correct taper angle. Generally the graduations on the scale marked degrees, is one-half of the whole taper angle, while taper per foot or percentage indicates the whole taper angle.

If a taper ring gage is used for checking the accuracy of the taper, Fig. 741,

Fig. 741

the male section should be given three lengthwise stripes of prussian blue about 120° apart and then carefully inserted into the ring gage with a slight twisting motion. If the surface being checked does not conform to the surface of the gage, the irregularity rubs the blue off and leaves a bright metallic ring indicating the high spot. If the gage bears only on one or two lines it indicates that the piece being tested is out of round.

STEEP TAPERS can be produced by swiveling the work head or by dressing the wheel at an angle. If the wheel is to be dressed at an angle it is absolutely necessary to have the diamond set on the exact centerline of the wheel. This is necessary, not only to insure dressing the correct angle on the wheel, but also to obtain a straight face on the wheel rather than one which is concave or convex.

The 60° point on a machine center is ground (Fig. 742) by placing it in the live

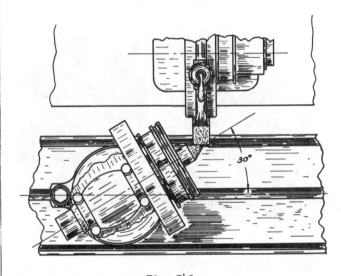

Fig. 742

spindle of the headstock and swiveling it through a 30° angle using the graduations on the base of the headstock. It is obvious that these graduations cannot be relied upon for extreme accuracy, therefore by using the flat center gage, Fig. 743, or the bell center gage, Fig. 744, as a check, the necessary adjustments can be made to insure accurate results.

The actual operation of grinding machine centers must be carried out with care as there is danger of burning the point because of the change in work speed due to the steep taper. This trouble may be overcome by using a slow work speed and small infeed, a flood of coolant, and starting the cut

at the point, ripping it back towards the shank.

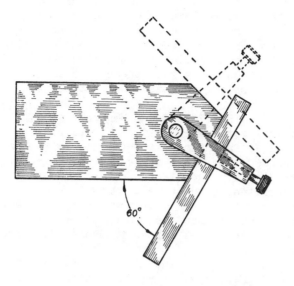

Fig. 743

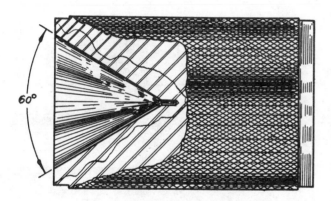

Fig. 744

WHEEL DRESSERS

On all external grinders some provision is made for dressing the wheel. The footstock is equipped with a suitable holder,

so arranged that it is adjustable and can be located on the centerline of the wheel spindle. Another type of diamond holder (Figs. 745 and 746) is a forged bracket having a keyed base to fit into the "T" slot of the work table and an adjustable upright which holds the diamond.

Fig. 745

Fig. 746

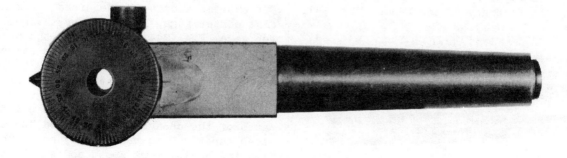

Fig. 747

Another type known as a micrometer adjustment wheel dresser (Fig. 747) consists of a hollow screw body surmounted by a dial graduated to read in thousandths of an inch, which passes through a threaded hole located on the axis of the footstock center. The diamond is inserted in the hollow screw and locked in place by a set screw. This dresser proves to be very efficient especially where the job consists of a number of pieces of the same size. To use it, after the first piece has been reduced to size, the point of the diamond is spotted on the grinding wheel so that the distance from the axis of the footstock center to the tip of the diamond is equal to the radius of the work. The micrometer screw is then locked in place insuring the correct sizing ability of the wheel.

SAFETY ITEMS

It is important when operating an external grinder to let the machine run for a few minutes when first starting it, to give it a chance to warm up. During this time let the coolant run on the wheel to balance it, because having stood over night the coolant drains to the bottom of the wheel causing it to be out of balance. Be sure to check the stops, feed trips, and levers to make certain that the wheel doesn't run into the machine and damage it. See that the driving pin between the drive plate and grinder dog is securely locked in place and proper adjusted. Keep a good stream of coolant running at the point of contact between the wheel and work as it helps to dissipate the heat, tends to give the work a better finish, and keeps the wheel clean. Keep your hands away from the moving wheel and work. If something goes wrong stop the machine by pulling the main switch and call an instructor.

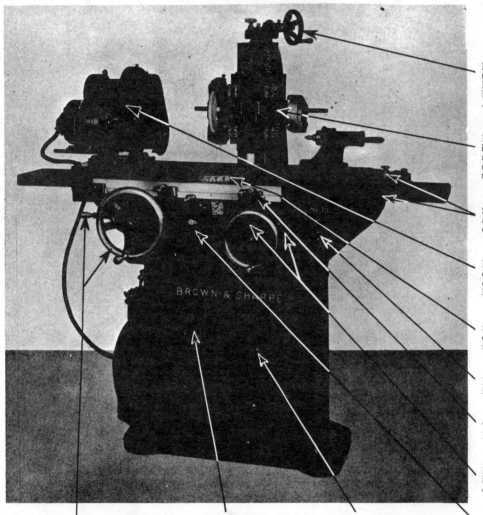

Elevating screw handwheel swivels to any convenient operating position. Bracket clamped in position by knob. Handwheel graduated to read to .0005"

Double-ended wheel spindle carried in sturdy slide. Has vertical adjustment of 8". Proper adjustment of spindle boxes automatically maintained by spring shoes.

Spring latch, and knob for fine adjustments of swivel table. Latch engages knob.

Motor-driven headstock has both dead-center and revolving-spindle drive. Swivels on graduated base. Lever operates quick-acting clutch.

Swivel table turns on stud to 90° either side of zero. Scale, graduated to degrees, indicates setting.

Main start-stop push button conveniently located. Controls both motors.

Table reversing dogs quickly set. Have fine thumbscrew adjustment.

Fine cross feed operated by small handwheel; engaged by knob on cross feed handwheel.

Table reversing lever. Can be operated by hand. Lock prevents engagement of longitudinal power travel when handwheel is in use.

Lever for selecting fast hand travel, slow hand travel, or automatic power travel of table; and longitudinal table handwheel.

Separate start-stop switch and overload relay reset for headstock motor. All electrical controls enclosed in base.

Reset for wheel spindle and table motor overload relay.

Fig. 748. Universal and Tool Grinding Machine

UNIVERSAL GRINDERS

Besides the external grinder just discussed another grinder which is rapidly growing as a utility, is the Universal Tool Grinder (Fig. 748). This machine is one which can, within its capacity, do the work of other grinding machines provided the necessary attachments are available. Because of its adaptability, some shops use nothing but universal grinders, other shops use them for work which might interfere with continu-ous operation of specialized machines. This machine is truly universal; the headstock is adjustable both horizontally and vertically, the wheel head can be swiveled in a horizontal plane and can be raised or lowered, and the table can be traversed and swiveled.

CENTERLESS GRINDERS

The Centerless grinder (Fig. 749) is a specialized machine which was developed for the rapid production of cylindrical, external

Fig. 749. Centerless Grinder

1. Handwheel for operating regulating wheel truing attachment.
2. Regulating wheel truing attachment.
3. Regulating wheel.
4. Ball crank for micrometer adjustment of regulating wheel.
5. Hand lever for adjusting pressure on regulating wheel.
6. Spar for adjusting regulating wheel and work rest slides.
7. Speed change levers controlling regulating wheel speed.
8. Lever for clamping regulating wheel slide to lower work-rest slide.
9. Lever for clamping lower work-rest slide in position.
10. Upper work-rest slide for transverse adjustment.
11. Work-rest.
12. Grinding wheel.
13. Lever for hydraulic control of grinding wheel truing attachment.
14. Grinding wheel truing attachment.
15. Lever for operation of profile attachment.
16. Valve handles for controlling coolant supply to diamond and grinding wheel.

tapers, or external profile work (Fig. 750). In centerless grinding two wheels are employed; one, the cutting or grinding wheel, is used to remove the excess stock; the other, a "regulating wheel," which is used to control the speed of rotation of the work and rate of feed. The work is supported on a work slide or rest.

This machine has many distinct advantages over other grinders in that the work does not have to be center drilled, thereby saving the time required for that operation on the lathe; since the work does not have to be mounted on centers and since the grinding operation is almost continuous, the time required for loading and unloading is saved; heavier cuts can be taken and the amount of material to be removed can be less, therefore saving more time and adding to the life of the grinding wheel; the operation of the machine does not require the services of a skilled mechanic and since there are few moving parts in the machine, the upkeep cost is very low, and the output rate very high.

It is generally thought that because the external centerless grinder was designed for jobs of a large number of pieces, that it is not suited for tool room use, but actually much time and money can be saved where the job lot contains only a few pieces because the machine is simple to set up.

The actual grinding operation depends for the most part on the pressure exerted by the grinding wheel on the work and the operation of the work with respect to the wheel centers (Fig. 751). In operation, the pressure exerted by the grinding wheel on the work forces the work against the work rest and regulating wheel. The regulating wheel revolves in a direction the same as that of the grinding wheel and has a horizontal movement. It has speed of 12 to 300 R.P.M., and at the same time feeds the work through the machine. The rounding of the work depends on how high the work rests are above the

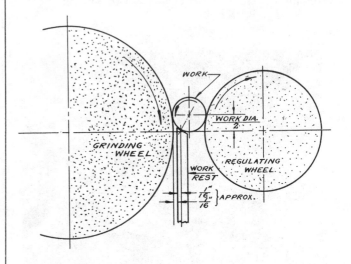

Fig. 751

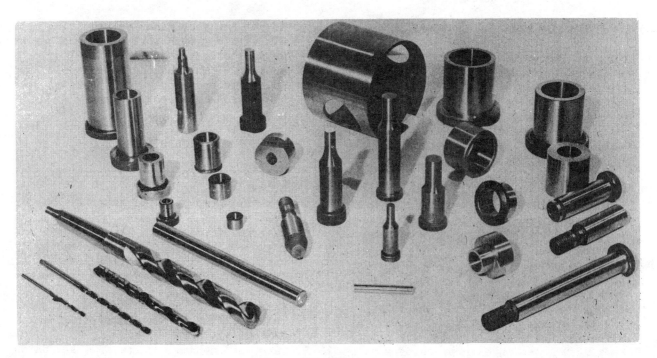

Fig. 750

centerlines of the wheels and the top angle
of the work rest.

METHODS OF FEEDING

Two common methods used for feeding
the work to the machine are the THROUGH FEED
and IN FEED. The through feed method is used
on straight cylindrical work whereas the in
feed method is used for shoulder, taper, or
profile work. See Figs. 752 and 753.

"The center of the work should be located approximately one-half the work diameter above the centerline of the wheels."

Fig. 752. Through Feed

SAFETY RULES FOR EXTERNAL GRINDERS

1. Under no circumstances attempt to operate
   an external grinder unless the wheel is
   guarded adequately.
2. When it is necessary to replace a large
   external grinding wheel, ask the instructor about methods of proper mounting and
   testing.
3. Before starting the work head, always test
   the work to see that it is between centers.
4. If the work must be tested for size while
   it is between centers, be sure to allow
   ample clearance between your hands and the
   grinding wheel.
5. Poorly adjusted drivers, loose dogs, defective center holes in the work, etc.,
   are a constant danger on the external
   grinder causing spinning of the work. Adjust the driver correctly and securely,
   fasten the drive dog tightly, and inspect
   the center holes in each piece of work before grinding it.

6. Be extremely careful in removing work
   from a collet head. Run the table back
   to a SAFETY STOP which will give ample
   hand clearance between the wheel and work.
   (ASK THE INSTRUCTOR ABOUT THIS).
7. If the work is heavy, shut the machine
   down when placing it between centers.
8. Avoid dressing the sides of a large grinding wheel, if possible, but if it is necessary for the job, ask the instructor to
   help. He will demonstrate the right
   method of dressing the wheel and how cuts
   can be taken so as to preserve the corners of the wheel.

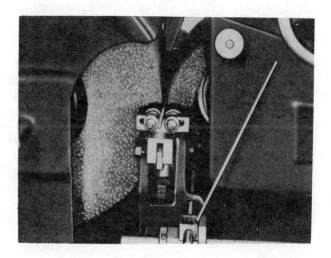

Fig. 753. In Feed

9. Do not put your hands on revolving materials with any open work in it such as keyways, slots, and flutes.
10. Be careful in handling sharp tools, such
    as drills, reamers, and cutters as severe
    lacerations may result.

QUESTIONS

1. What are the common shop terms for cylindrical grinding?
2. What is the purpose of external grinding?
3. What kind of jobs can be produced on the
   external grinder?
4. How does the work of the external grinder
   compare with that done on a lathe?
5. Into what three major groups are external
   grinders divided?
6. What companies manufacture the external
   grinders used in the Henry Ford Trade
   School?
7. Name the important parts of a standard
   external grinder.

8. What job operations can be done on the plain cylindrical grinder?
9. What movements are important in any cylindrical grinder?
10. Why are each of these movements important?
11. Explain how the job in external grinding is driven when supported between centers.
12. Generally speaking what should the length of the table traverse be?
13. Why is it important to permit the grinding wheel to over-run the end of the job?
14. What happens to the job if the grinding wheel completely over-runs the end of the work?
15. What does the speed of the table traverse depend upon?
16. How fast should the speed of the table traverse be?
17. How is long, slender work supported while being driven between centers?
18. What care should be exercised in the use of steady rests?
19. What provisions are made for feeding the grinding wheel of the work?
20. What depths of cuts is it possible to take with the automatic feed?
21. Why is it advisable to use the automatic feed rather than the hand feed?
22. What is the depth of a roughing cut using the automatic feed?
23. What speed and feed should be used for roughing out a job? For finish grinding a job?
24. Give the grades of bond, size, and kind of abrasive commonly used for external grinding wheels?
25. Why is it not possible to have a set rule for the rate of work speed?
26. What work speeds are commonly used?
27. What conditions might cause cracks, checks or burns on the surface of the job?
28. What conditions might cause tapering of the work?
29. How can tapering of the work be overcome?
30. What is meant by "plunge cut" grinding?
31. When, in particular, should plunge cut grinding be used?
32. Why should the job be checked for size before setting the job up?
33. Footstock centers which are cut away are called what kinds of centers?
34. What care must be exercised in attaching the grinder dog to the end of the job?
35. How should the diamond be used to make the wheel fast cutting?
36. When the wheel is first fed to the work what notice should be taken of the work contact?

37. Why should the work be checked after the first cut is taken?
38. How should the diamond be used to prepare the wheel for the finishing cut?
39. If the work has keyways or splines, how should it be prepared if steady rests are to be used?
40. How is it possible to grind a cylinder completely at one time without using a grinder dog?
41. If the work has several shoulders, what is the proper way to grind it? Why?
42. Why can micrometers not be used for gaging jobs requiring a high degree of accuracy?
43. What is an "amplifying comparator"?
44. Explain how to use the amplifier.
45. To what degree of accuracy can the ordinary amplifier be used?
46. How is it possible to produce taper work on the cylindrical grinder?
47. What tools are necessary for producing accurate taper work?
48. How is the taper of the job generally specified?
49. Explain how to use a taper ring gage for producing an accurate taper.
50. In producing steep tapers what care must be exercised in dressing the wheel?
51. Explain an efficient method of grinding machine centers.
52. How may the accuracy of the 60° conical point of the machine center be checked?
53. Name three kinds of diamond holders used on external grinders.
54. What advantage does the micrometer adjustment diamond holder have over the others?
55. Give six important safety rules for operating external cylindrical grinders.
56. What is meant by a "Universal Tool Grinder"?
57. What is the advantage of this machine?
58. What are its disadvantages?
59. What kinds of work can be done on the Centerless grinder?
60. What is the principle of operation of the centerless external grinder?
61. What advantages does the centerless grinder have over the plain and universal external grinders?
62. Upon what two factors does the actual grinding operation depend in centerless grinding?
63. In what direction does the regulating wheel rotate with respect to the grinding wheel?
64. Upon what two factors does the rapidity of the rounding of the work depend?
65. What is meant by "through feed"? "In feed"?
66. What is meant by "profile work"?

SURFACE GRINDERS

     Surface grinding is the act of producing and finishing flat surfaces by means of a grinding machine employing a revolving abrasive wheel.

     Surface grinding machines are divided into two major groups according to the shape of the table and how it moves. They are, the "planer type" where the table is rectangular in shape and traverses under the wheel, and the "rotary type," where the table is circular in shape and rotates under the wheel.

PLANER TYPE   EXAMPLE

1. Horizontal spindle using the O.D. or periphery of the wheel.   (#3B Abrasive)
2. Horizontal spindle using the rim of a recessed, or cupped wheel. (Gardner Surface)
3. Vertical spindle using the rim of a recessed, or cupped wheel. (6 X 18 Norton Surface)

ROTARY TYPE   EXAMPLE

1. Horizontal spindle using the O.D. or periphery of the wheel.   (Heald Rotary #22)
2. Vertical spindle using the rim of a recessed, or cupped wheel.   (#18 Blanchard)

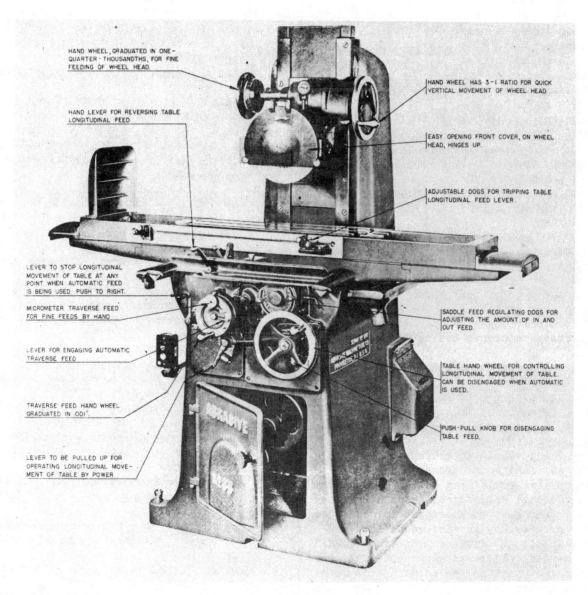

Fig. 754. Planer Type Grinder

Fundamentally any of the above mentioned machines consists of a spindle with a grinding wheel mounted on it, and a table or magnetic chuck for holding the work for the purpose of presenting it to the wheel. Each machine has its particular advantages but for the purpose of general tool room work this discussion deals principally with the horizontal planer type of surface grinder (Figs. 754 and 755) which uses the periphery of a disk wheel.

SURFACE GRINDER WHEELS

Grinding wheels of various shapes are used in the horizontal surface grinder, among the more common being the disk, gage, and thin rubber slotting wheels. The sizes of the wheels may vary from very small wheels of the internal class, which may be used with a high speed attachment, to wheels 10" in diameter having a 1" face. The wheels most commonly used on surface grinders are:

1 Handwheel, graduated to half-thousandths of an inch, provides easy and accurate vertical adjustment of spindle head.

2 Wheel spindle runs in phosphor bronze front box and ball bearing rear support. Front box has means of compensation for wear. Heavy wheel guard with removable cover gives ample protection.

3 Table reversing mechanism operated manually through lever shown, or automatically through second lever located on same shaft and tripped by adjustable table dogs.

4 Adjustable dogs permit stopping power movements automatically at any desired point in each direction of cross feed.

5 Knob and trip lever start and stop both longitudinal and transverse power movements. Turning knob to right starts power feed; depressing lever stops feed instantly.

6 Start-stop push button and overload relay reset conveniently located.

7 Large base compartment has shelf for mounting driving motor. On overhead-driven machine, provides handy storage space for Attachments and accessories.

8 Longitudinal table handwheel conveniently located and easily operated. Can be positively disengaged when power travel is used.

9 Adjustable stops provide for any cross feed from .01" to .09" (or zero feed) at either end of table travel. Knob in center disengages cross feed mechanism for manual operation.

10 Graduated handwheel permits rapid and accurate transverse adjustment.

Fig. 755. Planer Type Grinder

46-G; 46-H; 46-I; 46-J; 46-K.

60-H; 60-I; 60-J; 60-K.

Such materials as stone, glass, copper, or tungsten carbide require grinding wheels of special grain, material, and structure.

MOUNTING SURFACE GRINDER WHEELS

The following items are to be considered in mounting a grinding wheel on the horizontal planer type of surface grinder from the time the wheel is received at the crib until the machine is running:

1. Sound the wheel for cracks. Hold the wheel by the bore and tap it with a non-metallic object; if the wheel is not cracked it will give off a dull ringing sound; if the wheel is cracked it will give a dull thud.
2. Make sure that the wheel has blotting paper gaskets on both sides around the hole. (Most wheels arrive from the manufacturer with paper washers attached.)
3. Inspect the wheel flanges to make sure they are of the safety type and proper size.
4. Place the wheel on the spindle. The wheel should slide on the spindle, without bind or too much play (.003 to .005 clearance) up against the inner safety flange.
5. Put the outer safety flange on.
6. Put the spindle nut on securely.
7. Put the wheel guards in place and tighten them.
8. Make sure that the wheel clears the housing by turning it over by hand.
9. Start the machine and as the starter button is pressed step to one side, letting the machine run for at least one full minute before working with it.
10. Truing the wheel. Don't try to remove all of the runout with one pass of the diamond. Remove it a little at a time. The wheel speed has been built into the machine by the manufacturer. For safety, don't try to increase it without getting competent advice.

From a safety standpoint of view it is very essential to sound the wheel for cracks, because they may not be seen. If the wheel is cracked, and placed on the machine, the centrifugal force when the machine is started up will cause the wheel to burst, thus endangering the operator or near-by fellow workers.

Blotting paper or rubber gaskets should be placed between the safety flanges

80-0   120-P

$\frac{1}{32}$ to $\frac{3}{32}$ rubber slitting wheels.

and the wheel to evenly distribute the pressures around the wheel when the nut is tightened.

SAFETY FLANGES

Safety flanges are wheel flanges which act on the wheel between the spindle and nut. They should be at least one-third the diameter of the wheel and be relieved or undercut on the wheel side so that they bear on the wheel only at their outer edges. This bearing surface should be parallel with the side that the nut bears against to insure even holding or gripping of the wheel (Fig. 756).

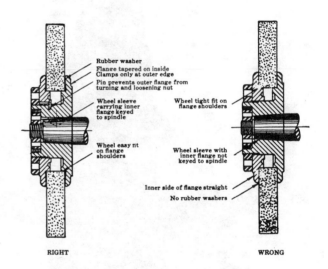

Fig. 756

Never force the wheel on the spindle. If it goes on snugly, the lead bushing should be scraped so as to be from .003 to .005 larger than the spindle, which enables the flanges to straighten the wheel up without putting an internal strain on it.

The nut threads on the spindle in a direction opposite to that in which the wheel rotates, so that the resistance offered by the work to the wheel tends to tighten the nut.

When starting the machine, especially after a new wheel has been mounted or first thing in the morning after the machine has been standing all night, step to one side as the starting button is pressed and let the machine run for at least one full minute. New wheels are apt to be out of balance due to being moisture logged on one side or the wheel spindle may have excessive end play in

it which upon starting would break the wheel. For that reason be clear of it in case it does break.

DRESSING SURFACE GRINDER WHEELS

In the case of surface grinding where the work is ground dry it is very necessary to keep the wheel sharp and true if the grinding operation is to be done efficiently. The area of contact between the wheel and work is much greater than it is in cylindrical grinding, which means that a great many more cutting tools are acting on the work at the same time. This means that the heat generated due to the added cutting tools must be kept at a minimum by keeping the wheel clean and sharp.

A dull or dirty wheel causes the work to burn or have hard cutting action. Hard cutting action means more strain on the wheel and machine parts causing the wheel and machine parts to heat up and operate inefficiently.

To dress a wheel it is merely necessary to remove sufficient material from the face of the wheel to sharpen it. Don't keep feeding the wheel down and hack away at it with the diamond. Grinding wheels are costly tools and should be shown due consideration. Some grinding wheel manufacturing companies claim that about two-thirds of a grinding wheel is wasted due to improper dressing, truing, or handling, so use it accordingly.

HOLDING WORK

Most surface grinders employ a magnetic chuck (Fig. 757) for holding the work, but it may be held by clamping it directly to the table or by clamping it in a vise and

Fig. 757

bolting it to the table. Work may be held in conjunction with the magnetic chuck by using a "V" block, angle plate, sine bar, or special fixtures.

The magnetic chuck holds the work by exerting a magnetic force on it. The magnetic poles of the chuck are placed close together so that it is possible to hold very

small pieces of work. Frequently, however, the work is too small for the chuck to hold it, so that long pieces of iron or steel, having a height less than that of the work, are placed along side to act as blocks thus retaining the work on the chuck. It must be remembered that only magnetic materials such as iron or steel will actually adhere to the chuck. When grinding such materials as bronze, brass, fibre, or certain kinds of stainless steel, the pieces must be held in place on the chuck by bars of iron or steel.

The accuracy of the work, such as parallelism, squareness, etc., depends on the accuracy of the holding face of the magnetic chuck. For that reason the holding surface of the chuck must be kept smooth and flat. As soon as the chuck shows nicks, scratches, or dents, or if the chuck has been removed for any purpose, it should be reground, in place, with care. Do this only on the advice of the instructor or foreman.

Generally the chuck is equipped with a back rail, the purpose of which is to support the work parallel to the table travel. This is removed when the chuck is reground, so that when it is replaced it must also be ground in place to restore its accuracy.

Few jobs are ever done on a surface grinder in the tool room, which do not have to be kept parallel within reasonably close limits. The most frequent cause of difficulty in obtaining parallelism is a dirty or poorly kept work surface on the chuck or the back rail not being parallel with the table travel. On thick work, if difficulty is experienced in keeping the opposite faces parallel, it may be overcome by reversing the position of the work on the chuck, putting it in nearly the same location, but without disturbing the wheel setting.

Thin work is especially hard to keep parallel because it warps so easily. The use of a free cutting wheel and light cuts taken alternately from each side do much to eliminate the warping and hence make it easy to keep the work parallel within reasonable limits. Another method is to place a thin parallel under each end of the work, taking a light cut or series of light cuts alternately from each side. When using this method, the work should be properly blocked on the table to keep it from sliding.

WITNESS MARKS

Occasionally the surface of a job is required to be ground "to clean up." By this is meant that the surface is to be ground so that a very small portion of the original

surface can be seen. This is known as leaving "witness marks" on the work.

    The purpose of leaving these witness
marks is to testify that only a small amount
of stock was removed from the job.   In some
jobs this means that the wheel must be spotted on the work and then traversed over the
entire surface to make sure that too much
stock is not removed. If the surface is
quite regular and has a good finish the edges
may be given a coat of blue vitriol or copper
sulphate to show that too much stock was not
removed.

ANGLE PLATE WORK

    90 degree angle plates (Fig. 758) are
used on the surface grinder for supporting
the work while two adjacent faces are ground
at 90° to each other.

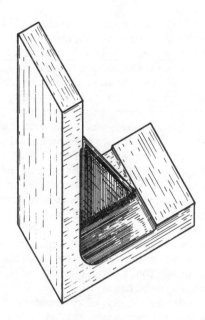

Fig. 758

    Obviously the accuracy of the work
will depend on the accuracy of the plate. If
the two adjacent faces of the plate are
ground at exactly 90° and if the edges of the
plate are maintained parallel and at 90° to
the adjacent faces, then it is possible to
grind two sides of the work at 90° with a
third side, thus cutting down on the work
set-up time.

    ADJUSTABLE ANGLE PLATES (Fig. 759),
precision made, can be had which make it possible to grind intricate angles in the work.
Some of these plates, known as "Sine Angle
Plates," and which have provisions for setting them with a micrometer, or Johansson

gage blocks, are quite suitable for grinding
angles which have exacting limits. The
"Sine bar" is frequently used for grinding
angles where an adjustable angle plate is not
available.

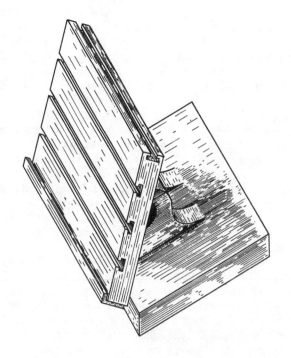

Fig. 759

Fig. 760

GRINDING WORK SQUARE

    To grind the work square after two

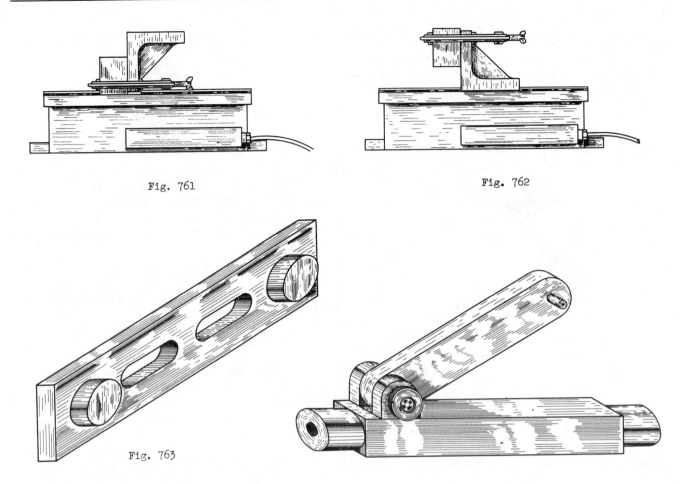

Fig. 761   Fig. 762

Fig. 763

Fig. 764

sides have been ground parallel, place the
side to be ground, down on the magnetic chuck.
Place a thin parallel or rule alongside of
the work, then place the 90° angle plate, top
edge down on the rule, bringing the outside
face of the plate up against the finished side
of the work (Fig. 761). Clamp the work and
plate together in this position, so that when
the angle plate is turned right side up on
the magnetic chuck (Fig. 762) the work face
to be ground will be exposed to the grinding
wheel. Make sure that the clamp has been
placed far enough below the face to be ground
and that the screw end is located inside the
angle plate to reduce the overhang and so as
not to interfere with the grinding operation.

ANGULAR SURFACES can be readily
ground on a surface grinder by dressing the
wheel to produce the required angle, by using
a sine bar (Figs. 763, 764, and 765) or adjustable angle plate and setting the work up
to the required angle or by using an adjustable magnetic chuck.

The grinding wheel should not be
dressed at an angle except as a last resort,
because it decreases the wheel life. If the
wheel must be dressed at an angle make sure
it is dressed correctly, that is, dress the
wheel so that it will produce the angle required on the work and not the complement of
it. For dressing the wheel at an angle,
either a standard angle block or sine bar may

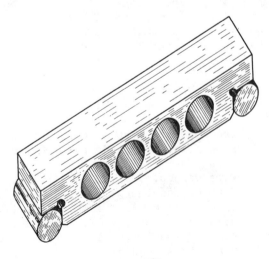

Fig. 765

be used to guide a sliding flanged block
which holds a diamond as shown in (Fig. 766).
After the angle block or sine bar has been
set up it is located so that the point of the
diamond is exactly on the vertical centerline
of the wheel spindle. See (Fig. 767).

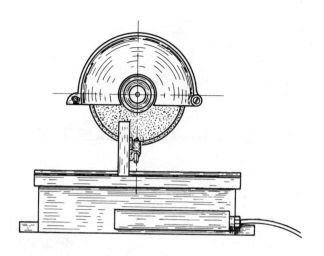

Fig. 767

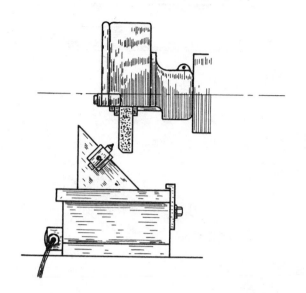

Fig. 766

the "V" maintained in the exact center of the
block.

JOB ANALYSIS

The following outline is a typical
analysis of the procedure for grinding two
opposite flats central on the OD of a shaft.

When the work is to be
set on an angle the common practice is to use a standard angle
block or sine bar (Fig. 768) in
conjunction with a right angle
plate. With some sine bars all
that is necessary is to stack
gage blocks or set a planer gage
at 5 or 10 times the sine of the
angle, depending on the length
of the sine bar; then place the
work on the sine bar squarely
against the 90° angle plate and
clamp it in place.   Other sine
bars require the addition of the
radius of plugs or thickness of
the base of the sine bar be added to 5 or 10 times the sine of
the angle.

V-blocks are used for
holding cylindrical work while
grinding flats, slots, radii, etc.   The
blocks used for this purpose are kept square,
the sides parallel, the ends parallel, and

Fig. 768

Type of job. . . . . shaft as per sketch, (Fig. 769(a)).
Type of machine . . . . Horizontal planer type.
Type of steel. . . . ."RR".
Heat treatment . . . ."Q".
Kind of wheel . . . . 60 - K.
Operations required . . . . Grind flats to size and central with OD.
Tools required . . . 4" V-block, 10" height gage, 1-2" mics, Ideal indicator.

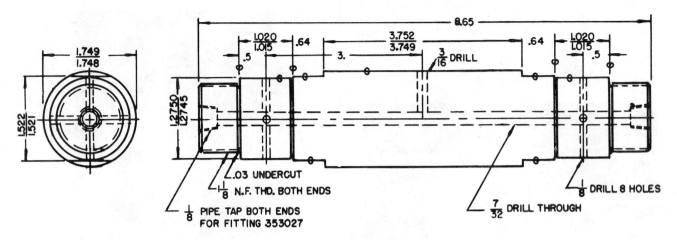

DET—56 SHAFT
630 MACHINE STEEL
STK. 2. DIA.   WT. 47 LBS.
HEAT TREATMENT "P"
ROCKWELL 64 65
GRIND

Fig. 769(a)

PROCEDURE

1. Check flats for sufficient grind stock.
2. Place the job in the V-block using a 10" height gage and Ideal indicator so that the rough flats of the job are parallel with the sides of the block.
3. Using the height gage and indicator, check the cylindrical surface of the job projecting from the V-block for being central.

This is accomplished by indicating the sides of the job as shown in Fig. 769(b) turning the V-block from side to side and observing if the reading is the same.

4. Check unground flats for being central, observing which side is high and whether or not it is within the grind stock, as shown in Fig. 769(c).
5. Place the job on the chuck with the low side up, and spot the wheel to clean up

Fig. 769(b)

Fig. 769(c)

this side of the flat leaving witness
marks.

6. Reverse the V-block to bring the unground
   side of the job up, and take the same cut
   from this face as shown in Fig. 769(d).

Fig. 769(d)

7. Check the job for size noting by what
   amount it is oversize.

8. Feed the grinding wheel down an amount
   equal to one-half the remaining stock and
   grind both faces; reversing V-block as in
   step 6. This should bring the job to the
   required size.

9. Check the flats for size and being central; then have it checked by the inspector.

GROOVE AND SLOT GRINDING

Certain classes of work on the surface grinder call for considerable skill and
patience on the part of the operator. Some
of the more common ones being groove or slot
grinding and radius grinding, especially
where they have to be maintained within accurate limits relative to another surface
(Fig. 770).

Probably the most important factor is
selection of the proper wheel. Where the
groove is wide enough, a gage wheel of medium
hard bond and from 46 to 60 grain size should
be used for roughing. If a gage wheel cannot
be secured a disk wheel having the same characteristics can be used but the sides must be
undercut or relieved to avoid tapering on the
top of the slot or groove.

Fig. 770

"Bell mouthing" of a slot or groove
is a condition where the ends of the slot become gradually wider than the center. On the
surface grinder this condition is generally
caused by spindle end play or pressure on the
wheel; the work forces the end play or wheel
in one direction, then as the wheel clears
the slot the work pressure decreases permitting the spindle to occupy a normal running
position. This condition can be overcome by
doing most of the grinding in the central
section of the groove or slot and only occasionally running the wheel off the ends.

To grind sharp corners on the bottom
of the slot, a vitrified wheel of 80-0 or
120-P should be used after the slot has been
ground to size.

Form grinding can be done on the surface grinder by dressing the wheel to the desired shape. In the case of convex or concave radii a special diamond holder, known as

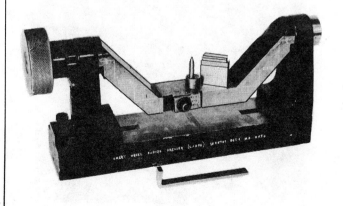

Fig. 771

a radius cradle, (Fig. 771) permits the diamond to be accurately set to the required radius.

Fig. 772

TO USE THE RADIUS CRADLE

Suppose that it is necessary to grind a 3/8" convex (male) radius. The wheel would have to be dressed concave as shown in Fig. 773. Adding 3/8" to "C" gives the distance which the diamond point is to be located above the gaging surface of the cradle; "C" being the distance from the axis of rotation to the gaging surface of the cradle.

Fig. 773

Stack "Jo" blocks to this distance, place the blocks on the gaging surface and adjust the diamond gage to the blocks and lock it in place; insert the diamond holder in the cradle and adjust it to the diamond gage, lock it in place then remove the diamond gage and the cradle is ready to use. The same procedure is followed for setting the diamond to dress the wheel convex (Fig. 774), except that the radius is subtracted from "C".

Fig. 774

It is very important when dressing the wheel for an angle or radius to have the diamond point set on the vertical centerline of the wheel as shown in Fig. 772, otherwise the wheel will be dressed to the wrong form.

The table of the horizontal planer surface grinder traverses very easily, and may be pushed back and forth by leaning against the end of it. This is a desirable quality of the table in so far as surface finish is concerned, but becomes dangerous when removing work from the magnetic chuck. Many ugly hand wounds are suffered by operators of this machine because they move the table to the right of the wheel, then take hold of the work and give it a pull in the direction of the wheel, so that if the table moves or when the work lets go the operator's hand or arm comes in contact with the revolving wheel resulting in a nasty burn and cut.

AFTER THE WORK TABLE HAS BEEN MOVED TO THE RIGHT OF THE WHEEL AND THE MAGNETIC CHUCK TURNED OFF, THE OPERATOR SHOULD REMOVE THE WORK FROM THE CHUCK BY PULLING IT IN A

DIRECTION PERPENDICULAR TO THE LONGITUDINAL
TRAVEL OF THE TABLE

The HORIZONTAL ROTARY grinder (Fig. 775) commonly known as a "Ring grinder" consists of a horizontal wheel spindle having a reciprocating motion similar to that of the shaper ram and a revolving magnetic chuck table supported by columns at the front of the machine. The work table can be raised or lowered and has provisions for tilting the table for concave or convex grinding. The machine is equipped with a coolant supply tank and pump for wet grinding and since it uses the periphery of the wheel is capable of producing a good finish

This machine is used for the production of flat, concave, or convex surfaces which makes it readily adaptable for either tool room or production purposes.

The VERTICAL SPINDLE ROTARY grinder (Fig. 776), consists of a cylindrical wheel mounted on a vertical spindle and supported on a vertical column. This vertical column provides a means of raising or lowering the wheel. The work table consists of a revolving magnetic chuck supported on ways or slides, which provide a means of moving the work to and from the wheel. When the work and wheel are engaged, the magnetic chuck rotates in a clockwise direction but the table is locked in place on the ways of the machine. The machine does not give as good a surface finish as the horizontal rotary grinder but has a use in the tool room as well as on production.

The work is placed on the magnetic chuck (see Fig. 777), the table is then moved in to bring the center of the chuck under the outer edge of the wheel where it is locked in place. The wheel head is then lowered very gradually until sparks indicate contact between the wheel and work, then the power feed is thrown in. If one of the

Fig. 775. Horizontal Rotary Surface Grinder

1. Knob for fine adjustment of work to wheel.
2. Handwheel for adjusting chuck to give concave and convex surfaces.
3. Handwheel for raising and lowering work chuck.
4. Start and stop buttons for rotation of magnetic chuck.
5. Diamond mechanism for truing wheel.
6. Off and on switch for magnetism to chuck.

7. Magnetic chuck for holding work.
8. Grinding wheel.
9. Safety dog for wheel slide.
10. Lever for adjusting speed of wheel slide.
11. Dogs for adjusting length of stroke of wheel slide.
12. Dog for quick return of wheel slide.
13. Wheel slide reverse lever.
14. Lever for adjusting speed of wheel slide and wheel.

Fig. 776. Vertical Spindle Rotary Grinder

1. Heavy Welded Steel Guards
2. Water Cocks
3. 25 H.P. Induction Motor
4. Ball Bearing Wheel Head
5. Air Outlet
6. Air Inlet

7. Wheel Dresser
8. Ammeter
9. Feed Variator
10. Oil Flow Indicator
11. Feed Dial 1 Turn = .100"
12. Feed Wheel 1 Turn = .025"

13. Oil Filter
14. Feed and Head Elevating
15. Control Cabinet
16. Chuck Speed Box With Oil Pump
17. Chuck Speed Control
18. Pump Control

19. Wheel Control
20. Table Traverse Control
21. Chuck Rotation Control
22. Chuck Switch
23. Base
24. One Piece Steel Magnetic Chuck

pieces has been previously ground to size at one spot and coated with blue vitriol and placed on the chuck, it can be used as a sizing block. The wheel can then be fed into the work until light scratches appear on the vitriolized surface indicating that the work is to size. When the correct size is reached the down feed is stopped, the wheel head raised, the work table moved out and the job removed.

The more essential points for the efficient operation of a Blanchard grinder are given below:

1. The selection of the proper grade and grain of wheel for the job.

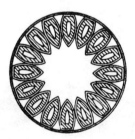

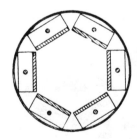

Fig. 777. Methods of Placing Various Types of Work on the Magnetic Chucks of Vertical Spindle Grinding Machines.

2. The selection of the proper wheel feeds and chuck speeds to keep wheel free cutting and as nearly as self-sharpening as possible.
3. The proper loading and blocking of the work, Fig. 777.
4. The condition of the working face of the chuck.
5. The condition of the working faces of the work.
6. The proper and judicious use of the wheel dresser when required.

The Vertical spindle surface grinder, such as a Blanchard grinder, etc., ordinarily uses a cylinder type wheel mounted in a suitable adaptor as illustrated in Fig. 778. Take notice that this wheel is reinforced with wire bands which must be removed as the wheel wears down.

Another type of wheel used on a machine of this kind is known as a segment wheel. It consists of a number of wheel segments mounted in a special wheel chuck. A wheel of this type has the advantage of removing stock faster on broad surfaces and may be changed with less loss of time than a cylinder type wheel. Both types of wheels are useful and each have advantages and disadvantages not possessed by the other. A machine of the Vertical spindle type is very sensitive as to wheel grade and grain and it is important that a wheel be used that is free cutting. Much may be done to overcome glazing of the wheel face by adjusting the speed of the chuck rotation and the wheel feed; it is better to slightly overfeed than to underfeed the wheel to the work as overfeeding tends to keep new sharp grains in contact with the job.

Where the excess stock on a large number of pieces of work does not exceed one-eighth of an inch, the vertical spindle grinder does a much faster job and leaves a much better finish than the planer, shaper, or mill.

(a)   (b)   (c)   (d)   (e)

Fig. 778. (a) 18-in. Segment Wheel; (b) 18-in. Sectored Wheel; (c) 18-in. Cylinder Wheel; (d) Ring Adaptor; (e) Wheel Cemented in Ring Adaptor.

TOOLS COMMONLY USED ON SURFACE GRINDERS

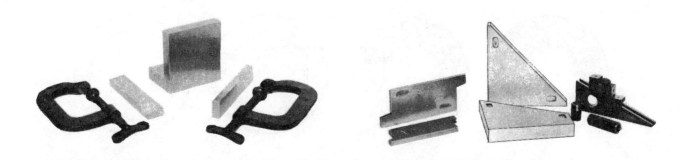

Fig. 779. 3" Angle Plate; H. & G. Parallels; 3" "C" Clamps.

Fig. 780. (left) Master Angle Blocks; (center) Planer Gage; (right) Adjustable Parallels.

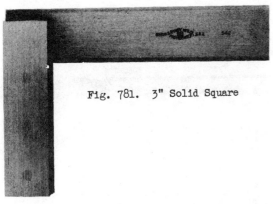

Fig. 781. 3" Solid Square

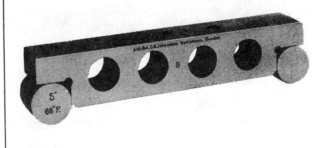

Fig. 782. 5" Jo Sine Bar

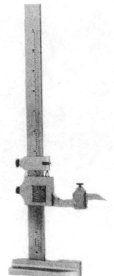

Fig. 783. 10" Height Gage

Fig. 784. Last Word and Ideal Indicator

Fig. 785. 3" V-Blocks

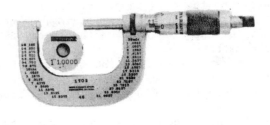

Fig. 786. 1-2" Mics

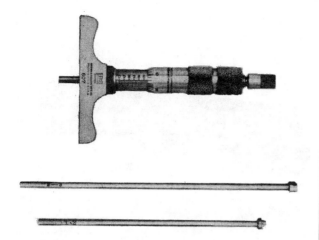

Fig. 788. Johansson Straight Edge

Fig. 787. Depth Mic.

Fig. 789. Johansson Gage Blocks

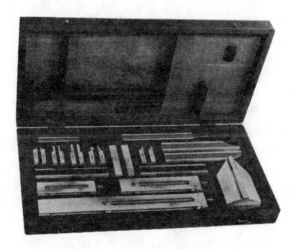

Fig. 790. Johansson Accessory Set

SAFETY RULES FOR SURFACE GRINDERS

1. Test all wheels for cracks or defects before installing them on the spindle.
2. When starting a machine with a new wheel, stand safely to one side until you are sure that the wheel is sound.
3. Whenever there is danger of injury in any way to yourself, in loading, unloading, or checking work on the chuck, STOP the machine.
4. In removing work from the magnetic chuck, crank the table clear of the grinding wheel and then pull the stock AWAY from the wheel.
5. Small thin pieces of stock, extra long stock, or stock with small contact surfaces should be thoroughly blocked on the chuck.
6. Do not use an unguarded wheel at any time if it has a diameter greater than 1-1/2 inches.
7. Wear goggles when using a rubber wheel on a surface grinder.
8. Do not wipe a chuck with a towel--use an oil brush.

9. Shut off the machine while checking or setting up a job.
10. Use properly undercut safety washers in mounting a wheel.
11. Exhaust hoods are supplied as a safeguard for the health of grinder operators. See that they are properly adjusted at all times and that they are not abused.

The one and two inch micrometer and solid square are the most common precision tools used on the surface grinder. The 10" height gage and Ideal or Last Word indicator is frequently used in connection with a planer gage or gage blocks. The planer gage and adjustable parallels are frequently used for measuring the widths of slots instead of the vernier caliper because they are easier to set and give a more positive measurement. A straight edge is frequently used on the surface grinder to test the flatness or straightness of a surface.

QUESTIONS

1. What is meant by surface grinding?

2. What is the difference between the planer and rotary types of surface grinder?

3. What companies manufacture the surface grinders used in the Henry Ford Trade School?

4. Name the principal parts of the surface grinder shown in Fig. 754.

5. What wheel shapes are commonly used on the surface grinder?

6. What are the grain and grade sizes of grinding wheels commonly used on surface grinders?

7. How should the wheel be inspected for cracks?

8. What is the purpose of blotting paper gaskets on the sides of the wheel?

9. What is meant by "safety flanges"?

10. Why is it necessary to step to one side when starting the machine up after it has stood idle all night or after a new wheel has been installed?

11. Why is it necessary to keep surface grinder wheels sharp and clean?

12. What is the result of continued use of a dull or dirty wheel?

13. Give five common methods of holding work on the surface grinder.

14. Explain the correct method of holding small work on a surface grinder?

15. How are non-magnetic materials held on the magnetic chuck?

16. What is the purpose of the "back rail" on the magnetic chuck?

17. What is the purpose of a 90° angle plate?

18. In what condition should the angle plate be kept for accurate work?

19. How many faces can be ground square with each other at the same time by use of the 90° angle plate?

20. From Figs. 761 and 762 explain how to set up a job for squaring by use of an angle plate.

21. Give two reasons for work being out of parallel in surface grinding.

22. Give an efficient method for keeping thick work parallel on the surface grinder.

23. How is thin work prevented from warping on the surface grinder?

24. What is meant by grinding the work to "clean up"?

25. Explain what "witness marks" are.

26. Briefly explain how to surface grind a job to leave witness marks on it.

27. When is the V-block used for holding work on the surface grinder?

28. Name four factors governing the accuracy of a V-block.

29. Briefly explain the procedure in grinding two opposite flats central on a cylindrical shaft.

30. Give two methods for grinding angular work.

31. Why should the grinding wheel NOT be dressed at an angle unless absolutely necessary?

32. When dressing the wheel at an angle what is an important thing to remember?

33. How is the wheel dressed at the desired angle?

34. What tools are commonly used for setting the work at an angle?

35. What is a "sine" bar?

36. Explain how the sine bar is used to set the work at an angle.

37. What two surface grinding operations call for considerable skill and patience on the part of the operator?

38. What shape, grain size, and bond of wheel should be used for slot grinding?

39. How can an ordinary disk wheel be used for grinding?

40. What is meant by "bell-mouthing" of the slot?

41. How can bell-mouthing of the slot be prevented?

42. How is it possible to maintain sharp corners in the bottom of a slot?

43. How is it possible to grind radii on the surface grinder?

44. What is the purpose of a radius cradle?

45. Explain briefly how a radius cradle is used.

46. What is the important thing to remember when dressing a radius or angle on a wheel?

47. What is the correct method of removing work from the magnetic chuck of a surface grinder?

48. What machine is frequently referred to as a "ring" grinder?

49. What is the principle of operation of the ring grinder?

50. What work operations can be performed on the ring grinder?

51. What is the principle of operation of the vertical spindle rotary surface grinder?

52. Describe the work table of the vertical spindle rotary surface grinder and tell how it operates.

53. After the work is loaded on the magnetic chuck how is it brought into contact with the grinding wheel?

54. How is sizing of the work accomplished on the vertical spindle rotary surface grinder?

55. Give six important points for the efficient operation of a vertical spindle rotary surface grinder.

56. What types of wheels are used on the vertical spindle rotary surface grinder?

57. How may glazing of the wheel be prevented on the vertical spindle rotary surface grinder?

58. What is safety rule #6 for surface grinders?

59. What is the purpose of exhaust hoods on grinders?

60. What precision tools are commonly used for checking work on the surface grinders?

INTERNAL GRINDING

Internal grinding is the act of grinding straight cylindrical, tapered or formed holes to accurate size. It is done on the Universal grinder (Figs. 791 and 792) or on machines especially designed for that purpose. Fig. 793 shows a collection of work commonly done on the internal grinders.

Machines for doing internal grinding are divided into three groups depending on how the work is held and operated. The Work Rotating type uses a three or four jaw chuck,

Fig. 791. Universal Internal Grinder

1. **Guard.** Covers spindle nose and hinges up for access to chuck, faceplate, or collet.
2. **Wheel head** on graduated swiveling base.
3. **Wheel head slide,** moves in and out.
4. **Fixture** for tripping diamond truing device.
5. **Fine adjustment** for upper table swivel.
6. **Table traverse locking lever.**
7. **Screw** for adjusting upper table swivel.
8. **Knob** for engaging wheel head slide power cross feed.
9. **Handwheel** for operating wheel head slide. Is graduated in .0005".
10. **Pilot wheel** for traversing table by hand.
11. **Crank** for setting length of stroke of table.
12. **Change gear handle** for changing speed of table traverse.
13. **Belt tension adjusting lever.**
14. **Screw** for positioning table stroke.
15. **Work head base.** Graduated in degrees for cutting steep tapers.
16. **Lever** for starting and stopping work head.

face plate or special fixture for holding the work. The Centerless internal uses a set of rollers which hold the work and give it a revolving motion. The Cylinder grinder holds the work in a fixed, non-rotating position on a reciprocating table and depends on the amount of eccentric wheel spindle travel to generate the correct size of the hole.

THE WORK ROTATING TYPE is the one commonly used in tool and die rooms. A three or four jaw chuck or drive plate screws on the spindle nose of the work head. The work head is mounted on the work table which in some cases moves back and forth. On others the wheel head moves back and forth, the work table being in a fixed position.

Fig. 792. Plain Internal Grinder

1. Diamond for truing wheel.
2. Grinding wheel head.
3. Motor for driving wheel head.
4. Fine adjustment for feeding wheel to work.
5. Cross slide hand wheel for feeding wheel into work.
6. Dogs for controlling length of stroke of table and grinding wheel.
7. Knob for adjusting engagement of quick return of wheel head.
8. Table speed control lever adjusts from 0 to 44 F.P.M.
9. Foot treadle lifts dog and lets table and wheel head go to rest position.
10. Ball lever reverses table in direction it is thrown.
11. Hand wheel for controlling movement of the table.
12. Lever operates collet for holding work.
13. Spindle lock knob convenient when removing fixtures.
14. Plate graduated for setting work head to grind tapers.
15. Knob for swiveling and lining up work head.
16. Lever for starting and stopping work head.
17. Guard around chuck and spindle nose. Lifts up.
18. Grinding wheel guard. Works automatically.

Fig. 793. Samples of Work Done on the
Internal Grinder in Henry Ford Trade School.

Since the work done on internal
grinders in tool rooms is generally ground
dry, these machines have to be built to much
closer specifications than other grinders,
because it is harder to protect the vital
parts from the ever present abrasive dust and
because the wheel and work speeds are much
faster. For the same reasons internal grinders should be kept well lubricated.

In these grinders the wheel and work
centerlines are in the same horizontal plane,
but because the wheel is smaller than the
hole they are not in the same vertical plane;
for this reason the grinding wheel must contact the work on the near or far side of the
hole depending on the construction of the machine. Fig. 794 shows the relation of the
internal grinding wheel to the work.

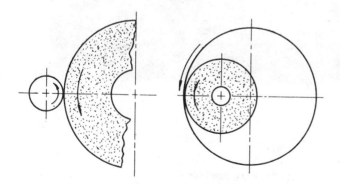

Fig. 794. (left) External; (right) Internal

In chucking a job in a four jaw
chuck, care must be exercised to get the work
running true. Fig. 795 shows a job in the
chuck and the points at which it must be indicated to get it running true. The dial indicator is placed first at point "A" and the
chuck jaws adjusted to give a minimum of runout to the work at that point; then the dial

indicator is moved to point "B". At this
point the work must be forced into line by
tapping it with a mallet. This procedure
should be repeated until the job runs true
within the required limits.

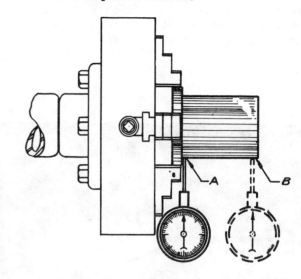

Fig. 795

The following outline is a typical
analysis of the procedure in internal grinding.

Type of job . Bushing as per sketch (Fig. 796)
Type of machine . . . . .Internal grinder
Type of steel . . . ."RR"
Heat treatment . . . ."Q"
Kind of grinding wheel. 60-K
Operations required . . Rough and finish grind
                              ID as per sketch
Method of holding . . . . 4 jaw chuck

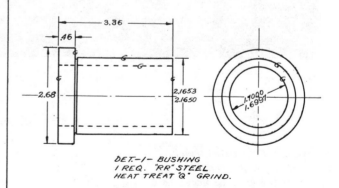

DET.-1- BUSHING
1 REQ. "RR" STEEL
HEAT TREAT "Q" GRIND.

Fig. 796

PROCEDURE

1. Check ID for sufficient grind stock.
2. Mount a four jaw chuck on head stock and
   chuck up the work according to the discussion on chucking.

3. Select the proper quill and grinding
   wheel, mount them on the wheel head and
   dress the wheel. The proper quill to use
   is one which is as short and strong as
   possible, consistant with the length and
   diameter of hole to be ground.
4. Adjust the machine for length of stroke
   and set the table to grind straight.
5. Take a trial clean up cut removing as little stock as possible.
6. Check the hole for straightness and make
   any necessary table adjustments.
7. With the wheel cutting straight, rough out
   the hole to within .001" or .002" of the
   required size.
8. Dress the wheel for finishing and finish
   grind the hole to size.
9. Check the hole with the proper size gage,
   noting any bell-mouthing or out of round-
   ness. Have the first piece inspected.

METHODS OF HOLDING WORK

Besides holding work in a three or
four jaw chuck, it may also be held in a collet; by clamping it directly to a face plate;
by holding the work in a V-block and clamping
the V-block to a face plate; by clamping the
work to a 90° angle plate then clamping the
angle plate to a face plate; or in case of a
large number of pieces of the same kind it
may be held in special fixtures or adaptors.
Work may also be held by a chuck and a steady
rest as shown in Fig. 797. At times it is
necessary to grind the hole in a spindle like
piece of work which necessitates the use of a
steady rest, and tying the end of the job to
the headstock center and drive plate. Fig.
798 shows such a set up. Page 259 illustrates a series of more common set ups and

operations as performed on an internal grinder.

Fig. 798

When holding thin-walled bushings in
a chuck the jaws should be tightened only sufficient to hold the work, otherwise the pressure exerted by the jaws will spring and distort the hole. In like manner when clamping
work to a face plate, be sure that the work
is firmly seated and that the pressure is
evenly distributed and only sufficient to
hold the work firmly.

Fig. 799 shows the correct and incorrect methods of clamping work to the face
plate. On the right-hand side note that the
clamp is not parallel to the face plate because the rest block is too high; note that
the bolt is too far away from the job; note
that the parallel, upon which the job depends
for end clearance of the grinding wheel, is
set too far in, and will interfere with the
grinding operation. The bolt, clamp, rest

Fig. 797

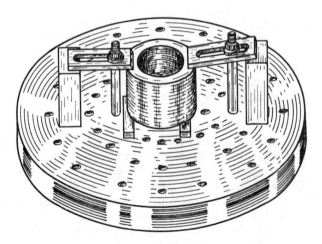

Fig. 799

block, and parallel are correct on the left-hand side.

Out-of-roundness of the work may be due to the work being improperly supported, overheating of the work, loose work head spindle, improper clamping, etc.

Generally speaking, the operator of an internal grinder is required to do one of two things, that is either internal cylindrical grinding or internal taper grinding. Internal cylindrical grinding can only be performed if the wheel and work heads are coordinated so that the axis of rotation of each are moved in parallel planes. This condition can be brought about by adjusting the work head on its swiveled base or the work table so that the horizontal centerline of the work will be parallel to the back and forth movement of the work table or wheel head.

EXTREME TAPER is produced in the work by swiveling the work head to one-half the taper angle by means of the graduations on the circular base of the head. Slight tapers may be produced by adjusting the table to the correct T.P.F. These graduations are obviously not accurate enough for precision work, so that a taper plug gage of the correct taper per foot is used for checking the work after a clean up cut is taken. The plug gage is given a light coat of Prussian blue and with a twisting motion is inserted in the hole to be checked. If the two tapers are not identical, a bright metallic line or surface will show on the plug gage which will indicate whether the taper is too much or not enough. If the head has been swiveled too much the brightened surface will appear on the small end of the plug gage. If the head has not been swiveled enough the brightened surface will appear on the large end of the plug gage.

When using plug gages for checking work be careful that the plug doesn't "freeze" in the hole. "Freeze" is the term applied to a condition where the plug is held fast by the work and is brought about by the fact that the heat of grinding causes the work to expand, so that when a cold plug gage is inserted, the work contracts thus locking the plug gage in the work.

Where a number of pieces are being ground this freezing may be prevented by keeping the plug and the work at the same approximate temperature. Leave the plug gage in the hole of the last piece ground until it is necessary to check the hole in the next piece. This keeps the plug gage warm and helps to prevent freezing.

The diameter of the wheel for internal grinding is based on securing the stoutest possible quill for maximum support, consistent with the size of the hole. Generally speaking the diameter of the wheel should not exceed two-thirds of the diameter of the hole. It must be remembered that as the size of the wheel increases, the diameter of the hole being constant, the greater the area of contact becomes between the wheel and work thus increasing the heating and probable distortion of the work.

The manufacturers of internal grinding machines build the limits of the wheel speed into the machine, and are such that speeds from 4000 to 6500 S.F.P.M. can be obtained. These speeds, of course, depend on the size of the wheel.

Various kinds of diamond or tungsten carbide dressers are available for mechanically or hand dressing the wheel. Occasionally grinding wheels are dressed by the operator holding a piece of silicon carbide in his hand, and passing it along the periphery of the wheel so that the wheel will contact the work properly.

The amount of stock to be removed from a hole to bring it up to size, depends on the diameter and length of the hole but is generally from .004" to .012". More stock than this, means a longer time to complete the job since the hole is smaller, and greater pressure and lighter cuts have to be taken.

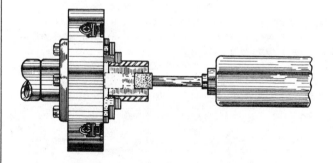

Fig. 800

"BELL-MOUTHING" of the work is a condition where the ends of the hole flare out or are increasingly larger than the required diameter. This condition is brought about by permitting the wheel to overrun the ends of the hole, wheel too hard, excessive wheel pressure, or wheels too short. It can be prevented by setting the length of stroke so that at the extremity of its travel, one-quarter to one-half the wheel face will be uncovered by the work, as shown in Fig. 800, and correcting other factors as listed above.

The work should be checked for size

and straightness after a clean-up cut has been taken, but be sure to move the work and wheel far enough apart to enable the wheel guard to swing down and cover the wheel, (Fig. 801) otherwise serious cuts and burns are apt to be received from the running wheel.

Wheel guard swings downward to protect operator, but does not interfere with fixtures. Can be adjusted for all sizes of wheelheads

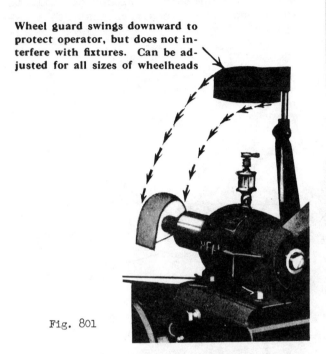

Fig. 801

BLIND HOLES can be ground on the internal grinder provided an undercut of sufficient width is made to give the end of the wheel a little over-run of the hole.

The INTERNAL CENTERLESS, Fig. 802 is especially designed for big runs of work of the same kind. It is fully automatic and assures that the hole will be concentric with the outside diameter of the work. It is capable of producing straight or taper, blind or through holes, interrupted holes and holes having a shoulder.

In this machine the work is driven and its speed controlled by a regulating roller which causes the work to revolve in the same direction as the grinding wheel. To support the work a second roller is mounted below the work and may be adjusted for varying distances from the work center. A third roller, known as a pressure roll, is supported on a swing bracket and holds the work in contact with the other two rollers and moves in and out to allow for loading and unloading the machine. The work head moves back and forth while the wheel head is fixed longitudinally. This makes for greater rigidity for the wheel, which keeps vibration reduced to a minimum.

The CYLINDER GRINDER, Fig. 803 commonly known as a "Planetary Grinder," was designed for the expressed purpose of doing internal grinding on work which was too cumbersome to be revolved on a work head. It is used to a good advantage in railroad shops or shops engaging in heavy machinery operation.

This machine has a reciprocating table to which the work is clamped in a fixed position. The grinding wheel is mounted on a spindle which, while revolving, travels in an adjustable circular path, the size of the path determining the diameter of the hole.

All of the internal grinders considered here, use Vitrified wheels of from 1/4" to 2-1/2" in diameter and from 3/8" to 2" face. The more common grades and grain sizes of aluminous oxide wheels used for internal grinding are; 46-I, 60-I, 46-J, 60-J, 60-L, and 120-P.

SAFETY RULES FOR INTERNAL GRINDING

1. Internal grinding wheels are provided with safety guards that enclose the wheel while the work is being checked. See that this guard is working at all times. Do not work without it in operation.
2. Mount all internal grinding wheels with either paper or brass washers; brass preferred for small wheels.
3. See that the machine is shut down when setting up a job, or while making any necessary changes to guards, stops, or other machine attachments.
4. Wheel breakage during facing operations is large. Keep the wheel dressed sharply and use goggles when doing work of this nature.
5. Do not use wheels in excess of 2-1/2 inches diameter on the internal grinder without a special permit from the instructor.
6. Do not try to wipe out the hole while the work head is in motion. This is a bad practice resulting in finger and hand injuries.
7. See that all heavy work is properly and securely clamped; use standard planer jacks and grinder clamps. Work off-center should be counterbalanced.
8. When using thin rubber wheels or extra hard wheels on the internal grinder, wear goggles as an extra protection from wheel breakage.
9. Any class of work that is too long to allow the proper use of the safety wheel guard while checking the work is a source of danger. Shut the machine down to check work of this nature.

QUESTIONS

1. What is meant by "Internal" grinding?
2. On what kinds of machines is internal grinding done?
3. What companies manufacture the internal grinders used in Henry Ford Trade School?
4. Into what three groups are internal grinders divided?
5. What kind of internal grinders are commonly used for tool room work?
6. How is work commonly held in the work rotating type of internal grinder?
7. Name the parts of the internal grinder which are pertinent to its operation.
8. Why must the internal grinder be built to closer specifications than most other machines?

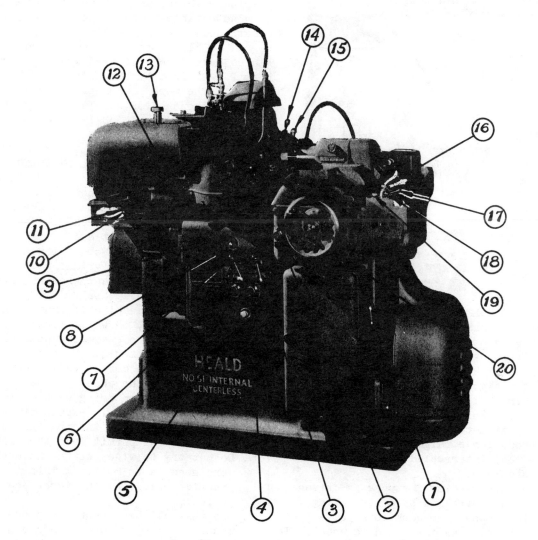

Fig. 802. Internal Centerless

1. Safety switch controlling all motors.
2. Wheel wear compensation knob.
3. Table traverse lever.
4. Knob controlling truing speed of table.
5. Delay valve controlling time of cross slide retraction.
6. Knob controlling finishing speed of table.
7. Throttle controlling table speed.
8. Hand lever for table runout.
9. Adjustment for amount of wheel retraction located here.
10. Trip screw for table runout when dressing wheel.
11. Trip screw for table runout when work is finished.
12. Traverse adjustment for regulating roll.
13. Knob for adjusting pressure roll.
14. Hand lever for operating wheel truing device.
15. Knob for micrometer adjustment of truing device.
16. Pick-up feed adjustment.
17. Lever to shut off all feed on wheel head cross slide.
18. Fine feed adjustment of wheel slide.
19. Coarse feed adjustment of wheel slide.
20. Handwheel for manual operation of wheel slide.

9. Why must the spindle bearings of the internal grinder be kept especially well lubricated?

10. In the grinding operation what is the relation of the grinding wheel with respect to the work?

11. Explain, exactly, how to chuck up a job in a four jaw chuck on the internal grinder.

12. Give six methods for holding work on the internal grinder besides the four jaw chuck and collets.

13. What care must be exercised in the chucking of thin walled bushings?

14. Explain exactly, the correct method of clamping work to a face plate on the internal grinder.

15. What are the most frequent causes of out-of-roundness of the work in internal grinding?

16. What two work operations is the operator of the internal grinder most frequently called upon to do?

17. What conditions must prevail for the production of internal cylinders on the internal grinder?

18. How can this condition be brought about?

19. How is extreme taper produced on the internal grinder?

20. How are slight tapers produced on the internal grinder?

21. Explain how to produce accurate taper work on the internal grinder.

Fig. 803. Cylinder Grinding Machine

1. Grinding wheel.
2. Screw for moving table crosswise of saddle.
3. Adjustable stops for length and position of stroke.
4. Lever for controlling speed of table reciprocation.
5. Shaft for adjusting height of knee.
6. Lever for engaging table reciprocation.
7. Crank for hand control of table to or from the grinding wheel.
8. Lever for starting and stopping grinding wheel spindle.
9. Handwheel for turning spindle by hand.
10. Lever for adjusting the depth of cut in steps of .0005" on the diameter of the work while the spindle is rotating at high speed.
11. Hole sizing mechanism. Operated by turning a knob for small adjustments and by turning a crank when going from a small hole to a large hole.
12. Grinding wheel spindle. Comes in six different sizes for different work requirements.

22. Explain the use of the taper plug gage in producing an accurate taper.
23. What is meant by the term "freeze" when checking a hole with a plug gage?

24. How can this freezing be prevented?
25. Upon what two factors does the selection of the diameter of the grinding wheel for internal grinding depend?

TOOLS COMMONLY USED ON INTERNAL GRINDERS

Fig. 804.   10' H & G face plate, 3" 90° Angle plate, 4" V-block.

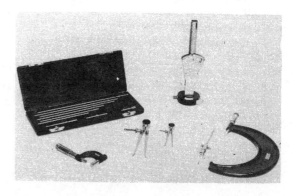

Fig. 805.   Inside mics, Tube mics, Inside calipers, #3 Jo gun.

Fig. 806.   Telescope Gages: AA, A, B, C.

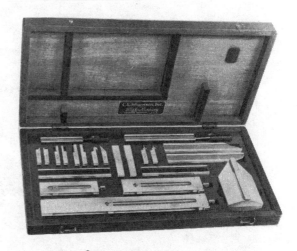

Fig. 807.   Johansson Accessories

Fig. 808.   Johansson Gage Blocks

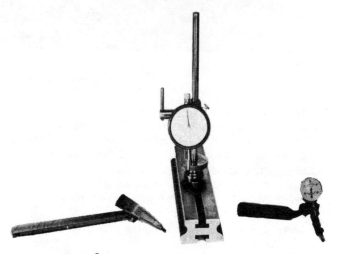

Fig. 809.   Ideal Indicator, Dial Indicator, Last Word Indicator

26. What are the limits of speed in F.P.M. for internal grinding wheel?
27. What is meant by the term "bell-mouthing"?

28. What is the most frequent cause of bell-mouthing?
29. How can bell-mouthing be prevented?

COMMON SET-UPS AND OPERATIONS ON INTERNAL GRINDERS

Fig. 810. Work Clamped to Face Plate

Fig. 811. Angle Plate Job on Face Plate

Fig. 812.. V-block Job on Face Plate

Fig. 813. Setting a 3" Jo Gun Using Accessories

Fig. 814. Checking the ID of a Bushing with a 3" Jo Gun.

Fig. 815. Using Inside Mic and Extension Handle to Check ID of Bushing.

30. When checking work on the internal grinder how is the operator protected from the rotating grinding wheel?
31. What condition is necessary for the production of blind holes on the internal grinder?
32. What is the purpose of the Internal Centerless grinder?
33. How is the work held in an internal centerless?
34. For what kinds of work is the internal centerless suited?
35. By what name is the cylinder grinder commonly known?
36. What is the purpose of the cylinder grinder?
37. For what branches of industry is the cylinder grinder especially suited?
38. How is the work held in the cylinder grinder?
39. How is the cylinder grinder adjusted to produce the correct sized hole?
40. Wheels used in internal grinding are made from what kind of bond?
41. What grades of bond and abrasive sizes are commonly used in internal grinding wheels?
42. How is the operator of a tool room internal grinder protected from the rotating grinding wheel?
43. What care should be exercised in mounting an internal grinding wheel?
44. At what two times in particular should the operator of an internal grinder wear goggles?

MISCELLANEOUS GRINDERS

A Floor or Bench grinder (Fig. 816) is usually an electric motor mounted on a suitable base and having the rotor shaft extended from each side with a grinding wheel mounted on each end, for sharpening tool bits, planer tools, boring tools, drills, etc.

Some floor grinders for heavy duty work have a wheel shaft mounted in heavy bearings which are an integral part of a heavy cast base. The wheel is driven by a belt connecting the wheel shaft pulley to a motor mounted on the base of the machine.

For grinding cutting tools these machines should be equipped with fine grain wheels of a silicate or vitrified bond. For rough grinding, snagging or heavy work, the machine should be equipped with coarse free cutting wheels.

All floor or bench grinders should be equipped with adjustable eye shields made from fire glass or safety glass and should never be used without wearing goggles.

One end of a Floor or Bench grinder may be equipped with suitable sewed canvas forms, the periphery of which have been coated with an abrasive to be used as a buffer. Buffing or polishing jacks are used for the purpose of securing a good finish without respect to surface accuracy. In using a buffer care must be taken to keep the work from getting caught in the wheel and being pulled into the machine or serious injury might result.

Fig. 816

GRINDING SUGGESTIONS

Well known grinding machine and grinding wheel manufacturers have made some very good suggestions relative to safety and economy of operation of grinding equipment. It will pay the student in grinding to pay particular attention to the more pertinent of these suggestions and if generally adopted should go a long way toward the elimination of machine abuses and accidents. A few of these suggestions follow:

1. Handle all wheels carefully.
2. Sound the wheel for cracks before putting it on the machine, especially in the case of cylindrical, and surface grinders.
3. Make sure the blotting paper or rubber gaskets are used between the wheel and wheel flanges.

4. Where a choice of machines is available use one that is rigid.

5. The wheel should fit freely on the spindle with about .003" to .005" play.

6. Use relieved or safety flanges which are at least one-third of the wheel diameter.

7. The spindle nut should be set up tight enough to properly hold the wheel.

8. Keep all wheel guards tight and in place.

9. After installing a wheel on a surface or cylindrical grinder turn it over by hand to make sure that it clears the guards and housing.

10. Keep the spindles and bearings well oiled to avoid overheating of the spindle and possible wheel breakage.

11. When starting to work in the morning, press the starter button on the machine and at the same time step to one side letting the machine run for at least one minute before engaging the wheel with the work.

12. When grinding long slender work on cylindrical grinders use steady rests to support the work and keep the shoes properly adjusted to the work. Make sure the driving pin for the grinder dog is firmly located in the arm on the drive plate.

13. When grinding wheels are removed from the machine handle them carefully and store them in a dry place.

14. When engaging the wheel with the work avoid excessive pressure; it is ruinous to the wheel, work supports, and spindle bearings.

15. In any grinding machine it is necessary for the work to be properly supported and that the operator keep his hands away from the revolving wheel and work.

16. When placing work on the magnetic chuck of surface grinders the table should be moved far enough away from the wheel so that the operator can locate the work without injury to his hands from the rotating wheel.

17. Since the table of a Horizontal Rotary grinder moves along its ways very easily, the work should be removed from the magnetic chuck by sliding it off in a direction perpendicular to the table traverse.

18. Close and easy fitting goggles should be worn on any dry grinding operation where there is danger of flying grit getting into the operator's eyes.

19. Exhaust hoods on dry grinders are provided for the benefit of the operator and should be kept in place and free from obstructions.

20. Magnetic chucks are expensive pieces of equipment. Keep them clean. Keep them free from scratches and nicks. Don't leave the machine for any length of time with the switch to the chuck turned on.

The thread grinder is a machine developed for the precision grinding of threads, and is becoming a necessity where accuracy and production are required or threaded work (both external Fig. 817 and internal Fig. 818) of various types and different materials.

The output is controlled by the automatic features of these machines. When it is necessary to grind only a few parts, the thread grinder can easily be set to new work and operated manually.

Besides greater accuracy, the thread grinder has other practical advantages. Exceptionally hard materials, including tough steel that is heat treated and are difficult to thread by other methods, can readily be threaded by these grinders. Threads are often ground directly from the solid. This

Fig. 817. External Precision Thread Grinder

1. Lead Pick-up
2. Work Drive and Lead Screw Housing
3. Work Head
4. Grinding Wheel
5. Coolant Valve
6. Tailstock
7. Helix Angle Graduation
8. Wheel Spindle Motor
9. Signal Lights Indicating Dressing of Wheel
10. Machine Table Slide
11. Electrical Compartment
12. Table Control Dogs
13. Manual Table Reverse Control Knob
14. Automatic Cycle Starting Lever
15. Size Setting Hand Wheel
16. Opening to Adjust Depth of Initial Grinding Cut
17. Manual Dresser Slide Adjustment
18. Work Drive Motor
19. Control Panel

naturally means a saving in time. Whether grinding fine threads from the solid or coarse threads previously roughed and given heat treatment, a tolerance on the pitch diameter of plus or minus .0002" can ordinarily be held.

The thread grinder uses a high precision ball bearing spindle to rotate the grinding wheel, the accuracy in size and finish of the ground piece depending to a considerable extent on the exactness with which the spindle operates.

On the external thread grinder shown in Fig. 817 for external threads the grinding wheel has a diameter of 18" and a thickness of 3/8", the internal thread grinder, Fig. 818, uses a grinding wheel from 3/4" to 4" in diameter. The grinding wheel is automatically kept dressed by the machine.

Fig. 818. Internal Precision Thread Grinder

1. Work Drive and Lead Screw Housing
2. Work Head Slide
3. Control for Right or Left-Hand Thread and Multiple Index
4. Work Piece
5. Grinding Wheel
6. Lights Indicating Cycle, Machine and Dresser Operation
7. Work Spindle Jog Control-Right and Left
8. Knob and Switch to Control Automatic Dressing
9. Electrical Compartment
10. Wheel Slide
11. Size Control Hand Wheel Assembly
12. Operator's Control Panel
13. Lead Pick-up and Automatic Backlash Compensation Control

Chapter 26

ROUTING OF BENCH TOOL WORK

The purpose of this group of sheets is to show the routing of Bench Work through the various departments of the Tool Room.

The term "Bench Work" is applied to the type of work done at the bench, consisting of planning, laying out, chipping, hack sawing, filing, threading, drilling and reaming of holes, scraping, lapping, checking each piece after each operation, and assembly of parts such as drill jigs, dies, gages, etc.

Orders coming into the Tool Room are received at the Clearing House. Here all the stock for the job, except that which is furnished by other departments and specified on the work order, is ordered on standard requisitions from the steel, tool, and miscellaneous stock rooms.

The job is then routed to the department which is to do the work and the foreman assigns it to a bench leader. A bench leader is a first-class toolmaker who has had sufficient training and experience to perform all the usual operations done in the tool room and to direct others in their performance. He is then in direct charge of the job and responsible for it. He plans the job carefully and decides how each detail will be made. Which operations will be done first and what machines are to be used must be considered carefully. A mistake in this might lead to serious results later as the job progresses. If special tools are necessary, such as jigs and fixtures, he must foresee this early enough to have them ready when needed. He must consider the most efficient and economical methods that he can use. A bench leader can often save many hours of time and a great deal of expense by having the work carefully planned, thus avoiding mistakes which might otherwise be made.

When the stock is received by the bench leader, it is checked with the blue print to make certain that the proper material and sizes have been sent. He then writes out a Tool Operation Card, routing the job to the required machine. This card is accompanied by a sketch or blue print showing the work that is to be done.

Since it is necessary for the bench leader to be able to locate any part of the job at any time, a complete record of each detail is kept on the back of the work order card. This record shows where each detail has been sent, the date sent, and the date returned.

He must follow up his orders so that the details are finished in the proper rotation, and are not allowed to lie around unnecessarily. As each detail is returned to the bench, it is checked to make sure that it is machined correctly according to the drawing.

Following the final assembly, the finished fixture is inspected for accuracy.

The bench leader then sends the finished fixture to the inspection bench for the final inspection.

The procedure of the details through the various departments is shown in the following sheets.

As an example of the operations and routing of a job, the expander shown in Fig. 819 is used.

The drafting room designs and makes a drawing of the expander, from which blue prints are made.

The order shown in Fig. 820 and the blue prints for the expander are sent from the work order department to the clearing house of the designated tool room, which in turn orders the material. The job is then sent to the bench foreman who assigns it to a bench leader.

When the work is sent to be machined, a record of the destination, date sent, date received, and time for the operation is kept on a Detail Work Card, as shown in Fig. 821.

A Tool Operation Card specifying the work to be done accompanies the stock and sketch or blue print to each machine. A new card is filled out for each new operation.

In this particular case, the Tool Operation Card is filled out as shown in Fig. 822 and sent with the blue print or sketch to the lathe.

After the job is returned from the Lathe Department, it is checked to determine whether the dimensions machined there are correct.

Details 1 and 2 are then sent to the Milling Machine Department, with a new Tool Operation Card specifying the operations to be done there.

After the milling is completed, both

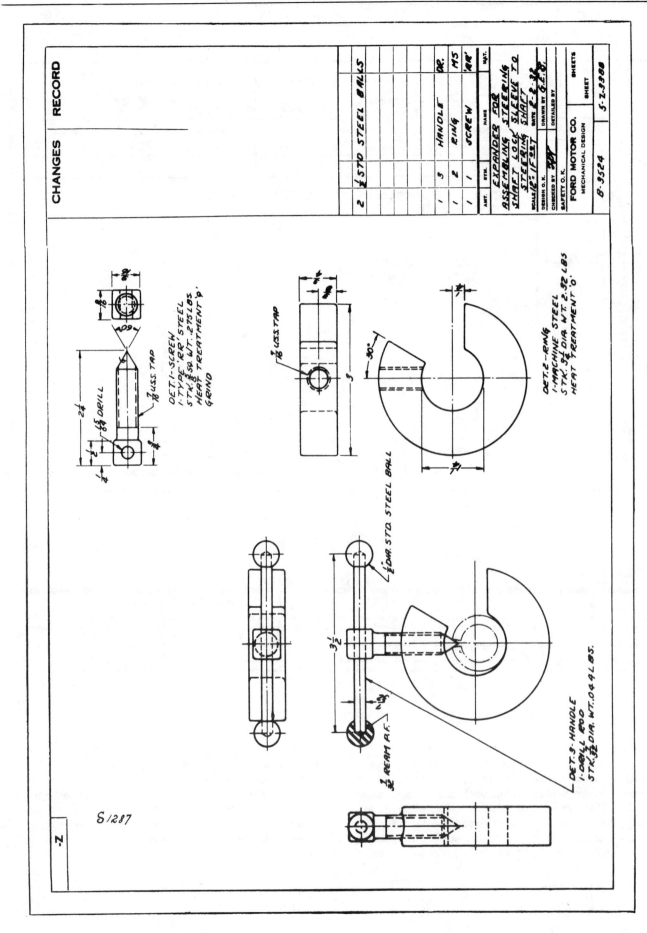

Fig. 819. Drawing of Expander

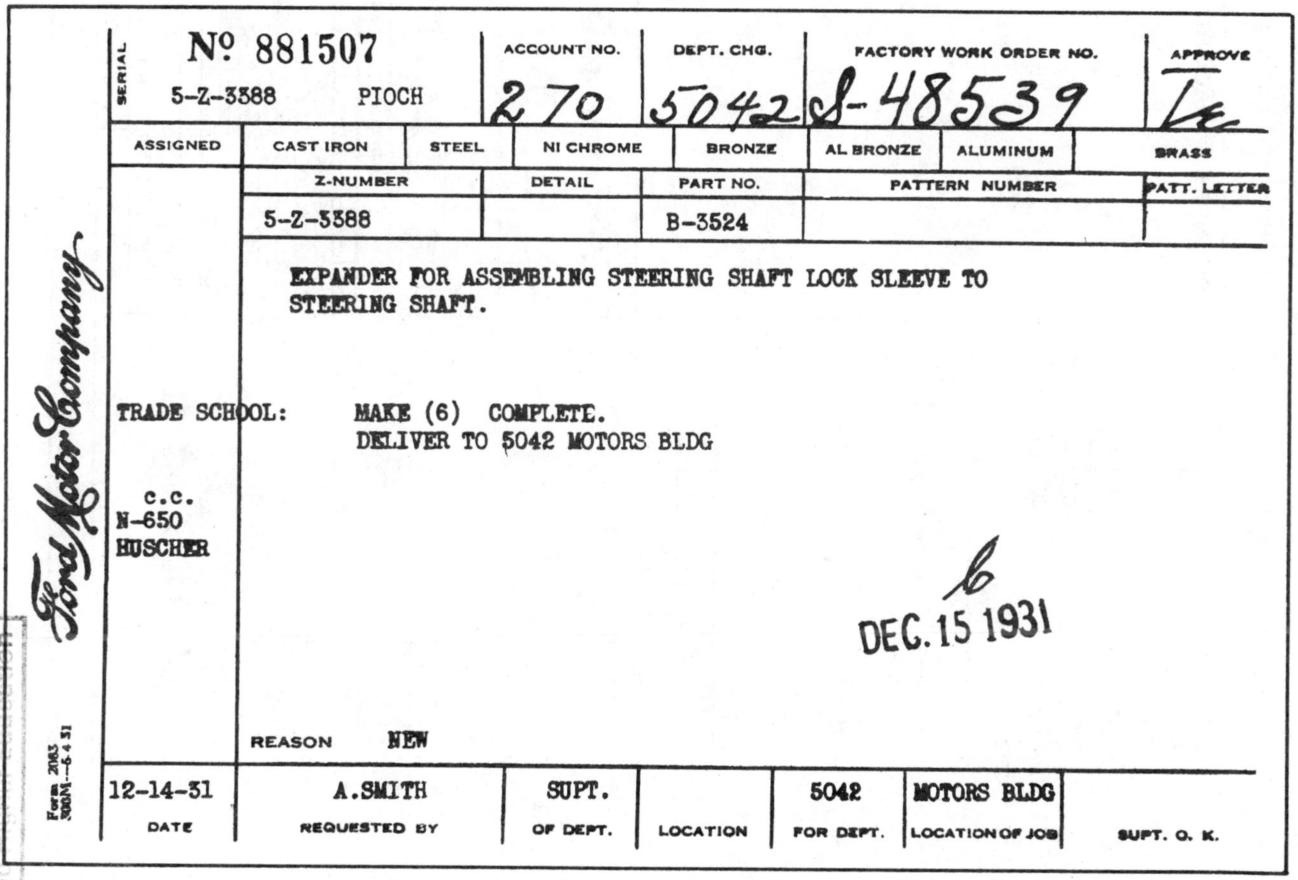

Ford Motor Company

SERIAL Nº 881507

5-Z-3388 PIOCH

ACCOUNT NO.	DEPT. CHG.	FACTORY WORK ORDER NO.	APPROVE
270	5042	S-48539	Le

ASSIGNED | CAST IRON | STEEL | NI CHROME | BRONZE | AL BRONZE | ALUMINUM | BRASS

Z-NUMBER	DETAIL	PART NO.	PATTERN NUMBER	PATT. LETTER
5-Z-3388		B-3524		

EXPANDER FOR ASSEMBLING STEERING SHAFT LOCK SLEEVE TO STEERING SHAFT.

TRADE SCHOOL: MAKE (6) COMPLETE.
DELIVER TO 5042 MOTORS BLDG

c.c.
N-650
HUSCHER

6
DEC. 15 1931

REASON NEW

Form 2063 300M—6-4-31

12-14-31	A.SMITH	SUPT.		5042	MOTORS BLDG	
DATE	REQUESTED BY	OF DEPT.	LOCATION	FOR DEPT.	LOCATION OF JOB	SUPT. O. K.

Fig. 820. Order Form

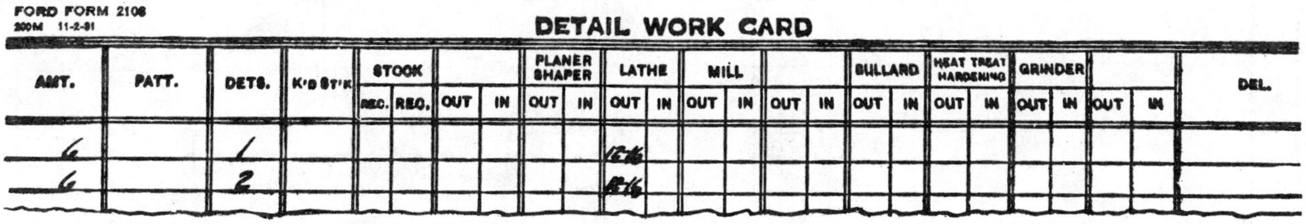

FORD FORM 2108
300M 11-2-31

DETAIL WORK CARD

AMT.	PATT.	DETS.	K'D ST'K	STOCK				PLANER SHAPER		LATHE		MILL				BULLARD		HEAT TREAT HARDENING		GRINDER				DEL.
				REC.	REQ.	OUT	IN	OUT	IN	OUT	IN	OUT	IN	OUT	IN	OUT	IN	OUT	IN	OUT	IN	OUT	IN	
6		1								12/16														
6		2								12/16														

Fig. 821. Work Card

details are again checked for correct dimensions.

The filing, drilling, tapping, and reaming operations are then completed at the bench, after which details 1 and 2 are sent to the Heat Treating Department to be hardened, tempered, and inspected.

They are then returned to the bench and detail 1 is sent to the Grinding Department.

When the grinding work is completed, detail 1 is returned to the bench, where it is again checked for size. The job is then ready for final assembly.

After the final assembly, the job is sent to the Inspection Department for final inspection, and from there to the clearing house for delivery to the department specified on the work order.

Form 3112 500M—4-15-36
Chg. Dept. *5042* **TOOL OPERATION CARD** Acct. No. *270*

Order No. *S-48539* | Part No. *3524* | Mach. No.

Z-No. *5-Z-3388* | Det. No. *1 and 2* | Date *2-22-38*

Sent by *Arnold* | To Dept *Lathe*

No. Pcs.	Patt. or Det.	REMARKS
6	1	} *Machine as per*
6	2	} *attached blue print.*

Foreman *Emptage* Dept. *T. S.*

Inspected by................................ Date...............

KEEP THIS CARD CLEAN AND RETURN WITH JOB

Fig. 822